# Trust, Complexity and Control

# Trust, Complexity and Control

## Confidence in a Convergent World

**Piotr Cofta**
Chief Researcher, Identity and Trust
BT, UK

John Wiley & Sons, Ltd

*Other Wiley Editorial Offices*

John Wiley & Sons Inc., 111 River Street, Hoboken, NJ 07030, USA

Jossey-Bass, 989 Market Street, San Francisco, CA 94103-1741, USA

Wiley-VCH Verlag GmbH, Boschstr. 12, D-69469 Weinheim, Germany

John Wiley & Sons Australia Ltd, 42 McDougall Street, Milton, Queensland 4064, Australia

John Wiley & Sons (Asia) Pte Ltd, 2 Clementi Loop #02-01, Jin Xing Distripark, Singapore 129809

John Wiley & Sons Canada Ltd, 22 Worcester Road, Etobicoke, Ontario, Canada M9W 1L1

Wiley also publishes its books in a variety of electronic formats. Some content that appears in print may
not be available in electronic books.

Anniversary Logo Design: Richard J. Pacifico

*Library of Congress Cataloging in Publication Data*

Cofta, Piotr.
   Trust, complexity and control confidence in a convergent world / Piotr Cofta.
      p.   cm.
   Includes index.
   ISBN 978-0-470-06130-5 (cloth)
   1. Computer security.   I. Title.
   QA76.9.A25C58 2007
   005.8—dc22                                           2007015560

*British Library Cataloguing in Publication Data*

A catalogue record for this book is available from the British Library

ISBN 978-0-470-06130-5 (HB)

Typeset in 9/11pt Palatino by Integra Software Services Pvt. Ltd, Pondicherry, India
Printed and bound in Great Britain by Antony Rowe Ltd, Chippenham, England.
This book is printed on acid-free paper responsibly manufactured from sustainable forestry in
which at least two trees are planted for each one used for paper production.

Do not trust people. They are capable of greatness.
(Stanislaw Lem, 'Holiday', 1963)

# Contents

# Foreword

There are very few areas where technological innovation and social science are such uncomfortable yet dependent bedfellows.

The concept of trust has a warm, comfortable, deep-rooted familiarity that you would expect from something that has played a major part in defining human evolution. We all know something about what it is to trust... and be trusted.

Technology all too often casts a long cold shadow of reality across the familiar warmth associated with being human. Technology deals in hard facts. Lacking supporting context, technology-driven spaces like the Internet struggle to support human endeavours without resorting to the endless quest for control, to offset the unquantifiable risk. As the new saying goes, 'On the Internet nobody knows that Alice is a dog!'

Given his background (university, Nokia, Medial Lab Europe) and the fact that Piotr is currently working as a Chief Researcher in the Mobility Research Lab of a global network operator, he is well placed to understand the generic limitations and needs of convergence. Even though from a user's point of view convergence is likely to remain a moving target, it provides an interesting backdrop to explore the relationship between trust, complexity and control.

Humanity evolved in a world where a desire to communicate (real time, person-to-person) generated a profitable business where the connectivity and the service were implicitly linked – the telephone conversation! That world, its focus on network architecture and the related business model are becoming less important as customers increasingly look to personalised and converged lifestyle solutions – not just the connectivity. Today, we still want our customers to talk – but more importantly to sign up, log on and consume content 24 hours a day, 7 days a week.

The technical challenge associated with convergence is to hide the underlying complexity from the user... who simply wants to get on and do it! One starting point is to attempt to mirror in the virtual world something of the natural human approach to trust in the real world. Our current research suggests that an improved understanding of confidence in a converged world has the potential to unlock some radical business models based on identity and risk rather than control.

Piotr's book, like the subject matter, is more than the sum of the individual parts. OK... it has all the presence of a well-researched, comprehensive, reference text that you would expect from an academic well established in his field. However, read from cover to cover, it is so much more. From the consistent reference to Alice's relationship with Bob to asides like actually buying 'Trust in a bottle', the author's obvious passion for exploring and demystifying this cross-over subject takes on the enthusiasm of a Grand Wizard's tutorial at times.

Looking beyond the original intent and necessary iteration, the book has as much to say about being human as it has to say about the technology of dealing with human interaction.

Richard Dennis CEng, MBCS, CITP
Head of BT's Mobility Research Centre

# Introduction

... in doubt a man of worth will trust to his own wisdom.

(J.R.R. Tolkien, *Lord of the Rings*)

This book is about Alice. Alice is a fictitious character who has been introduced by Bruce Schneier [Schneier1996]. Since then, Alice has been frequenting security-related books, together with the rest of the original cast: Carol, Dave, Trevor, etc. In this book we will look at Alice's uneasy (and usually unilateral) relationship with Bob, another fictitious character of the same provenance.

This book tells a simple story: Alice is dependent on Bob and she would like to determine whether she can be confident about Bob – that he will help her when needed. She can be confident either if she trusts Bob or if she can control him. Simple as that, the story gets complicated if we start considering why she should trust him, to what extent she can control him or how she should react if he changes his behaviour. Things get even more complicated if we place them far apart and let them communicate over the Internet – a picture so common today. To complicate things even more, we may introduce between them people and computers that will provide their opinions about Bob and about themselves. Some opinions will be true and honest, some will be deceitful and false. The more realistic we are, the more complicated the picture gets.

Despite all the obstacles we are piling up, we wish Alice all the best. We will succeed if we can advise Alice how to make the most of those situations – how to determine whether she should be confident about Bob or not. For that purpose she should understand herself (how her confidence is built) and she should understand the world around her (how her confidence can be influenced by others). We will provide her with the knowledge, but the decision is hers. We can only hope that she will trust her own wisdom.

Alice lives in a world that is saturated with technology and she participates in what is a continuous interplay between digital technology and society. In developing digital technology, we replicate what we know about ourselves, what we discover through social sciences, philosophy or psychology. There is nothing conceptually new that has not been explored before. Queues, messages, priorities, files, conflicts, services, etc.: the vocabulary of computer science surprisingly resembles the vocabulary of corporate management or the social behaviour of a group of people. It seems that what Alice knows of being confident about other people can be carried over to the technology-intensive environment.

However, once new technologies are implemented, they are changing and influencing Alice and her friends. Computers and networks not only change the way she works and play but also change the way she thinks of herself, and about the world around her. Communication and computing have been specifically challenging, demanding a re-evaluation of core existential assumptions of our identity and our relationships. For the first time there are systems of complexity that are comparative with living creatures, where from pure digital appearance Alice cannot easily tell the difference between a person and a machine. She know that she can be fooled into believing that what she sees or hears was a real person while in fact it was only the result of an extensive computational process. Being confident about something becomes increasingly harder, and old knowledge may be unwanted ballast, not a desired guide.

At the same time, the technology is becoming pervasive in every aspect of her life, combining computation and communication into a unifying experience. The communicating intelligence can be present everywhere, from digitally-enabled refrigerator to satellite-enabled car. How she relates to those devices and how she relates to others through those devices (and how they relate to her) is the question that must be addressed. Alice cannot ignore the role technology plays in her life and in the way she builds her confidence.

Confidence, the way it is presented here, is a self-referring sum of control and trust – a sum of two concepts that are of paramount importance when it comes to the relationship between technology and us. The method that has been chosen to explore this relationship is the interdisciplinary tour: from technology to sociology and back. By applying technical thinking to sociological terms and then by applying sociological thinking to technical terms, the book presents a unified approach to confidence; neither entirely social nor entirely technical, but demonstrating the synergy and interaction of both. Alice lives in an interdisciplinary world, so that an approach that is not interdisciplinary is unlikely to give the complete picture.

This book is personal, panoramic and propositional. It is personal because it presents the author's opinion on various aspects of the widely understood constructs of confidence, trust and control. This leads to agreement, disagreement or discussion with propositions from other sources: books, papers, presentations, etc. However, this is not an overview of the research area, as everything here is presented from the unifying perspective and through the lens of author's interest in the given part of the area. Whatever was deemed to be obvious has received the light treatment, everything that has been considered novel or particularly problematic resulted in a longer discussion.

The book is panoramic. It does not stop on a single subject to explore it in depth, but instead provides a view on the subject that is as wide as possible. It is interdisciplinary and freely links and cross-utilises research from different areas. Again, this may sometimes lead to uneven attention and superficial treatment of aspects that may deserve more thoughtful consideration. This is specifically visible if some areas have been deemed marginal to the main problem of the book.

Finally, this book is propositional: it is built on a single proposition (best laid out as the model of confidence) and seeks to explore the implications of this proposition on several areas of interests, across disciplines and across application areas. Whenever possible, there is a clear separation between an overview of existing works and the proposition (signalled quite often by the phrase 'the proposition is . . . '). The existing body of research is used extensively to support the proposition.

The book discusses trust quite extensively, even though it talks about confidence. There are several reasons for this – the fact that the construct of trust is harder to grasp, that it quite often encompasses confidence and control, the fact that trust is the only indispensable element of the relationship or the fact that relationships tend to move from control to trust [Lewicki1996].

Trust is annoying. It escapes clear definition (despite having 17 different ones [McKnight1996]), yet everybody can immediately recognise it – specifically recognise the lack of it. Trust is embedded in every statement we say [Habermas1986], yet when it becomes itself a subject of conversation it is usually trivialised and misused. Trust is recognised as an essential element of every social transaction (specifically commercial ones), but it can be hardly found in any written contract – and is even less likely to be enforced by law. Trust is a foundation of security, but we are unwilling to admit it. Trust escapes rational judgement yet everybody seems to be an expert in judging trust.

Trust should not be here. It seems to be superfluous, something that can be replaced by better design and the implementation of an ever-growing number of procedures, rules and regulations. The age of reason stresses contracts, cognition, understanding and reasoning, and trust does not seem to be found in any of them, being essentially orthogonal to reasoning [Faulkner2003]. Even though trust is enabled by emotions, it is not entirely driven by emotions – it belongs to the domain of reason, but in a way that is sometimes hard to grasp. The fact that we were not able to eliminate it demonstrates our dependence on it, but not our understanding of it.

The more we think about trust, the better we see that the question of trust is incorrectly positioned. Today, we do not need trust, but confidence. Trust is what (hopefully) happens between people, is not easily quantifiable, takes time and effort to build and can be easily lost. In the marketplace we do not really care whether others are providing us with goods or services because we trust them or because they are bound by regulations. Actually, we seem to prefer regulations. The volatility of our relationships decreases the need for trust: disposable customers cannot afford trust while disposable producers are unlikely to reciprocate.

Trust is perishable. People lament that there is not enough trust in our everyday life (even though there are dissenting opinions as well, e.g. [O'Neill2002], stressing our over-expectations). We can easily find (e.g. [Tyler1996a]) a wide palette of examples where trust is disappearing, loyalty is not reciprocated and formal contracts replace fiduciary obligations. The decline in trust can be attributed to many factors (some of them are discussed in the book), but one of the important factors is the decline in the 'shadow of the future', driven by the simultaneously individualistic and global nature of societies where individuals are less likely to ever meet again. The decline in trust is to a certain extent a self-fulfilling prophecy, as in the expectation of less trust people react with behaviour that reinforces less trust.

Trust is desired. There is a growing need to apply trust in order to achieve greater organisational flexibility and ability to manage during times of distress. Trust seems to be essential in the acceptance of decisions [Tyler1996b], increasing flexibility of organisations, building agile teams, supporting innovations, decreasing costs, improving quality, etc. Trust is increasingly a necessity in modern organisations, as our abilities to codify, monitor and enforce control are diminishing [Frankema2005]. It is also worth mentioning the economic and organisational factors: relationships that are built on trust are less expensive to manage (e.g. [Fukuyama1996]). Apparently modern organisations are wiling to utilise trust – if only they know of how to create and retain it.

Confidence can be achieved through trust or through control and we increasingly desire trust but implement control. Control is not as fashionable as trust, but it has its rightful place in the development of relationship. Specifically, while we are using computers, more control sounds like an attractive deal – maybe we will finally become masters of those machines, not only their helpless users?

Control has been with us for such a long time that we take it for granted. Organisations has been always built on control (e.g. [Pennings1987]) and modern societies have relied on it to protect their independence, to control their governments and to provide efficient creation and distribution of wealth. Better control meant survival, comfort and opportunity. It is only recently that the new approach, stressing trust, has become widely accepted.

It is maybe because we are so familiar with control or maybe because it has been working for us for such a long time that we are still willing to rely on it. It is maybe because the idea of a computer becoming the better informed partner that is able to command us may seem slightly insulting to our human pride (even though a look at a typical call centre will make us face this unpleasant reality).

Control alone is no longer working, but trust alone is not an alternative. We need both: separate but combined, different but inter-dependent. Above all, we need the ability and willingness to use our mind to grant trust where trust is due and to deploy control where control is necessary. Neither too little nor too much, but exactly as much as it is needed, to the best of our assessment.

Our relationship with our own technology is an interesting yet uneasy one. The Internet is being presented as a source and a vehicle of all the evil while noble institutions of the real world are supposed to uphold our morality. Strangely, it does not seem true any more. Looking at numbers what we can see is that being a responsible and decent trader on eBay results in the 8 per cent premium [Resnick2006], a margin worth considering by anyone who is into trading. However, if you happen to become an irresponsible spender and end up bankrupt, do not worry: banks (at least in late 2006) will still lend you money and will not even charge you more for it – just 0.5 per cent extra, well within the error margin of natural fluctuations of interest rates [Batchelor2006].

So, the rebellious Internet offers a strong incentive for (financially) decent behaviour while trusted institutions of the real world offer almost no disincentive for (financially) indecent one. If we consider Giddens' observation [Giddens1991] that we institutionalise our relationships in expectation that institutions will become guarantors of socially responsible (moral) behaviour, then we should think again, as it seems that it is the Internet that can teach us the lesson that our banks cannot. Maybe the Internet holds a lesson for us that we should learn? Maybe it is a technology-mediated trust that will offer the new way to build confidence?

The book talks about the confidence in mixed societies, where people and machines interact in a manner that makes it sometimes impossible to tell who (or what) is at the other end of a communication channel – not to mention that it is impossible to tell how the communication channel is constructed. Alice can be a person or a computer, Bob can be either and they may interact through whatever combination of social and technical means are available.

It is not exactly the situation of the future. We are quite rapidly getting to the position where the majority of our dealings are mediated in an electronic manner (e.g. over the Internet), with other people as well as with computer systems. Within such mixed societies our electronically-managed reputation may be more important than our real-life one (apparently with a different premium on both) and our personal web page may be designed for search bots rather than for humans (strangely, this is already a common design recommendation).

Convergence is in. The current trend in communication can be defined as convergence where different modes of communication and computation merge into the single meta-system that will eventually deliver the best service for the lowest price, any time and anywhere. Looking closer, there are several convergences to talk about, all happening at the same time [Mallett2005]: fixed and mobile, voice and data, media and network, etc.

The fixed and mobile convergence brings together the world of wires and cables with the one that has been based on airwaves. Voice and data convergence means that the communication increasingly interprets everything as a service – whether it is a voice communication, file transfer or music download. Communication converges with IT to form the computational-communicating mesh with intelligent devices that utilises the best of both, such as Internet-based TV or digital telephones. Finally, media converges with the network itself so that the content becomes a part of a communication.

This book builds on the expectation of three other convergences. They are not strictly related to communication, even though they are driven by the convergence that happens in the communication area. Those convergences are related to the growth in interdisciplinary understanding between research related to digital communication/computation and those that are related to social sciences.

The convergence of technical and social research into one interdisciplinary approach is the core of this book. It assumes that we can apply our knowledge of social sciences to the technology and that we can apply technology-based thinking to problems discussed within social sciences. In this process we can find methods and frameworks that benefit both areas.

The convergence of us and computers into one mixed society is yet another theme explored in this book. The theme has been already presented here, so that it is only worth mentioning that the main thread discussed in this book addresses the fact that we deal with other people through computer-based communication systems and we are not even sure that those that we are dealing with are human. Bob may be a man or a computer, Carol may be a friend or an autonomous agent. Even Alice does not have to be a person.

Finally, the book proposes the model of confidence that essentially assumes the convergence of two distinct concepts: trust and control. Saving detailed explanations for later, it is necessary to mention that even though trust and control work in different ways, they both are like two sides of the same coin, together forming the perception of confidence.

This book intentionally stays away from a moral judgement of trust, control and confidence. Even though it is widely accepted that trust is good (and therefore that distrust is bad), the book offers no opinion regarding those beliefs. We do not discuss whether trust is a virtue or a vice, whether it is good or bad to control, whether trustworthiness should be rewarded or punished.

The approach here is Alice's. The book discusses how she can properly assess whether she should be confident in Bob. The thing that we care about is her ability to make the correct decision – not her trust, not her ability to control Bob, not Bob's trustworthiness and not the happiness or welfare of the whole society. Further, no judgement is offered whether Alice should choose trust or control to achieve her confidence in Bob. It is again left up to Alice to decide. We cannot help her with it, but we can make sure that she is in possession of all the available information possible to make such a decision. We trust that she will trust her wisdom.

The book is structured into three parts. In Part I we explore how Alice is able to manage the unbearable complexity of life through trust and control. We will start by setting the scene and exploring the interleaving concepts of trust, control, confidence and complexity. From there we will introduce the model of confidence that will be used throughout the book. The comparative analysis of the model relates it to other models that can be found in the literature. The first part closes with the analysis of the darker side of the relationship – lack of confidence and distrust.

Throughout Part II Alice will face the challenges and perils of the digital communication. We will start from a general analysis of how digital media impairs Alice's ability to establish trust (but not necessary control) and how the omnipresent technology shifts our perception of the need for trust. From there we will see how Alice copes with her relationship in different digitally-enhanced communication settings. We close by postulating and then exploring the concept of trust-enhancing technologies: the family of digital technologies that should increase the ability to establish confidence through trust.

In the final part, we will look at the correlation between confidence in a world of convergent communication and e-commerce and Alice will become our model user and customer. We will start from general observations about the impact of convergent digital communication on confidence in e-commerce. From there we will look at interesting cases of e-businesses that build on confidence, to see how they manage to construct the complete confidence eco-system. We will then move on to address problems of authentication and privacy and to find a new role for trust management. Finally, we will re-enter the relationship between confidence and economy to re-define the relationship between confidence and security.

The majority of chapters presented in this book originated as research papers. However, as they are now liberated from formal restrictions, they have been expanded and enriched to provide more complete discussion on the subject, for the added benefit of the reader.

## Bibliography

[Batchelor2006] Charles Batchelor: Shaking off the Debt Stigma. *Financial Times*, July 8/9 2006.

[Faulkner2003] Paul Faulkner: The Epistemic Role of Trust, In R. Falcone et al. (eds) *AAMAS 2002 Workshop on Trust, Reputation,* and Security: Theories and Practice. LNAI 2631, pp. 30–38. 2003.

[Frankema2005] Katinka Bijlsma-Frankema and Ana Cristina Costa: Understanding the Trust-Control Nexus, *Int. Sociology*. 20(30): 259–282. 2005.

[Fukuyama1996] Francis Fukuyama: *Trust: The Social Virtues and the Creation of Prosperity*. New York: Touchstone Books. 1996.

[Giddens1991] Anthony Giddens: *Modernity and Self-identity: Self and Society in the Late Modern Age*. Cambridge: Polity Press. 1991.

[Habermas1986] Jürgen Habermas: *The Theory of Communicative Action*. Vol. 1. *Reason and the Rationalization of Society*. Cambridge: Polity Press. 1986.

[Lewicki1996] Roy J. Lewicki and Barnard Benedict Bunker: Developing and Maintaining Trust in Work Relationships. In Roderick M. Kramer and Tom R. Tyler: *Trust in Organizations*. London: Sage Publications. 1996.

[Mallett2005] Colin Mallett, Richard Dennis and K. Jit Singh: 2005: An Interesting Year for Convergence? *Hong Kong Mobility Roundtable*, 2–3. June. 2005.

[McKnight1996] D. Harrison McKnight and Norman L. Chervany: The Meanings of Trust. In University of Minnesota. http://www.misrc.umn.edu/wpaper/wp96-04.htm. 1996.

[O'Neill2002] Onora O'Neill: *A Question of Trust: The BBC Reith Lectures*. Cambridge: Cambridge University Press. 2002.

[Pennings1987] J. M. Pennings and J. Woiceshyn: A Typology of Organisational Control and its Metaphors. In *Research in the Sociology of Organizations*, 5: 73–104. 1987.

[Resnick2006] Paul Resnick: The Value of Reputation on eBay: A Controlled Experiment. *Experimental Economics*, 9(2): 79–101. Available at: http://www.si.umich.edu/~presnick/papers/postcards/PostcardsFinalPrePub.pdf. 2006.

[Schneier1996] Bruce Schneier: *Applied Cryptography: Protocols, Algorithms and Source Code*. Chichester: John Wiley & Sons, Ltd. 1996.

[Tyler1996a] Tom R. Tyler and Roderick M. Kramer: Whither Trust? In Roderick M. Kramer and Tom R. Tyler: *Trust in Organizations* London: Sage Publications. 1996.

[Tyler1996b] Tom R. Tyler and Peter Degoey: Trust in Organizational Authorities: The Influence of Motive Attributions on Willingness to Accept Decisions. In Roderick M. Kramer and Tom R. Tyler: *Trust in Organizations*. London: Sage Publications. 1996.

# Part I

## Confidence

### Trust and Control

# 1

# Setting the Scene

What is trust?
I know.
Tell me.
I knew . . .

(overheard)

## 1.1 Basic Concepts

Let us set the scene. There is Alice and Bob. We do not know what they think (unless they
decide to tell us, but then – should we trust them?), but we can observe how they behave.

We know that Alice would like the future to be good to her. This may include the successful
completion of what she is doing right now, her long-term plans or possibly the lack of
damaging events in the near future. If we place Alice at the pedestrian crossing, she may
simply want to cross the street safely – maybe not a particularly great achievement, but
sometimes one that may be hard to achieve.

When Alice leaves the pavement and starts crossing the street, she become vulnerable.
Every driver can easily hurt her, potentially in a way that will permanently and negatively
impair her life – with little physical consequences to the driver or to his car. She is aware of
this, and – seeing that the car driven by Bob is approaching – she must decide whether she
is confident enough that the driver will slow down and let her walk across.

This is the image we start with: Bob behind the wheel, Alice at the pavement, considering
whether she should cross the street or not. For that moment, the relationship between Alice
and Bob emerges. In a second it will be gone, but currently Bob becomes an inseparable part
of Alice's future. His actions (or inactions) may significantly impact on her life.

This book is about moments like this, moments that we experience throughout all our life:
vulnerable to decisions of others yet dependent on them. Banks considering mortgage application, managers deciding about promotion, government regulating taxes, doctor suggesting
a treatment . . . the list is very long. Should we be confident that their decisions will benefit
our future?

*Trust, Complexity and Control: Confidence in a Convergent World*   Piotr Cofta
© 2007 John Wiley & Sons, Ltd

This book concentrates on what can be called Alice's ***transactional assessment*** of confidence: the process that may eventually lead her to rely on Bob, in anticipation that Bob will support her goals. Such assessment is clearly transactional: Alice may be confident about Bob at certain moments in time (e.g. not on Friday night) and may do it only regarding certain subjects (e.g. not when it comes to classical philosophy). Her assessment of confidence within the scope of one transaction does not determine her assessment within other transactions. Her assessment is not necessary binary, as she may engage into the transaction at different levels.

Another interesting aspect of this relationship is Alice's willingness and ability to improve Bob, her long-term ***relational decision*** about confidence. We will see later that Bob may behave in a desired manner because this is in his nature or because he has been forced to behave. Further, his behaviour may be altered by the fact that Alice is confident about him. From the long-term perspective, Alice may be sometimes better off if she spends some effort on strengthening Bob's good nature, rather than using enforcement tools on him. In the future, it will save her hassle (and the cost) of dealing with the battery of instruments designed to coerce Bob.

As Alice and Bob are increasingly interacting through and with a help of digital media, another concept becomes important. ***Trust management*** can be understood as technical replication, processing and enforcement of confidence that already exists in the social world. For example, if Alice is confident about Bob, she may be willing to give him a token that can be verified later by an infrastructure managed by Alice. Here the token becomes a technical representation of a confidence that existed at certain moment between Alice and Bob. The notion of trust management will be particularly important if we assume that neither Alice not Bob must necessarily be humans.

## 1.2 Research Overview

Trust, control and confidence are the key elements of the book, with trust being the most fundamental one (and the hardest to define). Even though the phenomenon of confidence is interdisciplinary, research in confidence (trust and control) tends to follow the recognised structure of disciplines. The following is a short and non-exhaustive overview of different works related to trust and confidence, structured into two main groups. The first covers the widely understood social sciences (including philosophy, psychology, economy and management) while the second covers the technical approach to confidence, mostly within the context of computer science. Each group presents a varied set of approaches to confidence. The selection has been made on the basis of their importance for the discussion presented here and is not intended to provide complete coverage of the field.

### 1.2.1 Social Sciences

The ethics of trust and control has been a fundamental question for a long time. However, there is a general disagreement whether trust is morally right or wrong, following the disagreement about the foundations of human nature. Proponents of trust applaud it as a social virtue and point that without exercising trust, trust will never be reciprocated (e.g. Aristotle, Hume, Kant). Usually, they suggest reforming society by exercising individual virtues in order to maximise trust and minimise control. Opponents tend to think of trust as a vice and demonstrate the irrationality of trusting in light of visible high levels of betrayal (e.g. Plato, Machiavelli, Hobbes). This usually goes together with a call for increased control and a strong state as they see the certain opportunity for everyday trust being guaranteed by strong institutions. Whatever the proposition, the importance of trust (and to a certain extent the importance of control) to human life is clearly obvious.

The importance of trust and control as a complexity reductor [Luhmann1979] created the opportunity to discuss trust and control not from a moral, but from a rational position, treating them both as forms of social enablers. Specifically trust, through its low internal complexity and significant power to reduce complexity, is able to overcome our limitations of bounded rationality [Simon1957] and reduce the otherwise unbearable complexity of all foreseeable futures. Therefore, trust can be discussed as a psychological and social necessity that can be rationally analysed, weighted and explained. The fact that trust is present in all forms of communication [Habermas1986] reinforces the perception of trust as a fundamental social enabler.

Trust developed in early childhood is believed to be the foundation of a general disposition to trust, but even more importantly, such trust supports our notion of normality and lowers our existential anxiety [Giddens1988]. Interestingly, this rudimental form of trust seems to develop in parallel with the understanding of self-identity, thus creating at the same time the notion of self versus others and the notion of trust in specific others (usually caregivers), within the basic understanding of limitations of time and space.

The observation that there is a link between the level of social trust, welfare and development (e.g. [Fukuyama1996], but see also [Dasgupta2000] and [Luhmann1979]) has created significant interest in considering trust as a form of social capital and has led research into a closer understanding of the process of creation and distribution of such capital. Even though the original concept of 'high trust' and 'low trust' societies may not necessarily hold, it has been widely accepted that trust (or, to be more specific, trustworthiness that leads to a vested trust) decreases transactional cost, thus leading to greater efficiency in economical relationships. Even though it is not clear whether it is the network effect that facilitates trust or trust that allows the network of social relationships to develop [Hardin2002], the social benefit of both is clearly obvious.

Confidence can be interpreted as an unidirectional relationship (Alice is confident about Bob), but it can be also the two-directional one, where the mutual relationship between Alice and Bob evolves over time. Deutsch [Deutsch1973] differentiates between interpersonal trust (where one party is not aware that the other party trusts it) and mutual trust (where both parties are involved in the mutual relationship of trust). As the concept of reciprocity (that is available mostly in relationships) may significantly alter the dynamics of confidence building, the process of relationship was widely studied. The dynamics of relationship [Lewicki1996] suggests that the relationship develops through a three-stage process, starting with the control-driven stage and hopefully ending with the trust-based one.

Organisations, the scope of management sciences, have usually been associated with hierarchical control, driven by specialisation and efficiency. More recently, however, there has been a visible trend to discuss trust as an important element of organisation (e.g. [Solomon2001]), both internally (e.g. as a facilitator of innovations [Afuah2003] and as a supporter of organisational changes in times of internal conflicts [Webb1996]) and externally (e.g. in the virtual organisation [Handy1995]). The growth of outsourcing has raised questions of trust between cooperating parties, specifically in the context of different cultures (e.g. [Sako1992]). Similarly, the need for increased agility has created interest in the rapid creation of trust in temporary groups [Meyerson1996]. There is an underlying assumption that trust and confidence are attributable to organisations in a same way as to individuals, following the concept of intentional stance [Dennett1989].

Game theories use the concept of trust to explain phenomena that counter the instant economic rationality of utility maximisation. The ability of economic players to go beyond obvious self-interest, (potentially in expectation that the other party will reciprocate) became the basis of several economic games. The game of trust [Berg1995], allows trust and trustworthiness to be expressed in monetary terms, thus becoming a model solution in situations where trust should be measured. More recently, the game of distrust [Bohnet2005] has been used to explain some significant recent management disasters. Similarly, studies of the Prisoner's Dilemma (formalised by W. Tucker in [Poundstone1992]) are used to link trust with economic utility and demonstrate the rationality behind reciprocity.

## 1.2.2 Technology

Within the realm of technology, trust and control have usually been associated with reliability (e.g. [Powell2003]) and were not seen as a separate issue until the arrival of complex computer-controlled systems. Computer science had initially approached trust and control from the perspective of security. Recognising that trust is not controllable, the security developed an elaborate structure of control (e.g. [Bishop2005]), in an attempt to minimise elements of trust (e.g. [Anderson2001]). However, more recently, the recognition of the fundamental nature of trust has been addressed in initiatives such as trusted computing [Pearson2002], where individual devices are given assurance in their own configuration on the basis of a highly-protected, hardware-based root of trust. The need for a portable root of trust has also fuelled the creation and popularity of smart cards [Rankl2003].

In data communication, the understanding that trust precedes meaningful (and secure) communication has eventually led to the concept of trust management, the separate layer of interactions that lead to the creation (and maintenance) of trust relationships between communicating nodes, following e.g. business agreements, contractual dependence, personal relationship, etc. PGP [Zimmermann1994] has been exploring the area of peer-to-peer trust while PKI ([Adams2002] or [Perlman1999] for alternative models of trust enabled by PKI) proposed the multi-stage model of trust. More recently, WS-Trust [Anderson2005] has established itself as a standard within service-oriented architecture (SOA), the potential foundation of Web 2.0 while [Ishaya2004] offers trust management for virtual organisations. Grid computing (e.g. [Dragovic2003]) and pervasive computing environment (e.g. [LoPresti2005]) have brought different challenges to trust management.

The need to effectively manage distributed computing systems has led to constructs such as trusted domains (several computers trusting each other's authentication capabilities), trusted credentials (others' identities accepted without any further proof), trusted storage (storage space accessible only to selected users), trusted zones (privileged Internet address space) etc. In all these cases there is a notion of trust as essential yet different from actual cooperation (or communication), something that requires special management practices. Usually, the ability to manage trust is granted to system administrators or users, in the expectation that the technical structure of trust will reflect trust in respective social relationships.

Research on autonomous agents (e.g. [Falcone2006]) has liberated trust management from the need for an a priori trust, managed by the user or the administrator. Agents were vested with the ability to make and break the trust relationship (that can be more correctly called 'the relationship of confidence'), usually on the basis of past experience, through the process of learning, whether from direct interactions or from others' experience. Autonomous agents have brought the notion of imperfect trust (where trust is no longer a binary proposition), the problem of trust propagation [Josang2006] and reasoning. The new approach to trust has also – unfortunately – revealed new threats to trust, usually in the form of attacks on reputation [Dellarocas2004].

Interest in large systems (whether created by autonomous agents, ad-hoc networks or in any other way) required more specific instruments to discuss the reasoning about trust. To name a few, Josang [Josang2001] proposed the algebra of uncertain probabilities, introducing the notion of uncertainty to the probabilistic distribution of outcome. Formalisation of trust (e.g. [Demolombe2004], [Marx2001], [Huang2006], [Herrmann2006]) proposes logical primitives and schemes that can be used in reasoning about trust. The formalisation of reasoning has led to the creation of several formal systems and supporting tools. Marsh's [Marsh1994a] formal model of trust brings the concept of trust closer to the domain of computation while Grandisons's Sultan [Grandison2003] allows the capture, simulation and reasoning about trust-based relationships.

Both reasoning and transitivity require trust (confidence) to be qualified. The desire to measure trust (and confidence) generated significant amount of research. There are several works that link trust with probability either directly (where trust itself is perceived as probability – e.g. subjective probabilities [Gambetta2000]), or through models related to economics, e.g. Barber's [Barber1983] model of probability or Hardin's [Hardin2002] Bayesian model associated with economic payoff. Almost every model of trust introduced a different range of values that can be assigned to trust (see e.g. [Abdul-Rahman2005] for a review), sometimes with conflicting semantics.

From a more application-specific perspective, electronic commerce has used various metrics of trust to develop risk assessment, both for the seller and for the buyer. This has become an important focal point of several works [Rutter2001], [Kracher2005]. The commercial value of eBay's reputation system [Resnick2006] is widely known, and similar rating systems [Dellarocas2004] are used by other e-commerce sites. Collaborative filtering [O'Donovan2005] has been used to aid information search (following the concept that trust is a qualified reliance on information [Gerck2002]), but as more automated systems moved into the area [Page1998], collaborative filtering became the preferred solution for the recommendation. The needs of electronic commerce have stimulated the interdisciplinary approach to trust [McKnight2001], [Tan2000].

Another effect of the introduction of electronically-mediated communication is the development of research in user trust in digital devices, e.g. in a form of web features that facilitate the creation of perceived trust [Egger2000], trust in information systems [Li2004] or in improvements of trust between people while communicating through a digital channel.

## 1.3 Terminology

The hard part of talking about trust, confidence or control is to getting the terminology straight. Trust has already 17 different meanings [McKnight1996] so it is not possible to casually introduce the concept and expect that everybody will end up with the same understanding. Even worse: every definition will potentially be both accepted and contested. Attempts to deliver the structured yet rich ontology of trust (e.g. [Quinn2006]) may come in handy.

Definitions of trust (here these should properly be referred to as confidence) usually highlight such elements as Alice's expectations regarding Bob and Bob's supportive action, adding sometimes aspects of Alice's vulnerability and Bob's independence. Those definitions tend to describe different aspects of confidence and tend to fall into defining one or more of the following categories: propensity, beliefs, behaviour, decisions, relationship, etc. The set of definitions offered here structures the problem from the perspective of a book where we are interested in Alice being able to make the proper assessment of her confidence in Bob.

We will start from the initial scene again: Alice and Bob, each person thinking something and is behaving in some way (see Figure 1.1.).

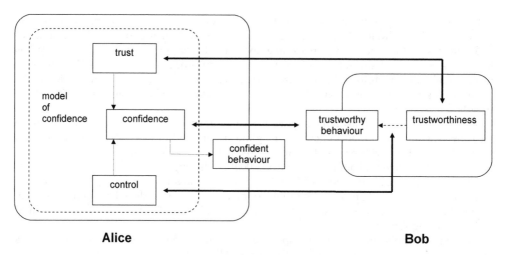

**Figure 1.1** Basic concepts

### 1.3.1 Alice

We are interested in Alice's assessment regarding her confidence in Bob, as this will presumably drive her behaviour. We are, however, not only interested in her decision alone: we would like to understand how Alice reached her decision – we would like to create a mental model of Alice, thinking process in an expectation that such a model will help us in understanding others (and will be potentially implemented in some technical devices).

We will define confidence as the belief that the other party (Bob) will behave in a way that is beneficial and desired for the first party (Alice), e.g. supporting Alice's goals, helping her or refraining from actions that may harm her. Confidence is Alice's internal state of mind, an expectation, and is not directly visible to the external observer. Only Alice can tell us how much confidence she has in Bob. In order to establish a link with measurable statistics, we will define confidence in terms of probability, but as confidence is bound to Alice's belief, such probability will be subjective – we are concerned here strictly with what Alice thinks.

> *Confidence is one's subjective probability of expectation that a certain desired event will happen (or that the undesired one will not happen), if the course of action is believed to depend on another agent.*

Confidence, apart from being a belief, can also be interpreted as a uni-directional relationship that is controlled by Alice. Alice can initiate, maintain, change the level of intensity or terminate this relationship through her own cognitive process (even though we expect that Alice will take into account certain reasonable evidence while doing it). Also in this sense confidence is subjective, as it is only Alice's cognition that makes her confident in Bob. We will see later that Alice can even misinterpret her knowledge only to continue such a relationship, no matter what Bob is doing.

We will use the following notation to describe confidence as a relationship (1.1):

$$conf_{Alice \to Bob}^{subject, time}$$                                    (1.1)

where:

| | | |
|---|---|---|
| *Alice* | – | the agent (person) that is confident |
| *Bob* | – | the agent (person) that Alice is confident about |
| *subject* | – | the frame of discourse; action or state that Alice is concerned about |
| *time* | – | temporal limits of discourse |

If this is not confusing, we can leave the subject and time blank, which should be interpreted as a current subject and current time (present in current considerations).

Alice's confidence is a belief, her cognitive state that emerged from certain mental process. We cannot learn about her confidence in other way than through her introspection, but we can reasonably guess how she thinks by inspecting our (and others) mental process. What we end up with is the *model of confidence* that attempts to explain how Alice (like all of us) thinks about confidence.

Her internal confidence should drive her *confident behaviour*, the externally and socially recognisable set of actions that clearly signals that she is confident about Bob. For example, she may without hesitation cross the street, thus demonstrating that she is confident that Bob will stop his car when needed. She may also cross the street for several other reasons, from being careless to being distressed. Therefore we must be careful to differentiate between confidence (as a belief) and confident behaviour. Alice may be confident about Bob but this may not be visible in her behaviour. For example, when she request a verification of certain work that Bob has done, she may be confident about the outcome, but she may be obliged to conduct such verification.

Further, she may express confident behaviour even though she is not confident about Bob. Deutsch [Deutsch1973] provides a long list of motivations that may cause Alice to behave in a confident manner, even though her beliefs may contravene our expectations regarding confidence. Alice may look confident because of despair (choosing it as the lesser of two evils); out of social conformity (if she does not look confident then she may fear social ostracism); due to her innocence (not knowing enough about a cruel world); she may follow her impulsiveness and emotions; she may consider confidence a social virtue that should be demonstrated; she may be masochistic – so that she defies our perception of her utility or finally she may like risk and gambling.

*Trust* is a part of Alice's belief that leads to confidence in Bob. Another part of the same process is *control*. We will discuss how they interact later throughout the book. Neither trust nor control is directly visible through Alice's actions, but they are both embedded in her confidence that can potentially be observed through her behaviour. Even though they are never directly visible, we will see that there are interactions where Alice's behaviour is mostly motivated by trust and interactions where trust and control go hand in hand – or where control prevails. We will not yet introduce definitions of trust or control as they can only be defined in relation to each other and to other elements of the model.

## 1.3.2 Bob

Let's look at Bob. He may be inclined to support Alice in her undertaking (e.g. by stopping the car) because of his intentions – his nature, so to speak. His internal intentions regarding Alice within the scope of this transaction will be called *trustworthiness*, in a slightly confusing manner (because trustworthiness, similarly to trust is overloaded with meanings).

Trustworthiness is a cognitive state that Bob can manage, with or without Alice. Bob can decide to do something or not, whether he is disposed to help, Alice in particular or whether he is ready to help everybody through certain actions or finally whether he would like to do it now but not later.

Again, what Bob does is not always what he thinks. In particular, Alice would like him to express *trustworthy behaviour* – simply to help her by providing what she wants (stopping the car, etc.). From our perspective this is what we can observe (as we do not have access to Bob's intentions). Bob's behaviour may not always be the result of his trustworthiness. He may exercise certain scenarios that may make Alice confident in him even though he does not intend to support her – the common salesman ploy. He may also suppress his trustworthy behaviour despite his trustworthiness, e.g. to scare Alice away. The most important thing for us is that Bob can be driven to expose trustworthy behaviour (which is beneficial for Alice) even if his trustworthiness is not up to it.

There is a potential gap between Bob's intentions (his trustworthiness) and his actions. The explanation for such a gap lies in the control element – Bob may not be willing to support Alice, but he can somehow be forced to do it. For example, Bob may stop and let Alice go not because he is of a particularly good spirit but because he is afraid of possible consequences of not stopping, e.g. in a form of a fine.

## 1.3.3 The Relationship

Let's now consider the complete picture of Alice and Bob (Figure 1.1) again. We can see that Alice can maintain certain levels of trust and control, collectively leading to her confidence. How does it match Bob?

We know that trust can have several interpretations. Specifically [Baier1986], we should distinguish between trust as behaviour (Alice is behaving in a way that signals her trust in Bob) and trust as belief (Alice believes that Bob is worth her trust, i.e. he is trustworthy). Here, trust is used in the latter meaning.

Bob keeps his internal state of trustworthiness, his intentions regarding this particular interaction. Whatever the reason, Bob's good intentions are potentially (but only potentially) visible through his trustworthy behaviour. As has been already stated, Bob can also be forced to behave in a trustworthy manner regardless of whether he is trustworthy or not. Alice can rely on one or more of her instruments of control to enforce desired behaviour. In this case Bob's trustworthy behaviour is an effect of both his trustworthiness and Alice's control. Note that Bob's behaviour relates to the overall confidence, not only to trustworthiness and it should properly be called 'confidence-signalling behaviour'. However, following the prevailing terminology that can be found in literature, we will stay with the less exact but more familiar term of trustworthy behaviour.

We are interested in Alice's transactional assessment about confidence and her decision to proceed (or not) depending on whether she is confident enough about Bob. The key element therefore is her ability to estimate Bob's intentions. If she over-estimates him, she is risking failure, but if she under-estimates him, she is missing an opportunity. What she needs is a correct estimate.

While looking at both the model and the relationship, we can see that Alice's estimate of trust in Bob should match his trustworthiness. We have provisionally defined trust as Alice's expectation that Bob will support her in the absence of any control element – in line with the way we defined Bob's trustworthiness. Alice's estimate of control over Bob should therefore match the difference between his trustworthiness and his trusting behaviour, i.e. it should explain Bob's behaviour that is not driven by his intentions. Consequently, Alice's confidence in Bob should become an estimate of his trustworthy behaviour.

Assuming that we can consistently assign certain numeric values to all elements of the model (preferably with identical semantics), we can express the relationship as follows (1.2.):

$$conf_{Alice \to Bob} = est_{Alice}(twb_{Bob})$$
$$trust_{Alice \to Bob} = est_{Alice}(tw_{Bob}) \quad\quad\quad (1.2)$$
$$ctrl_{Alice \to Bob} = est_{Alice}(twb_{Bob}) - est_{Alice}(tw_{Bob})$$

where:

$est_{Alice}$ – the function to produce Alice's estimate of its argument
$tw$ – trustworthiness
$twb$ – trustworthy behaviour

Note that we did not introduce the concept of 'controllability' similar to trustworthiness, i.e. we do not assume that Bob has a certain predilection (or resistance) to be controlled. We can deduct the existence of control only as a difference between trustworthy behaviour and trustworthiness.

Defining all elements of those formulas in terms of subjective probability, we have:

• Trust is an estimate of a subjective probability of certain desired course of actions in the absence of control.
• Control is the difference between the estimate of the subjective probability of trustworthy behaviour and trust.

Thus, trustworthiness is Bob's internal intention while Alice does not exercise any of her control instruments. It is 'Bob on his own', without any pressure from the outside. It is Bob's internal disposition to support Alice. By contrast, Bob's response to control is visible only if Alice exercises one of her instruments of control. Such a response is a result of Alice's action so that it does not exist as Bob's mental state before Alice uses her instruments. The amount of control does not depend on Bob so that is not part of his description. Alice, if not bound by complexity, can exercise any amount of control and Bob will eventually behave in the expected manner.

## 1.4 Agents Alice and Bob

Alice and Bob cannot just be anyone if we are to discuss the confidence between them. We do not expect a book to trust the table or a cup to control tea (even though, to a certain extent, those examples do not seem to be entirely irrational). Discussion about trust and control is restricted to entities that can hold beliefs and opinions and to entities that may or may not act the way they are designed: to cognitive agents and intentional agents.

This restriction does not exclude inanimate objects from the discussion. Even though a book cannot trust the table, it may be convenient for us to perceive them both as agents of assumed quality and then extend the notion of trust to them. Concepts such as trust, confidence or control are often applied to activities of non-human, inanimate or social objects. We can trust our car, be confident in the postal system or control the computer. The anthropomorphic intentional stance [Dennett1989] may come handy if entities are too complex to be fully understood. This is particularly true when we start dealing with complex digital systems that tend to behave in an unpredictable ways.

### 1.4.1 Cognitive Agent Alice

The belief of confidence requires at least the ability to have beliefs – internal states of mind. If we want to talk about Alice being confident about someone, then we must attribute to Alice the ability to hold beliefs. Certainly we can have beliefs and we can infer (through assumed psychological similarity) that other people can have beliefs. So, the concept of people being confident about other people seems to be justified. We can extend this concept to cognitive agents [Castelfranchi2000] to cover not only humans but also other entities that we attribute with the ability to have beliefs.

Moving away from natural cognitive agents (people) and simple systems (such as cars or elevators) towards more complex systems (such as software agents or 'the Internet') we enter a realm where entities seem to be able to reason, yet we are aware that such reasoning is not 'natural'. It might have been programmed into them or it might have been the outcome of a collective mindset. The borderline between what constitutes cognitive agent and what is not cannot be clearly outlined: it is our perception whether the cognitive explanation seems to be plausible and beneficial.

Note that if we decide to consider artificial Alice (e.g. a software agent that is supposed to mimic Alice's decisions), we are gaining the opportunity of a direct inspection of Alice's 'mental states' – something that we cannot do with a human. However, this will help us less than we expect – if we want to re-create Alice's reasoning in bits and bytes, we should first understand how the real Alice thinks. If we decide to call certain internal variables 'trust', 'confidence' or 'control', then we simply assign attributes that are convenient for us to states that are available to us, with no guarantee that they have any relevance to what humans are doing.

### 1.4.2 Intentional Agent Bob

Bob does not have to hold beliefs – unless we would like to consider the mutual relationship between Alice and Bob, beyond the uni-directional Alice's assessment of her confidence in Bob. However, Bob must be able to have intentions and express certain free will, i.e. Bob should be able to determine his goals and act upon those goals. If Bob is not attributed free will, then his actions cannot be explained by his intentions, he cannot be trustworthy and Alice has nothing to estimate – her estimates of his trustworthiness are void.

If Bob does not have intentions, then his behaviour should be explainable entirely through control, mostly by Alice's knowledge about Bob. If she knows him, then she can always predict his actions. This deterministic explanation of Bob's behaviour applies well if Bob is a relatively simple entity, but fails if Bob is a complex, even though a deterministic one. What is complex and what is simple depend only on Alice's perception of Bob.

Let's consider an example – Bob is a simple dice-tossing software program, that draws numbers from its pseudo-random number generator. From the perceptive of software developer, Bob's behaviour is completely deterministic, and (assuming a known starting point), the sequence of numbers is completely predictable. From Alice's perspective, however, Bob's behaviour is a mystery, as the generator provides a sequence that seems to be random. As Alice is unable to determine any visible rules in Bob's behaviour, she may even over-interpret Bob and attribute intentions, free will or even with a supernatural power [Levy2006] to him.

Let's consider another example where fully predictable Bob-computer suddenly significantly slows down. Certainly, Alice may attribute this behaviour to some technical problem, possibly incorrect settings, overheating or maybe some form of security breach. However, this requires from Alice expert knowledge. It may be easier for her to endow the computer with intentions and develop explanations that make use of such intentions (e.g. the computer does not like Mondays). Technical fault is interpreted as a sign of free will.

In both cases Alice interpreted Bob incorrectly (from the perspective of an expert), but in both cases Alice came up with an explanation that better suited her needs. Her intentional stance [Dennett1989] allows her to deal with Bob on known terms that are derived from human relationships. Surprisingly, Alice's strategy is rational: by altering her perception of Bob and by introducing intentions, she is able to decrease the perceived complexity of Bob so that she can reason about him. Without an intentional stance, the complexity of Bob would overwhelm her cognitive capabilities.

## 1.5 Confidence as a Strategy

There are potentially several reasons why Alice may want to be confident about Bob. We can distinguish two main lines of strategies that will be apparent throughout the book.

First, Alice can treat confidence as a *limiting factor* in her dealings with Bob, i.e. she may decide that she will not proceed unless she can attain sufficient confidence. In this case, Alice sets the required threshold value upfront and tests whether she is able to reach it. This is a luxury approach, as it assumes that Alice is able to decide whether to proceed with Bob or not – so that her action is not essential to her. Say, Alice may decide not to buy a new pair of shoes if a shop assistant does not inspire sufficient confidence – but she already has another pair.

Second, Alice may use her confidence as a *selecting factor*, so that she is willing to proceed with the one that inspires the highest confidence, regardless of the actual level. This can be somehow defined as a strategy of necessity, as in this case Alice does not know what the minimum level of confidence is (as she does not define any), but she uses confidence in a comparative way among contenders. Apparently, in order to employ this strategy, Alice should have a choice so that she can choose between e.g. Bob and Dave. Say, Alice is hungry and would like to choose a restaurant that she is most confident about. Apparently, she will choose a certain restaurant (she cannot afford to wait), it is only a question which one.

What if Alice must proceed but there is only Bob to choose from? Apparently, in this case she does not have to be bothered with any evaluation of confidence as she is not in a position to make a decision anyway. For her own peace of mind she may pretend that she is confident about Bob and for the benefit of her experience she may evaluate her confidence – but neither has any practical bearings. This is one of those cases where her confident behaviour is not driven by her confidence [Deutsch1965].

Bob can support Alice (by expressing trustworthy behaviour) for several reasons, possibly too numerous to be discussed here. He may feel an ethical obligation or he may be driven by greed. He may understand that his interest is aligned with Alice's [Hardin2002] or he may be afraid of a retaliation in a case of default. If Alice is interested in a one-time relationship, either will work for her. If Alice prefers a long-term relationship, she may be willing to explore Bob's motivation to see whether she may increase his trustworthiness.

## 1.6 No Confidence

There is an interesting question whether Alice can live without confidence (specifically without certain level of trust). It seems (e.g. [Kipnis1996]) that such a life is potentially possible but likely not desired. To live without confidence means to give up expectations regarding others or to rely entirely on faith rather than on confidence (so that no evidences are sought and no relevance to the real world is expected). The life without confidence may eventually lead to paranoid states where a person is unable to develop any confidence in an external world and thus either isolates herself in a controllable world or accepts the randomness of her future with no hope of establishing any expectations.

Looking at digital systems, we can see again that there is little difference: if the system is not confident about information that flows from another system, then the only solution is to ignore it. It may work well for isolated systems, but this is hardly a proposition for a converged, networked world.

It is possible to strike a balance between trust and control within the scope of confidence. Alice can offset her lack of trust with increased control over Bob or she may decide to trust Bob and relax her control. She may also try to reduce her dependence on Bob, shift it towards more trusted subjects (e.g. she may prefer to be dependent on Carol, but not on Bob), etc. In other words, even though life without any need for confidence seems to be an unlikely proposition, Alice has a certain flexibility how, whom, and to what extent she would like to trust and control.

Specifically, we can see the tendency to talk about trust but implement control – in the form of surveillance systems, access codes, complex authentication rules, audit trials, etc. This is discussed throughout the book, but again there is little difference between our social behaviour and the technology – in both cases we can see the desire to reduce the dependence of trust (so that only the minimalist 'root trust' is allowed) and build on control.

## 1.7 Pragmatics of Confidence

This book intentionally stays away from an ethical judgement of trust, control and confidence. It is widely accepted that trust is morally good, superior and desired (and therefore that distrust is bad, inferior and unwanted (e.g. [Kipnis1996]). Even though there are more diverse opinions about moral values of control, the book offers no judgement regarding those beliefs and there will be no further discussion about them (with the exception of distrust). The interest of several researchers seems to be in improving society, mostly by maximising average level of trustworthiness, with a certain disregard to the needs of the individual.

The book does not aim to show how to achieve the all-trusting society, or even the all-trusting individual. Instead, the pragmatic and personal approach is taken. We are interested here only in Alice's welfare that we understand as her ability to make a right decision about her confidence in Bob. What we maximise is not a general trustworthiness but Alice's ability to properly assess her confidence in Bob. The better she is able to judge Bob, the better it is for her. Whether she ends up being confident about him or not is secondary. It is also secondary whether she may end up trusting him or controlling him.

The way the problem is treated in this book borrows heavily from engineering. Confidence is viewed here as an optimisation of a decision-making process while facing bounded rationality. The main question is how Alice, having limited resources, can achieve the level of a rationally justified confidence in Bob, where her trust matches his trustworthiness and her confidence matches his trustworthy behaviour.

Neither trust nor control are considered particularly beneficial, morally superior or desired for an individual or for society. The overall benefit is seen in the ability of every Alice to properly assess every Bob, potentially through a proper balance of trust and control. There is an obvious and immediate benefit for Alice, but Bob and the whole society benefit from it as well. Strange as it is, unbounded trust and trustworthiness can possibly be detrimental to Alice, Bob and to all of us. We will explore this claim below.

Let's first assume that Alice is selfish and is driven only by her self-interest. She actually has two interests. One is to achieve what she desires, another is to make it at the minimum expense (so that she may e.g. pursue other goals as well). This calls for the low complexity of the desired relationship with Bob – something that can be delivered by trust rather than control.

Apparently, Alice benefits significantly from being able to determine whether she can trust Bob. She can achieve her goals easily if Bob is very trustworthy – he will help her just because this is the way he is and she will not have to spend her resources on controlling him. Due to the directional nature of the relationship, we do not assume anything more from Alice. She does not have to be trustworthy or reciprocate Bob's trustworthiness. She therefore has a strong incentive to become a free-rider, the person that benefits from other's trustworthiness without contributing anything.

Alice may not be selfish and may want to go beyond the one-time contact with Bob to invest in the long-time relationship. She can nurture Bob's trustworthiness with the expectation that in future she will benefit from it. For Alice, this is an investment proposition: she invests her time, effort (and possibly some other resources) expecting payback in the future. Even though it looks better, the endgame is the same: trustworthy Bob and Alice exploiting him.

Surprisingly, then, Bob's high trustworthiness is not beneficial for society, as it makes Alice live off Bob, against his will, thus decreasing his ability to contribute to the society. Alice may gradually become interested (and skilled) in exploiting others without contributing anything, thus becoming a ballast to the society. It is neither beneficial for Bob (he is actually exploited) nor for Alice (whose standards may deteriorate). So, Bob's unbound trustworthiness, if met by selfish Alice, leads to damage done to both (as well as to society).

Let's now consider situations where Alice is unable or unwilling to properly assess her confidence in Bob – we have decided to impair what we claim to be her core competence. As long as she is incidentally matching his trustworthiness and his trustworthy behaviour, no harm is done. However, if Alice over-estimates her confidence in Bob, she may easily end up pursuing the risky choice, while if she under-estimates Bob, she may end up not pursuing the potential one. Either case does not seem particularly interesting, so that again it seems that for Alice the best option is to be confident enough – not too much and not too little.

Society does not benefit from them either. Even though certain elements of optimism [Marsh1994b] are beneficial, undeserved optimism leads to the reverse exploitation – Alice being abused by Bob. Undeserved pessimism, in turn, leads quite easily to paranoia and states of social exclusion where Alice intentionally limits her contact with other members of society – not a great benefit either. So, by impairing Alice's ability to assess her confidence, we made her either an easy prey or a social outcast. Neither prospect looks particularly attractive.

Finally, let's consider that society is interested in maximising the total level of confidence among people (specifically increasing the level of trust, e.g. [Fukuyama1996]), in the expectation that this will lead e.g. to lower barriers and general improvements in our relationships. Such increased level of confidence may be beneficial for the development of trade, growth of civic societies, etc. Society will be satisfied with the increase in the total level even if some members suffer in this process.

The vision seems to be valid, with one caveat: it assumes that all people become trustworthy at the same time, so that all people benefit from this. Unfortunately, the missing part of such a vision is how those people should become trustworthy in a synchronous way. Such a situation is simply not sustainable: lapses in judgement, differences in competences, natural errors and fluctuations make people continuously more or less trustworthy and make others trust them less or more. The mechanism of confidence is statistical and subjective. Therefore, in the absence of a plausible explanation of how to make everybody equally and simultaneously trustworthy, we must assume the existence of people of different levels of trustworthiness. If this is the case, then again, Alice is better off if she is able to determine confidence that is due. If society supports trust, she may operate with high registers of trust, rather than with low registers, but her ability to determine the level of confidence is essential.

Actually, the total trustworthiness is not the prerequisite for market efficiency. It has been demonstrated for negotiated contracts [Braynov2002] that a market in which agents are trusted to the degree they deserve to be trusted (so that trust reflects trustworthiness) is as efficient as a market with complete trustworthiness. Complete trustworthiness is not a necessary condition for market efficiency, but the ability to properly assess other's trustworthiness apparently is. If Alice's trust matches Bob's trustworthiness, they are both individually better off – and society is better off as well. If she is unable to assess Bob correctly, results are socially unacceptable or may lead to failures.

As we can see, whether Alice is selfish or not, whether she wants to invest in the relationship or not, whether society is trusting or not, her core skill is the same: to be able to assess her confidence in Bob. Too much blind trust is damaging – the same as too little.

What if Bob is no longer a person? What if Bob is a computer – or what if Alice does not know whether Bob is a computer or not? Our confidence in computers (and in all other classes of devices) differs from our confidence in people. We have significant experience in dealing with people but we have very limited experience in dealing with devices. Our expectations are different: we do not expect the computer to act for our benefit, we want it (quite often in vain [Lacohee2006]) to do what it has been told to do. We trust them and we are confident about them in a way that is simultaneously similar and different to our relationship with people.

It is similar, because we are confident if we can trust or control. As devices are growing more complex, we tend to trust – mostly because we cannot control them [Dennett1989]. Similarly, we assess our trust and our confidence according to what we know about devices. Even though expectations and evidences are different, the mechanism seems to be similar.

It is different because this relationship is morally void. There are no ethical guidelines whether to trust the Internet or not. There is no moral ground in trusting a faulty computer in the expectation that it will reciprocate. Our relationship perhaps is equally emotional, but otherwise it is utilitarian and rational.

However, the core competence remains the same: to be able to assess whether we should be confident about the device. Pragmatic as it is, it encapsulates exactly what we always need.

# Bibliography

[Abdul-Rahman2005] Alfarez Abdul-Rahman: A Framework for Decentralised Trust Reasoning. PhD thesis. Available at: http://www.cs.ucl.ac.uk/staff/F.AbdulRahman/docs/thesis-final.pdf. 2005.

[Adams2002] Carlisle Adams and Steve Lloyd: *Understanding PKI: Concepts, Standards, and Deployment Considerations*. Sams Publishing. 2nd edn. 2002.

[Afuah2003] Allan Afuah: *Innovation Management: Strategies, Implementation and Profits*. Oxford: Oxford University Press. 2003.

[Anderson2001] Ross Anderson: *Security Engineering: A Guide to Building Dependable Distributed Systems*. Chichester: John Wiley & Sons, Ltd. 2001.

[Anderson2005] Steve Anderson et al.: Web Services Trust Language (WS-Trust). Available at: http://specs.xmlsoap.org/ws/2005/02/trust/WS-Trust.pdf. 2005.

[Baier1986] Annette Baier: Trust and Antitrust. *Ethics*, 96: 231–260. 1986. Reprinted in: *Moral Prejudices*. Cambridge: Cambridge University Press. 1986.

[Barber1983] Bernard Barber: *The Logic and Limits of Trust*. New Jersey: Rutgers University Press. 1983

[Berg1995] J. Berg, J. Dickhaut, and K. McCabe: Trust, Reciprocity, and Social History, *Games and Economic Behavior*, 10: 122–142. 1995.

[Bishop2005] Matt Bishop: *Introduction to Computer Security*. Reading. MA: Addison-Wesley. 2005.

[Bohnet2005] Iris Bohnet and Stephan Meier: Deciding to Distrust. KSG Working Paper No. RWP05-049. Available at: http://ssrn.com/abstract=839225. 2005.

[Braynov2002] Sviatoslav Braynov and Tuomas Sandholm: Contracting with Uncertain Level of Trust. *Computational Intelligence*, 18(4): 501–514. 2002.

[Castelfranchi2000] Cristiano Castelfranchi and Rino Falcone: Trust and Control: A Dialectic Link. *Applied Artificial Intelligence*, 14(8). Available at: http://www.istc.cnr.it/T3/download/Trust-and-control.pdf. 2000.

[Dasgupta2000] P. Dasgupta (2000): Trust as a Commodity, in Diego Gambetta (ed.): *Trust: Making and Breaking Cooperative Relations*, electronic edition, Department of Sociology, University of Oxford, Chapter 4, pp. 49–72. Available at: http://www.sociology.ox.ac.uk/papers/dasgupta49–72.pdf. 2000.

[Dellarocas2004] Chrysanthos Dellarocas: Building Trust Online: The Design of Robust Reputation Reporting Mechanisms for Online Trading Communities. In Wen-Chen Hu et al. (eds): *Advances in Security and Payment Methods for Mobile Commerce*, Chapter VII, pp. 95–113, Idea Group. 2004.

[Demolombe2004] Robert Demolombe: Reasoning About Trust: A Formal Logical Framework. In: *Proc. of Second Int. Conf. iTrust 2004, LNCS 2995*. 2004.

[Dennett1989] D.C. Dennett: *The Intentional Stance*. Cambridge, MA: MIT Press. 1989.

[Deutsch1965] Morton Deutsch and Robert M. Krauss: *Theories in Social Psychology*. New York: Basic Books Inc. 1965.

[Deutsch1973] Morton Deutsch: *The Resolution of Conflict: Constructive and Destructive Processes*. New Haven, CT: Yale University. 1973.

[Dragovic2003] Boris Dragovic, Steven Hand, Tim Harris, Evengelos Kotsovinos, and Andrew Twigg: Managing Trust and Reputation in the XenoServer Open Platform. In Proc. of iTrust 2003, in P. Nixon and S. Terzis (eds): *Trust Management 2003, LNCS 2692*, pp. 59–74. 2003.

[Egger2000] Florian N. Egger: From Interactions to Transactions: Designing the Trust Experience for Business-to-Consumer Electronic Commerce. PhD thesis, Eindhoven University of Technology (The Netherlands). ISBN 90-386-1778-X. 2000.

[Falcone2006] Rino Falcone et al. (eds): Ninth Int. Workshop on Trust in Agent Societies (TRUST). In *Proc. of Fifth Int. Conf. on Autonomous Agents and Multiagent Systems AAMAS-06*. Hakodate, Japan. 2006.

[Fukuyama1996] Francis Fukuyama: *Trust: The Social Virtues and the Creation of Prosperity*. New York: Touchstone Books. 1996.

[Gambetta2000] Diego Gambetta: Can We Trust Trust? In Diego Gambetta (ed.): *Trust: Making and Breaking Cooperative Relations*, electronic edition, Department of Sociology, University of Oxford, Chapter 13, pp. 213–237. Available at: http://www.sociology.ox.ac.uk/papers/gambetta213–237. 2000.

[Gerck2002] Ed Gerck: Trust as Qualified Reliance on Information. In *The COOK Report on Internet*, X(10), January 2002, ISSN 1071–6327. Available at: http://nma.com/papers/it-trustpart1.pdf. 2002.

[Giddens1988] Anthony Giddens: *The Consequences of Modernity*. Cambridge: Polity Press. 1988.

[Grandison2003] Tyrone Grandison: Trust Management for Internet Applications. PhD thesis, University of London. 2003.

[Habermas1986] Jürgen Habermas: *The Theory of Communicative Action*. Vol. 1. *Reason and the Rationalization of Society*. Cambridge: Polity Press. 1986.

[Handy1995] Charles B. Handy: Trust and the Virtual Organization. *Harvard Business Review*, May–June: 40–50. 1995.

[Hardin2002] Russel Hardin: *Trust and Trustworthiness*. New York: Russell Sage Foundation. 2002.

[Herrmann2006] Peter Herrmann: Temporal Logic-Based Specification and Verification of Trust Models, in K. Stolen et al. (eds): *iTrust 2006, LNCS 3986*, pp. 105–119. 2006.

[Huang2006] Jingweni Huang and Mark S. Fox: An Ontology of Trust: Formal Semantics and Transitivity. In *Proc. of Eight Int. Conf. on Electronic Commerce ICEC2006*, Canada, pp. 259–270. 2006.

[Ishaya2004] Tanko Ishaya and Darren P. Mundy: Trust Development and Management in Virtual Communities. In *Trust Management. Proc. 2nd Int. Conf. on Trust (iTrust '04)*. LNCS 2995.2004.

[Josang2001] Audun Josang: A Logic for Uncertain Probabilities. *Int. J. of Uncertainty, Fuzziness and Knowledge-Based Systems*, 9: 279–311. 2001.

[Josang2006] Audun Josang, Stephen Marsh and Simon Pope: Exploring Different Types of Trust Propagation. In K. Stolen et al. (eds): *iTrust 2006, LNCS 3986*, pp. 179–192, Springer-Verlag. 2006.

[Kipnis1996] David Kipnis: Trust and Technology. In Roderick M. Kramer and Tom R. Tyler: *Trust in Organizations*. London: Sage Publications. 1996.

[Kracher2005] Beverly Kracher et al.: A Foundation for Understanding Online Trust and Electronic Commerce. *Info, Comm and Ethics in Society*, 3: 131–141. Available at: http://www.troubador.co.uk/image/J.s/3-3-4-Kracher6651.pdf. 2005.

[Lacohee2006] Hazel Lacohee, Stephen Crane, and Andy Phippen: *Trustguide*. 2006. Available at: http://www.trustguide.org/Trustguide%20-%20 Final%20 Report.pdf

[Levien1998] Raphael L. Levien and Alex Aiken: Attack-Resistant Trust Metrics for Public Key Certification. *Proc. of 7th USENIX Security Symposium*, San Antonio, Texas, January. 1998.

[Levy2006] Steven Levy: *The Perfect Thing: How the Ipod Shuffles Commerce, Culture and Coolness*. New York: Ebury Press. 2006.

[Lewicki1996] Roy J. Lewicki and Barnard Benedict Bunker: Developing and Maintaining Trust in Work Relationships. In Roderick M. Kramer and Tom R. Tyler: *Trust in Organizations*. London: Sage Publications. 1996.

[Li2004] Xin Li, Joe S. Valacich and Traci J. Hess: Predicting User Trust in Information Systems: A Comparison of Competing Trust Models. In *Proc. of the 37th Hawaii Int. Conf. on System Sciences, HICSS2004*. 2004.

[LoPresti2005] Stéphane Lo Presti, Michael Butler, Michael Leuschel and Chris Booth: A Trust Analysis Methodology for Pervasive Computing Systems. In Rino Falcone, et al. (eds) *Trusting Agents for Trusting Electronic Societies LNCS 3577/2005*. Berlin: Springer-Verlag. 2005.

[Luhmann1979] Niklas Luhmann: *Trust and Power*. Chichester: John Wiley & Sons, Ltd 1979.

[Marsh1994a] Stephen P. Marsh: Formalising Trust as a Computational Concept. PhD thesis, University of Stirling. Available at: http://www.nr.no/~abie/Papers/TR133.pdf. 1994.

[Marsh1994b] Stephen Marsh: Optimism and Pessimism in Trust. Technical Report CSM-117. In *Proc. IBERAMIA94/CNAISE'94*. Ed. Ramirez, J. McGraw-Hill. 1994.

[Marx2001] Maarten Marx and Jan Treud: Trust Dynamics Formalized in Temporal Logic. In Y. Z. L. Chen (ed.) *Proc. of the Third Int. Conf. on Cognitive Science, ICCS 2001*, pp. 359–363. Beijing: USTC Press, 2001.

[McKnight1996] D. Harrison McKnight and Norman L. Chervany: The Meanings of Trust. In: University of Minnesota. Available at: http://www.misrc.umn.edu/wpaper/wp96-04.htm. 1996.

[McKnight2001] D. Harrison McKnight and Norman L. Chervany: Conceptualizing Trust: A Typology and E-Commerce Customer Relationships Model. In *Proc. of the 34th Hawaii Int. Conf. on System Sciences*. 2001.

[Meyerson1996] Debra Meyerson, Karl E. Weick and Roderick M. Kramer: Swift Trust and Temporary Groups. In Roderick M. Kramer and Tom R. Tyler: *Trust in Organizations: Frontiers of Theory and Research*. London: Sage Publications. 1996.

[O'Donovan2005] John O'Donovan and Barry Smyth: Trust in Recommender Systems. In *Proc. of IUI'05*, pp. 167–174. January. 2005.

[Page1998] Larry Page, Sergey Brin, R. Motwani, and T. Winograd: *The PageRank Citation Ranking: Bringing Order to the Web*. Stanford Digital Library Technologies Project. 1998.

[Pearson2002] Siani Pearson et al.: *Trusted Computing Platforms: TCPA Technology in Context*. Englewood Cliffs, NJ: Prentice-Hall. 2002.

[Perlman1999] Radia Perlman: An Overview of PKI Trust Models. *IEEE Network*. November/December. 1999.

[Poundstone1992] William Poundstone: *Prisoner's Dilemma*. New York: Doubleday. 1992.

[Powell2003] David Powell and Robert Stroud: Conceptual Model and Architecture of MAFTIA. Deliverable D21. Project IST-1999-11583. Available at: http://www.maftia.org/deliverables/D21.pdf. 2003.

[Quinn2006] Karl Quinn et al.: The Design, Generation, and Utilisation of a Semantically Rich Personalised Model of Trust. In K. Stolen et al. (eds): *iTrust 2006, LNCS 3986*, Berlin: Springer-Verlag, pp. 324–338. 2006.

[Rankl2003] Wolfgang Rankl and Wolfgang Effing: *Smart Card Handbook*, 3rd edn. Chichester: John Wiley & Sons, Ltd. 2003.

[Resnick2006] Paul Resnick: The Value of Reputation on eBay: A Controlled Experiment. *Experimental Economics*, 9(2): 79–101. Available at: http://www.si.umich.edu/~presnick/papers/postcards/PostcardsFinalPrePub.pdf. 2006.

[Rutter2001] Jason Rutter: From the Sociology of Trust Towards a Sociology of 'E-trust'. *Int. J. of New Product Development & Innovation Management*, 2(4): 371–385. 2001.

[Sako1992] Mari Sako: *Price, Quality and Trust: Inter-firm Relations in Britain and Japan*. Cambridge: Cambridge University Press. 1992.

[Simon1957] Herbert Simon: A Behavioral Model of Rational Choice. In: *Models of Man*. Chichester: John Wiley. 1957.

[Solomon2001] Robert C. Solomon and Fernando Flores: *Building Trust in Business, Politics, Relationships and Life*. Oxford: Oxford University Press, 2001.

[Tan2000] Yao-Hua Tan and Walter Thoen: Formal Aspects of a Generic Model of Trust for Electronic Commerce. In: *Proc. of the 33rd Hawaii Int. Conf. on System Sciences*. 2000.

[Webb1996] Eugene J. Webb: Trust and Crisis. In: Roderick M. Kramer and Tom R. Tyler: *Trust in Organizations*. London: Sage Publications. 1996.

[Zimmermann1994] Phil Zimmermann (ed.): *PGP User's Guide*. Cambridge, MA: MIT Press 1994.

# 2

# The Model of Confidence

The best material model of a cat is another, or preferably the same, cat.

(Norbert Wiener, *Philosophy of Science* (1945))

## 2.1 Introduction

Alice's transactional assessment of her confidence in Bob is an exercise that she is performing in her mind – quite often fast and possibly partly subconsciously. We do not have access to details of this process (probably even Alice cannot tell us everything), but we can speculate about its nature and its details. This model is an outcome of such speculative works. It does not mean that Alice comes to her assessment exactly through the process described by this model (this is actually quite unlikely), but only that the model is a sufficient explanation of this process.

The main purpose of the model proposed here is to serve as a tool for a better understanding of the human process of the decision regarding confidence. Another goal of this model is to bridge the gap between the sociological and psychological perception of trust (confidence) and the perception of confidence as a measurable property of digital systems. The model can be used to explain phenomena of confidence in different application areas related to the Internet or other forms of digitally-mediated interaction such as e-commerce, e-government, e-health, etc.

The model uses the complexity as both the underlying driver for confidence and as the limiting factor. Within the model, complexity is exchanged for the better assessment of confidence (the one with lower uncertainty) and such assessment is built through trust or through control, on the basis of available evidence.

This chapter starts with the review of fundamental concepts (discussed earlier). A short discussion of trust and control as well as a discussion of the relationship between complexity and confidence follows. The model is presented next, including the identification of entities, evidence and enablers as well as some formalisation of the concept. The extended discussion of trust and control closes this chapter.

*Trust, Complexity and Control: Confidence in a Convergent World*   Piotr Cofta
© 2007 John Wiley & Sons, Ltd

## 2.2 Fundamentals

The starting point for the model comes from Luhmann's observation [Luhmann1979] that trust arises from our inherent inability to handle complexity, specifically the enormous complexity represented by all the possible variants of the future. Illustrating this concept, let's consider Bob that has certain amount of independence – he is an intentional agent. A human is an excellent example, but social groups, complex technical systems [Dennett1989] or animals may be considered as well. Bob's independence is demonstrated by externally perceived uncertainty about the future action that can be taken by him (sometimes referred to as 'free will').

### 2.2.1 Confidence

Let's further assume that Alice (a person) is pursuing her goals and – in order to reach those goals – she needs certain courses of action in the future. She may be after something very simple, like safely crossing the street or something complicated, like receiving her degree. Alice may be unwilling to proceed unless she has certain confidence that the beneficial course of action will happen – the driver will slow drive, her tuition has been paid, etc. Her goal is contingent on Bob's actions – Bob can make or break her ability to achieve her goal. Bob can suddenly accelerate while driving his car or Bob may embezzle money that Alice has been saving.

### 2.2.2 Trust and Control

Alice may have some reasons to expect that Bob will actually act in her favour. For example, Alice may have her ways to influence Bob (she may cross the street next to a policeman) or she knows Bob well enough to predict the course of his actions (she has chosen the reputable bank). If Alice can reasonably enforce or expect a favourable action of Bob, then Alice has *control* over Bob, usually using specific *instruments* of control (a policeman, the bank's reputation). On the other hand, if the behaviour of Bob cannot be reasonably controlled but depends only on his intentions, Alice must resort to trust, trusting that actions of Bob will be favourable.

This leads to an interesting relationship between trust and control, observed also e.g. by Castelfranchi [Castelfranchi2000]. *Trust is in fact a deficiency of control* that expresses itself as a desire to progress despite the inability to control. Symmetrically, *control is a deficiency of trust*. This resonates with Cummings [Cummings1996] who identifies trust as a cognitive leap of faith beyond reason. Deutsch [Deutsch1962] has also noticed the complementary relationship between both. This leads to an additional observation that even though trust cannot be analysed in the same terms as control, both must share certain common characteristics.

Separating decisions regarding the outcome of some future-looking actions between the part of the confidence that is control-based and the part that is trust-based, we can assume that our confidence is driven by the sum of confidence from both sources: control and trust. We do not define 'sum' in arithmetical terms – in fact, we do not define any particular method to 'add' trust, control or confidence. However, we can observe that trust can be a substitute for control and control can be used as a substitute for trust.

### 2.2.3 Transactional Horizon

Alice's success within a transaction is likely to be contingent not only on Bob, but also on other entities. For example, Alice's safe passage depends on the weather, street lights, policeman, other drivers, time of day, etc. Unfortunately, Alice cannot include everyone and everything in her reasoning – it would be too complex.

Therefore Alice decides on the transactional *horizon* of what she wants to include in her reasoning. The extent of such a horizon depends on her *disposable complexity*, her ability to manage the number of entities and relationships between them.

### 2.2.4 Evidence and Enablers

Let's now finally have a look at what the foundation of Alice's knowledge is in both areas of control and trust. Usually Alice is in possession of certain *evidence* that can be used to support her decision. Even though no amount of evidence is sufficient to guarantee the future, Alice is always able to reason on the basis of available evidence – some evidence seems to be better than none at all.

In order to process, Alice builds on four *enablers*. She must be reasonably confident about the *availability* of evidence. Similarly Alice must be reasonably confident about Bob's *identity* – otherwise she cannot consistently infer evidence about Bob. As she is assessing Bob's intentions, she should be confident about Bob's ability to have intentions – that Bob is *similar* to her perception of what he is. Finally, she must be confident that others who deliver opinions about Bob are *honest* – that they tend to tell her the truth.

## 2.3 Two Trusts

The model presented here proposes that trust and control together contribute to the perception of confidence, driven and limited by complexity. The concept of 'two trusts' (i.e. trust and control) is not a new one, even though the names presented here may be different from what can be found elsewhere. The understanding that one should mix trust in intentions with a certain reinforcing structure provided by the environment is clearly visible throughout the literature.

What is called 'trust' here can be related to the intrinsic properties of a person while 'control' roughly relates to contextual properties (see [Riegelsberger2005] for an overview). What we call here 'control' can be also known as control trust, reliance trust, assurance-based trust, calculus-based trust, guarded trust, deterrence-based trust or cognitive trust. In contrast, 'trust' is known as party trust, affective trust, identification-based trust, relational trust or extended trust.

This proliferation of terminologies has been one of the drivers to introduce the construct of confidence in the place traditionally taken by trust and to restrict trust to what is related to the intentions of a person. The construct of control is therefore used to express what is enforced on a person. Even though it may lead to certain initial complications, the clarity provided by this justifies such a terminological decision.

The relationship between trust and control (expressed directly in those terms, no longer hidden in different flavours of 'two trusts') has also been widely studied. Apart from the general observation of a substitutive relationship between them, there is no general agreement about the exact nature of the relationship (see [Frankema2005]). Opinions range from oppositions [Child2001], dualism [Das1998] to duality [Mollering2005], [Farrell2004] where they are not mutually exclusive, even though the existence of control may render trust irrelevant.

The proposition here is as follows. Trust and control both contribute to the confidence (the substitutive approach), but while control is reducible to trust, trust cannot be reduced to control, due to the instrumentalisation of control. We are less concerned here with the influence of institutions, that are seen here mostly as instruments of control, not as a means to preserve a common identity.

## 2.4 Complexity and Confidence

While we are concerned with Alice in a rich context of a transaction, we are here mostly interested in her confidence in Bob. We recognise (and Alice recognises it as well) that Bob is the main source of her uncertainty regarding the future, so that the extent she is confident in him will be essential to her transactional confidence.

One of the main propositions of this book is that Alice can exchange her affordable complexity for the better assessment (lower uncertainty) of her confidence in Bob. Note that we do not propose here that she can actually 'build' or 'increase' her confidence by spending additional complexity, something that may seem to be contradictory to everyday expectations where e.g. promises of increased control are supposed to lead to greater confidence.

### 2.4.1 Examples

Let's take two examples. In the first, Alice is considering a transaction with an Internet merchant that she has never dealt with before. Her confidence in a merchant is rather low, but she has decided to spend some of her complexity (e.g. time, resources) to ask her friends about this merchant. Her friends reassure her that the merchant has a good reputation, Alice's confidence grows and she is willing to engage in the transaction. Additional complexity seems to increase the confidence.

In the second example, Alice is looking at the web page of a reputable company (this is actually a real case). The page states that the communication between Alice and the site is secured. Alice's confidence is therefore high and she may be willing to engage in the transaction. However, Alice decides to spend some of her affordable complexity and check what her browser says about the connection. To her amazement, the browser claims that the connection is not secured. Here, Alice's confidence decreases because she has afforded additional complexity.

### 2.4.2 Proposition

The proposition here is that for every entity there is (potentially unattainable for Alice) an optimum level of confidence that is appropriate for it, i.e. the level where Alice's confidence exactly matches its trustworthy behaviour and where Alice's trust matches its trustworthiness. Within the scope of the transaction Alice may have problems in determining this level, as she may have insufficient information. By affording additional complexity, she can expand her horizon beyond what is narrowly available and she can gather more information, thus allowing her to adjust her confidence.

Alice can simply include in her horizon what is directly relevant: Bob, the world, etc. However, the more powerful method that Alice is using to expand her horizon is to use instruments that involve other entities: other's opinions about Bob, others acting as enforcers or guarantors, etc. Those agents become a part of the transaction in addition to Bob, i.e. Alice is effectively re-defining what she means by the transaction. Adding those agents is associated with a growing complexity, as Alice must manage them and she must assess her confidence in them (potentially on the basis of yet another agent). However, by expanding the scope of transaction Alice is able to benefit from a more complete and holistic view of what she can expect.

Considering the second example, Alice has moved from the obvious (what the website has said) and expanded the scope to add her browser. Alice may not stop here: she may add her bank that probably provides certain insurance, she may consult her friend to learn how risky it is to deal over the unprotected link, etc. In this process, she is building the relatively complex horizon of a transaction that includes not only her and the website but also her browser, her bank, her friend, etc. While she is still concerned about her confidence in the website, her reasoning includes several other entities.

If Alice does not have much complexity to spend, we can easily see that her assessment may be off target: it may be both wrong (under- or over-estimated) and uncertain – e.g. Alice will build it only on the impression that is available to her within the narrow scope of the transaction. What is, however, the situation if Alice's affordable complexity is unlimited and she is wiling to spend it on the transaction? She may expand the scope to include other entities and ultimately she may gain all available information about Bob and all control about Bob she can possibly has. Her knowledge of all possible interactions will give her exact and certain understanding to what extent she should be confident in Bob, but nothing more.

## 2.5 The Model

The model itself consists of four main parts. While the core proposition has been presented above, the detailed description follows. The model can be briefly explained as follows:

1. The question that is considered here, is to what extent Alice should be confident that a course of actions within a transaction will be beneficial for her.
2. While our primary interest is in Alice's relationship with Bob, Alice must recognise that there are several entities that the success is contingent upon.
3. Alice assumes that certain enablers are in place: identity, similarity, honesty and availability.
4. Alice considers four kinds of entities that may influence the outcome: world, domains, agents and transactions.
5. Depending on her affordable complexity, Alice includes in her transactional horizon a selection of entities. Adding each entity increases the complexity, but may decrease the uncertainty of her assessment.
6. Having decided on entities, Alice buildis the model of confidence, stressing the relationship between those entities.
7. Alice uses available evidence to assess confidence in each entity, separately assessing trust and control.
8. On that basis, Alice assesses her confidence in the horizon of a transaction.

### 2.5.1 Entities

Alice structures her world into abstract entities of certain types. It is the defining element of an entity that Alice is willing to express her confidence in such an entity, e.g. she can be confident about Bob, about doctors, about the world or about recent payment. Even though the exact structuring depends on Alice, the proposition is that Alice is dealing with entities of four different kinds:

1. *Agent* is the smallest entity that is attributable with intentions and free will. Examples of agents are individuals (e.g. Bob), computers, companies, governments, shops, doctors, software, etc. Agents can be both the subject of evidence and sources of evidence. The abstraction of agents captures the semi-permanent characteristics of them. Confidence in

agents is of major to Alice concern as agents are perceived to be highly independent in their intentions.

2. *Domain* is the collective abstraction of a group of agents that share certain common characteristics, e.g. all agents that are doctors, all agents that engage in commerce, all computers, etc. Each domain allows Alice to express confidence about a particular behavioural profile of agents (e.g. 'I am confident in doctors') and replaces the missing knowledge about a particular agent if such an agent is believed to share the characteristics of a domain. One agent may belong to several domains and one transaction may include several domains as well (e.g. 'electronic commerce' and 'governed by EU law').

3. *World*. This top-level abstraction encapsulates all agents and is used as a default to substitute for unknown information of any kind. The confidence in the world is often referred to as an 'attitude' or 'propensity to trust' and can be captured by statements such as 'people are good', 'the world is helpful', etc.

4. *Transaction*. This entity introduces the temporal aspect as it encapsulates transient elements that are associated with an individual transaction. Specifically, the behaviour of various agents at the time of transaction, the value of a transaction, particularities of transaction, etc. are captured here. Note that transaction-entity differs from the horizon of a transaction.

The ability to generalise and abstract confidence allows one to build confidence in more abstract entities on the basis of experience with more specific ones. Specifically, evidence relevant to transaction (e.g. positive or negative outcome of such transaction) can gradually accumulate as confidence in the world.

The opposite is not true – confidence in the world (or in the domain) cannot be directly used to build confidence about an agent. However, such confidence substitutes for and complements the evidence of an individual, working as a default, thus decreasing the need for confidence in individual agents.

## 2.5.2 The Complexity-scope Exchange

Alice's choices regarding her horizon are restricted by her limited disposable complexity. The amount of such complexity can depend on several factors, including physical and psychological predispositions, current cognitive capabilities, tiredness, etc. For example, young, healthy, educated Alice may be willing to accept much more complexity then the old and sick one. Similarly Alice-organisation may be willing to deal with much higher complexity than Alice-individual.

Alice's disposable complexity can be traded to improve the assessment of the confidence by expanding Alice's horizon of the transaction. Specifically, such expansion may include additional entities. Figure 2.1. shows an example of the exchange process. Having only limited affordable complexity, Alice has initially decided to include in the scope of her transaction only four entities (note that this is the reasonable minimum): world, Bob, domain (electronic commerce, EC) and particulars of this transaction (t1). Having realised that she has some complexity to spare, she decided to extend the scope and include more entities: Carol (who e.g. knows Bob), her bank (that can guarantee the transaction), her past experience with Bob (t2), etc. Note that there are likely several entities that are left behind the horizon.

We do not require from Alice any attempt to reach any particular level of confidence. Alice might have decided on a certain minimum level of confidence and may refuse to progress if such a level is not attainable. However, Alice may be also in a position where she is able to make a decision that she is most confident with, regardless of the actual level of confidence. Whatever her situation is, the only thing she can trade her complexity for is additional information that comes from extending the horizon of a transaction.

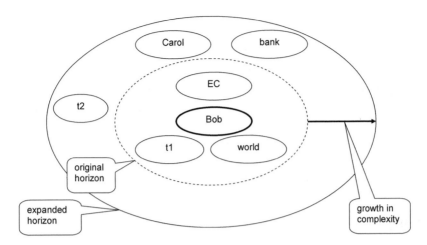

**Figure 2.1**  The complexity–horizon exchange

Alice is free to include whatever entity she likes. However, if her choice is not relevant to the transaction, she is wasting some of her affordable complexity and is not gaining in the quality of assessment. Alice can also artificially increase or decrease her confidence in Bob by selectively extending the scope of the transaction – something that can make her happy but not necessary wise.

What should drive Alice when she is choosing her horizon is the premise of a reduction of uncertainty. Even though no entity can provide an absolutely certain assessment of Bob, each can provide an assessment at its level of certainty. If Alice wants to proceed wisely and to get the best assessment, she should pick those entities that can help her reach the lowest possible uncertainty (or the highest possible certainty).

When choosing an entity, Alice bears in mind that trusting Bob is usually cheaper exactly because Alice does not have to deal with instruments that control Bob (in addition to Bob), but only deals with Bob. If trust can be further assessed on the basis of first-hand experience (so that Alice can provide all the necessary evidence), this option looks lean compared to anything else.

However, this is not always true. If Alice can trust Carol and Carol promises to control Bob, then controlling Bob may be a cheaper option. For example, if Alice has little experience with Bob but Carol is her lifelong friend who knows how to control people, then Alice may find trusting Carol and controlling Bob both cheaper and more efficient.

### 2.5.3 The Assessment of Confidence

Let us now consider how confidence can be assessed, i.e. how Alice can determine her level of confidence. The proposition is that within the horizon of the transaction, Alice can determine her overall confidence by processing her level of confidence from all the different entities: world, domain, agents and transactions, according to her perceived relationship between them.

Figure 2.2. shows the simplified relationship between confidences coming from different types of entities, assuming that there are only four entities within the horizon and that there are no relationships between them.

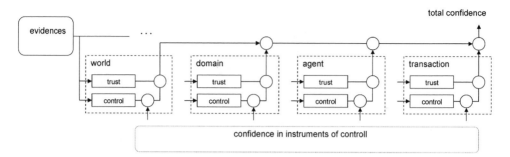

**Figure 2.2**   Confidence across different entities

## Two Paths

The proposition is that there are two conceptual assessment paths: one for trust and another for control. Evidence of trust is delivered to trust processing that assesses it and generates the level of trust associated with it. Evidence of control is delivered to control processing, but the outcome from the control processing is further influenced by the confidence in the instrument of control (Figure 2.2 and Figure 2.3).

Alice has two choices: she can either directly use trust or she can use control with the associated confidence that she has in the efficiency of such instruments, provided that those instruments are included in her horizon.

In either case, trust is the prerequisite of action: either Alice trusts Bob or Alice is confident (which recursively requires trust) in someone (something) else to control Bob – and she must include such an entity within the scope of the transaction. In fact, trust is the only irreducible element of the model: at a certain level, control requires trust.

## Sources of Evidence

Trust and control requires evidence. There is an understanding that trust cannot appear in the absence of evidence – the situation that is referred to typically as faith. However, neither trust nor control require the complete set of evidence [Arion1994].

Evidence is not judged identically. The significance of each evidence is attenuated depending on the perceived honesty of the source of such evidence. First-hand evidence can be also weighted by the confidence that the agent has in its own honesty. Honesty of the agent is one of the enablers and will be discussed later. It is worth noting that the confidence in the agent's honesty can be achieved by the recursive usage of the model, within a different area of discourse.

Evidence of trust and control are different, even though they may come from the same source. There are three classes of evidence related to trust: continuity, competence and motivation. There are three classes of evidence regarding control: influence, knowledge and reassurance.

## The Building Block

The graphical representation of the confidence building block (Figure 2.3.) can be used to further illustrate the proposition. This representation does not directly capture the recursive nature of the evaluation.

In order to graphically represent the recursive and iterative nature of the confidence, within the horizon of a transaction, Figure 2.4 demonstrates how several building blocks can be

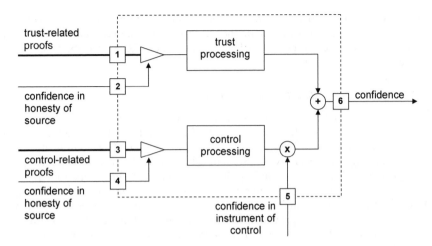

**Figure 2.3**   Basic building block

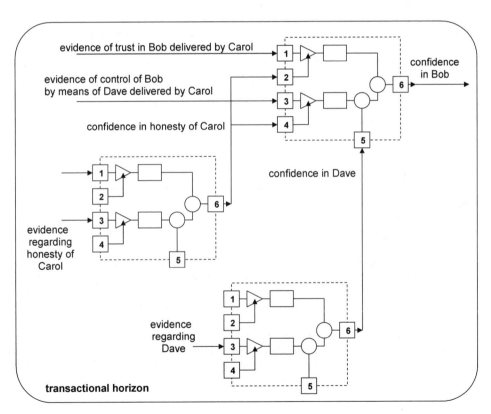

**Figure 2.4**   Building blocks within a horizon

combined to develop confidence in more complex (and more realistic) cases. Here, within her horizon, Alice must assess her confidence in three entities: Bob, Carol and Dave. While Bob seems to be critical to success, Carol is the main source of evidence about him and Dave is an instrument to control him.

### 2.5.4 Assessment as an Optimisation Process

The assessment of confidence can be also modelled as an optimisation process. When choosing entities for the inclusion in her horizon, Alice is trading complexity for the reduction of uncertainty. Every entity comes with a cost that determines how much of her complexity Alice must spend on maintaining such an entity. If Alice selects herself, the cost may be negligible, Bob may come cheaply but e.g. Dave may require significant effort to make him work for her.

Every entity also comes with an offer to increase Alice's certainty about her transitional confidence (possibly with regard to another entity). The ability to increase certainty may come e.g. from experience, the proven ability to deal with a situation, etc. While maintaining the spent complexity within the limit of her disposable complexity, Alice must maximise the certainty that she receives.

Further, every entity offers its own assessment of confidence. Depending on the position on a horizon, such confidence may be about Bob, about controlling Bob, about being honest, etc. Even though Alice may want confidence, actual values that are offered by entities should not motivate her decisions regarding their inclusion within a scope. It is only the maximisation of certainty that she should want. She is therefore trying to optimise her decision in such a way that for the minimum amount of disposable complexity she will gain the maximum possible certainty about the level of confidence.

As the process is NP-complete, Alice is not likely to actually proceed exactly according to the description provided above and come up with the best solution. From this perspective, the 'true' level of confidence is not available to Alice. Alice is likely either to stay with simplistic and sub-optimal solutions or possibly apply simple heuristics to try to identify potential improvements.

## 2.6 Formalisation

Complete formalisation of the model is beyond the scope of this book and may in fact be unnecessary. First, the model is used only to facilitate our understanding of how Alice might reason about confidence, not to imply that she is in fact reasoning exactly the way that is described here. Within the constraints proposed by the model, Alice has freedom to choose the reasoning that suits her and this makes the overly-prescriptive model futile. Further, the complete formal model must include as its component the formalisation of a significant part of the theory of Alice's mind, a formidable task in itself and still far from completion. Taking that into account, the formalisation below concentrates on clarifying those elements of the model that may benefit from it. In addition, the formalisation introduces the notation that will be used throughout the book.

Throughout this description, the operator $\oplus$ (as well as the operator $\Sigma$) and the operator $\otimes$ all have the intuitive meaning of addition and multiplication, respectively. However, this does not imply that actual semantics of those operators or any numeric metrics of confidence. Alice's world $W$ can be modelled as a directed labelled network of entities and evidence (2.1):

$$W = \{A, E\} \tag{2.1}$$

where $A$ (the use of letter $A$ follows the common perception of entities being mostly agents) is a set of nodes of the network and can be further defined as (2.2):

$$A = \{Alice, Bob, \ldots, a_1, \ldots x\} \tag{2.2}$$

Each entity is a pair of its identifying name and a complexity that is associated with dealing with such an entity (2.3):

$$a_i = (name, cplx) \tag{2.3}$$

Pieces of evidence (that serve as edges of the network) are defined as a set (2.4). We will later introduce various subsets of this set.

$$E = \{e_1, \ldots, e_n\} \tag{2.4}$$

Each evidence is further described as a five-tuple (2.5), ignoring for clarity other potential attributes such as time, location etc.:

$$e_i = \left(a_k, a_l, subject, conf_{a_k \to a_l}^{subject}, cert_{a_k \to a_l}^{subject}\right) \tag{2.5}$$

where

| | |
|---|---|
| $a_k$ | entity that holds an evidence |
| $a_l$ | entity that an evidence is about |
| $subject$ | label that is used to distinguish between different classes of evidence |
| $conf_{a_k \to a_l}^{subject}$ | level of confidence in an evidence |
| $cert_{a_k \to a_l}^{subject}$ | level of certainly about confidence in an evidence |

We can now define the horizon $H$ as a subset of the world $W$. Alice builds her transactional horizon by removing irrelevant entities and evidence from the world. Alice may exercise quite a lot of freedom in doing this. Alice is supposed to end up in a directed forest (rather than a network) where certain entities are designated as roots, i.e. there is no outgoing evidence provided by such entities.

We define two functions on $H$. The first one, $cplx(H)$ provides the overall complexity of dealing with $H$, possibly by adding up complexities of dealing with all entities that are included in $H$. The second one, $cert(H)$ provides the certainty of the assessment that can be delivered by $H$. The proposition for the formula for evaluating the certainty will be discussed later. The complexity-certainty exchange can be then formulated as an exercise of finding such $H$ that at least satisfies (2.6):

$$cplx(H) \leq disposable\_complexity \tag{2.6}$$

The problem of finding the optimum can be therefore formulated as finding such $H$ that satisfies (2.7), considering all possible horizons $H_i$ within $W$:

$$\begin{cases} cplx(H) = \min(cplx(Hi); Hi \subset W) \\ cert(H) = \max(cert(Hi); Hi \subset W) \end{cases} \tag{2.7}$$

Alice's confidence in the horizon can be expressed as a sum of her confidences in all the top-level root entities (i.e. entities that within the scope of a horizon do not hold any further evidence) as shown in (2.8). It is proposed that the level of certainty should be calculated using the identical formula. Note, however, that certainty and confidence are not calculated simultaneously:

$$conf_{Alice \to H}^{subject} = \sum_{i=1}^{n} conf_{Alice \to a_i}^{subject}; \, a_i \in H, E_{a_i}^{subject} \cap H = \Phi \qquad (2.8)$$

where:

$E_x^{subject}$ set of evidence regarding the subject that is held by $x$

Alice's confidence in a particular entity, e.g. in Bob can be described as in (2.9). Note that the confidence in instruments is different from the confidence in Bob: while Bob is expected to behave in a manner that is beneficial to Alice, instruments are only expected to be able to control Bob:

$$conf_{Alice \to Bob}^{subject} = trust_{Alice \to Bob}^{subject}(Et) \oplus \sum_{i=1}^{n} ctrl_{Alice \to Bob}^{subject}(Ec \,|\, y_i) \otimes conf_{Alice \to y_i}^{control} \qquad (2.9)$$

where:

$trust \, (Et)$     level of trust from available evidence of trust $Et$

$ctrl \, (Ec \,|\, y_i)$     level of control while using the control instrument $y_i$; assessed from available evidence of control $Ec$

The attenuation (weighting) of evidence coming from different sources can be captured as (2.10):

$$Et = \{et_x^{subject} \otimes conf_{Alice \to x}^{truth}\}$$
$$\qquad \qquad \qquad \qquad \qquad \qquad \qquad \qquad \qquad \qquad (2.10)$$
$$Ec = \{ec_x^{subject} \otimes conf_{Alice \to x}^{truth}\}$$

where

$et_x^{subject}$     evidence related to trust in subject that is coming from the entity $x$; $et_x^{subject} \in E_x^{subject}$

$ec_x^{subject}$     evidence related to control in subject that is coming from the entity $x$; $ec_x^{subject} \in E_x^{subject}$

$conf_{Alice \to x}^{truth}$     level of Alice's confidence in honesty of $x$

## 2.7 Evidence

A piece of evidence (opinion, fact, signal) in this model can be defined as an arbitrary piece of information that is provided by an agent about an entity. The concept that a belief of confidence is in fact a game of signals (evidence) is proposed e.g. in [Bacharach2001] while the similar structuring of evidence is provided in [Mui2003]. The formation of confidence in the absence of any evidence (which is psychologically possible, yet practically unlikely) is excluded from this analysis. This ability to engage in collecting evidence is the foundation of the model and is reflected in its most fundamental enabler.

### 2.7.1 The Trust Path

There are several ways of categorising evidences of trust. For example, McKnight [McKnight2001] suggests that evidence related to trust can be broadly divided into evidence of competence, benevolence, predictability and integrity. In another approach, Hardin [Hardin2002] suggest looking for evidence that the other party's interest is aligned with ours while Barber [Barber1983] seeks evidence of continuity, competence and fiduciary obligations. Mishra [Mishra1996] proposes four dimensions of trust: competence, openness and honesty, concern and reliability, suggesting that the total level of trust is a multiplication of trust in all four dimensions.

This proposition follows Barber's line of thought (and is affected by the other works discussed here) and classifies evidence of trust into three categories: continuity, competence, and motivation.

### Continuity

Continuity deals with the perception that the relationship between Alice and Bob will go beyond the current transaction. Continuity is governed by the perception of shared norms that Bob is willing to adhere to and Alice is approving. Even though continuity cannot be guaranteed (as it can be interrupted e.g. by an unexpected accident), Alice is willing to build on the perception of continuity, in understanding that the vision of common future will increase Bob's trustworthiness.

Continuity is sometimes called 'the shadow of the future' [Tyler1996] to stress its relationship with expectations about the future. The expectation of the continuity of the relationship is fundamental to the creation of trust and is captured e.g. by the phenomenon of reciprocity [Lindskold1978]. If one party cannot reasonably assume that the other party will exist and that the relationship between them may potentially remain the same, trust may suffer significant loss [Keser2003].

Types of evidence that support continuity vary. In some cases the evidence of sustainable legal or moral order may be necessary while in other cases the evidence of the expected existence of the other party in the future will be sufficient, specifically in small groups where the probability of frequent encounters is high. Similarly, the investment made by one party (e.g. a large shop) can be used as an evidence that the party is willing to trade in the future. The perception of shared principles (moral, organisational, etc.) can be also used.

### Competence

The competent party contributes to the creation of trust exactly because it is able to positively influence the future. The lack of competence, abilities or means to support Alice's cause means that that party, even though willing, is unable to help us.

Competence is the category of evidences where the specialisation of modern life is visible. While continuity and (to lesser extent) motivation are usually associated with moral virtue and refer to the 'whole person', competence can be restricted to the particular discipline, e.g. competence in teaching does not translate to competence in driving a car. Even competence in such fundamentals as an ability to tell the truth about others is regarded as a separate area of expertise – one can be an expert driver but have problems articulating one's opinion about other people.

Competence is the common category in several models of trust building, even though actual evidence can be hidden under different names. For example, in case of e-commerce, technical excellence of a website can be taken as an evidence of competence, on the assumption that the entity that is able to professionally manage the website is also able to professionally manage the relationship conducted through this site. Also, certificates issued by professional organisations can serve as evidence of competence within a profession.

## Motivation

Trustworthiness is an intention that cannot be enforced. Therefore what drives Bob is his internal motivation, his willingness to help Alice. Evidence of motivation should demonstrate that Bob has a reason to actually act for her benefit. Motivation to a certain extent can be deduced from norms that bind Bob or from his desires that drive him.

The most popular evidence of the motivation is the encapsulation of Alice's interest in Bob's interest [Hardin2002], so that it can be demonstrated that Bob actually benefits from helping Alice. For example, the fact that the banker benefits from the transaction can be used as evidence that he will be motivated to handle the transaction well.

### 2.7.2 The Control Path

Control is closely related to the preparedness and readiness to act, possibly rationally and in the pre-planned manner. In order to assess her confidence in Bob, Alice must factor in not only Bob's controllability, but also her own confidence in instruments that are at her disposal, not to mention the effectiveness of such instruments.

Alice's actions related to control assume the dynamic nature of the relationship so that they can best be classified by their relation to time. Knowledge is needed for Alice to plan ahead; influence alters Bob's behaviour when necessary and finally assurance works by affecting Bob's future. Knowledge works because Bob is concerned with his past, influence with his present, and assurance because of his concern about his future.

## Knowledge

Control is associated with rational planning and execution and this requires knowledge. What Alice needs is truth about Bob's past – how he really behaved. Alice can gain knowledge about Bob's behaviour either passively (e.g. by observing him directly or through personal contacts) or actively, by requesting information from others.

In either case (whether passively or actively) Alice is employing certain instruments that record and report the behaviour of Bob. In direct observation she acts as her own instrument which provides her with limited but trustworthy information. If she solicits information from others, those 'others' are acting as her instruments. If, for example, she discovers Bob's reputation, the centre that has provided the reputation becomes her instrument.

The way knowledge works as control is twofold. First, it improves Alice's understanding of what can really be expected from Bob, i.e. how large the gap is between her estimate of his behaviour and his actual behaviour. Such knowledge can be used to estimate his future expected behaviour. This way of using knowledge does not affect Bob – it only works to improve what Alice knows.

However, if Bob knows that Alice (or others) are collecting information about him, the act of information-gathering becomes the instrument to control Bob's behaviour. If he is concerned about his reputation, he will try to live up to it. Even though it may not change his trustworthiness, it may affect his trustworthy behaviour. Regardless whether Bob is influenced by information-gathering process or not, Alice is better off with her knowledge. In the worst case she can use the more accurate estimate of his trustworthy behaviour, in the best case, Bob will correct himself to support her.

## Influence

Alice may try to influence Bob on or before the transaction so that his behaviour alters. Alice is supposed to apply certain power, either directly or indirectly to force Bob to perform the desired action. It does not have to be a physical force – if Bob positively respond to persuasion or to changes in his environment, then he can be directed towards the expected behaviour. Sometimes the withholding of power can act as an influencing factor. Alice indeed has a wide palette of tools to choose from.

This instrument works in two slightly different ways. First, Bob may be willing to alter his behaviour due to the expectation (threat) of an influence. Even though an instrument of influence itself stays outside of the transaction (i.e. it does not contribute to what Alice considers a success – it is still Bob that she is after), Alice must include such an instrument in her horizon and must make sure that it is real and that Bob is aware of it.

Alternatively, Alice may include an instrument in the transaction, so that an instrument can actively force Bob within a transaction. While re-defining the transaction (from 'you and me' to 'you, me and my guard'), Alice also re-defines her perception of a success: she should be satisfied not only in Bob helping her but in seeing Bob forced to help her.

## Assurance

Assurance works with the future the way knowledge works with the past. It works because Bob is concerned about what will happen to him after the transaction. Note that such consideration is not present in trust, as trust does not cater for (nor expect) the negative outcome – that's why the breach of trust is always psychologically painful.

In case of assurance, Alice is relying on control instruments to shape the way Bob thinks about his future. She is using the instrument that can apply future reward or punishment, depending whether Bob has supported her or not. The most typical example can be the use of a legal system to enforce the contract. Here, Bob will be forced to behave in a particular way even if he does not want to, under the threat of being charged, sentenced or maybe even jailed. Another, more positive example is the corporate bonus plan: by behaving according to expectations of the management, an employee will be financially rewarded.

The assurance system works in two different ways. First, if Bob is aware of it, then it alters the way Bob behaves. In this sense, the assurance instrument remains beyond the scope of the transaction, even though Bob's behaviour is guided by expected reward or punishment. It is the expectation of the punishment that Alice is counting on, not the punishment itself.

Another way of looking at assurance is to include the instrument within the scope of the transaction as an alternative to Bob. If Carol represents the legal system and she is included in the scope, then the transaction is no longer the two-party one between Alice and Bob, but there are three parties involved. The perception of success also changes: Alice is satisfied not only if Bob helps her, but also if she can sue him for not helping her (and potentially receive some compensation).

## 2.8 Recursion

Recursion is an attractive property of this model: even though the basic building block is relatively simple, it can be replicated and inter-connected to model assessments in complex transactions. The recursive nature of the model leads in two different directions. First, the confidence in honesty is used within the model to qualify evidence. We can see that it leads us to yet another level of confidence assessment where the context is not about a particular service but about the ability to tell the truth. This will be discussed later as one of the key enablers.

Second, the confidence in instruments of control is used to qualify confidence that comes through the control path (see also e.g. [Tan2000]). The reasoning behind recursion is very simple – if we want to use certain instrument of control, then we should better be sure that such an instrument works – that it will respond to our needs and that it will work for us.

We can combine here trust with control, exactly in the same way as in the main building block of the model. Specifically, we can control instruments of control – but this requires yet another recursive layer of trust and control. There is an interesting question of how deep should we go. If we employ the watcher to gain control, who is watching the watcher? Who is watching those who are watching the watcher? – etc. Even though we can expand the model indefinitely (within limits set by our resources), this does not seem reasonable, as each layer of control adds complexity.

This recursive iteration may be terminated in several ways. The simplest one is to end up with an element of control that is fully trusted. For example, trusted computing [Pearson2002] conveniently assumes that certain parts of hardware can be trusted without being controlled. Other solutions may loop the chain of control back to us – following the assumption that we can trust ourselves. Elections in democratic societies serve this purpose, providing the control loop that should reaffirm our confidence in the government.

## 2.9 Enablers

Alice's desire is to assess her confidence in Bob within a horizon of the transaction and with regard to a given subject – she wants Bob to behave in a particular way, to help her. Her reasoning about Bob is therefore focused on the specific subject and resides within what we can call the 'frame of discourse', the potentially rich set of entities, evidence, rules, states, etc. associated with a subject.

While concentrating on her main subject, Alice can rely on several supporting processes that facilitate her ability to reason about Bob. She needs sufficient amount of evidence, she would like to understand whether such evidence are true, whether Bob actually exists or whether he is a person. Such enablers indirectly affect the way Alice is performing her assessment of confidence. If one or more of enablers are not satisfied, Alice loses her ability to determine confidence – the process will fail.

The proposition is that there are four enablers – similarity, honesty, identity and availability – that form four separate frames of discourse below the one defined by the main subject. Those frames can be viewed as stacked one upon the other (in the order presented in Figure 2.5). Each frame is concerned with a separate question, and – interestingly – all those questions refer to Alice's confidence in certain properties of entities. Similarity verifies how confident Alice is that Bob's intentions can be correctly assessed, honesty addresses the question about Alice's confidence in Bob's trustworthiness (which is a prerequisite to accepting Bob as a source of evidence), etc.

The honesty enabler is directly referred to in the model in a form of a $conf^{truth}$ factor that influences the weighting of evidences. Other enablers are not directly visible, but all are

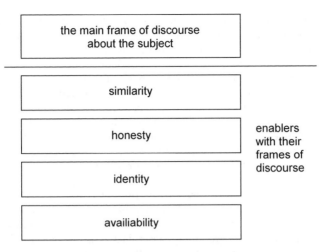

**Figure 2.5** Frames of discourse with enablers

necessary for the model to operate. For example, similarity is implied by the model: Alice can trust only those who can have intentions (like herself) – so that if Bob is dissimilar, then trust cannot be extended to him (while control can).

When Alice is reasoning about enablers, she can use the same model that has been presented here, i.e. she can use trust and control to resolve whether she is confident about the subject of the given frame of discourse. This may seems like an impossible task, as each reasoning about confidence requires Alice to reason about confidence! However, there is a way out.

Alice should reason about enablers in a bottom-up manner: if she is not confident about evidence, then she should withhold her opinion about identity, etc. Availability of evidence is therefore the basis of confidence about identity (in accordance e.g. with [Giddens1991]), identity allows evidence to be gathered about honesty, etc. However, practice shows that this is not always the case: shared identity can be used to demonstrate similarity while expectations of similarity (e.g. similar background or interest) support the perception of honesty [Dunn2005]. If Alice is not careful, she may easily end up in a self-reinforcing situations, with a positive feedback loop: if Bob is similar, then he is believed to be honest so that because he is honest, Alice will accept from him evidence about his similarity, thus indirectly driving her perception of his honesty even further.

Alice is expected to keep frames of discourse separate, without logical loops, so that she never uses the potential outcome of the reasoning as the input to the same reasoning, whether directly or indirectly. For example, if reasoning about her confidence in the main frame $(conf^{subject}_{Alice \rightarrow Bob})$, she may be considering also her confidence in Carol's honesty $(conf^{truth}_{Alice \rightarrow Carol})$ and her confidence in Dave's identity $(conf^{identity}_{Alice \rightarrow Dave})$, but not her confidence in Bob's similarity $(conf^{similar}_{Alice \rightarrow Bob})$. She must be confident that Bob is similar (or dissimilar to a known extent) prior to thinking of him helping her. For example, if Bob is a dog, Alice cannot expect him to help her with her computer (but she may expect help in tracing a lost ball).

The foundation and beginning of reasoning should be Alice herself. She must be confident that she knows enough about herself, that she knows herself, that she understands herself and that she is honest with herself before thinking about others. If, for example, she does not understand herself, she may have problems establishing her confidence in Bob's similarity – she may not have anything to compare Bob to. If she is not honest with herself, she may be willing to lie to herself about facts, making evidence useless. Starting with herself, Alice can gradually establish her confidence in others using her own first-hand experience and later also opinions provided by others – but then she should first establish her confidence in their honesty. This calls for the stepwise approach, driven by introduction, exploration and incremental widening of the group that Alice is confident about.

It is also possible to let Alice resolve all enablers simultaneously without any reference or link to herself, only on the basis of observations, e.g. allowing Alice to interact with a group of strangers. On the ground of logic alone (e.g. while resolved only on the level of honesty), this situation may end up in an unsolvable paradox: Alice will not be able to determine who is telling the truth. However, if she is resolving the problem on all levels at the same time, she may decide on the basis of greater availability or perceived similarity [Liu2004]. Similarly, the perception of strong identity in inter-personal contacts may increase the perception of honesty [Goffee2005].

The concept of separate layers of discourse is presented e.g. in [Huang2006], with two layers: one that refers to the main frame and another that refers to the frame of truth (honesty). Several interesting social phenomena are related to enablers. Gossip (that seems to stabilise society and contribute to 'social grooming' [Prietula2001]), the perception of identity and honesty, empathy [Feng2004] (that contributes to the perception of similarity) and emotions [Noteboom2005] (that facilitate experimentation, affects risk-taking and general accessibility) rightfully belong here.

### 2.9.1 Similarity

The main problem that is addressed by similarity is whether Bob is capable of intentions (whether Bob is an intentional agent) and whether his perception of norms, obligations, intentions and communication agrees with Alice's. Without discussing inter-cultural problems [Kramer2004], [Hofstede2006], even the simple act of differentiation between Bob-person and Bob-company has a significant impact on how Alice feels towards Bob and whether she can develop confidence in Bob.

The problem of similarity also applies to relationships. For example, (e.g. [Fitness2001]) in the communal relationship, Alice and Bob are expected to care about each other's long-term welfare while in the exchange relationship they are bound only for a duration of exchange. If Alice and Bob have dissimilar perception of their relationship, the formation of confidence may be problematic and the destruction of the relationship is likely.

Certain clues that allow Alice to determine whether Bob is a person are of value in creating trust [Riegelsberger2003], even if they do not directly provide information about trust itself. Empathy, the ability to accurately predict others' feelings seems to contribute to trust [Feng2004] by attenuating the perception of similarity. The simple act of Bob sending his picture to Alice by e-mail increases her willingness to develop the relationship, potentially towards trust, by the process of re-embedding [Giddens1988]. Note, however, that the fact that Alice is good at dealing with people face-to-face may actually prevent her from developing trust on-line [Feng2004], as the similarity is perceived differently in direct contacts and through digital media.

The role of the similarity enabler is to provide the estimate of a 'semantic gap' [Abdul-Rahman2005] that can potentially exist between the way Alice and Bob understand and interpret evidence. The larger the gap, the harder it is to build trust, as evidence may be misleading or inconclusive, so that evidence from similar agents should bear more weight [Lopes2006].

Similarity and the semantic gap are not discussed in the model – we conveniently assume that all agents are similar to Alice in their reliance on reasoning, social protocols, etc. (which is a very reasonable assumption for e.g. a set of autonomous agents, but may be problematic for a multi-cultural society). If there is any concern about the semantic gap, it may be captured in the lower value in Alice's confidence in truth.

## 2.9.2 Honesty

Alice's confidence in others' honesty is strongly and visibly present throughout the model in a form of weighting element for all evidence. The particular frame of discourse focuses on the question of truth: $conf^{truth}_{Alice \rightarrow x}$. The role of the honesty enabler is to allow Alice to accept opinions that come from others, in the form of a qualified reliance on information [Gerck2002].

The ability to tell the truth is the core competence of agents and it seems that no confidence can be established if we do not believe that the truth has been disclosed. Unfortunately, the construct of truth is also vague and encompasses several concepts. We may assume, for the purpose of the model, that by 'truth' we understand here information that is timely, correct and complete to the best understanding of the agent that discloses such information. This factors out the concept of objective truth but factors in the equally complicated concept of subjective truth. However, we cannot ask for more as the agent can hardly achieve an absolute reassurance about information it discloses.

Confidence in truth may be quite hard to establish. In digital systems, confidence in truth can sometimes be assumed a priori, without any proof. For example, when the mobile operator is issuing SIM, it is confident that the card (to be exact: any device that responds correctly to the cryptographic challenge) reports what is true (e.g. the content of a phone book) to the best of its understanding. This assumption can be challenged in principle (as the card cannot even reasonably check what physical form it has), but it holds in practice.

If Alice assumes that certain entities may not tell the truth, she may want to identify and exclude 'liars' from her horizon. This, however, leads to challenges associated with the construct of truth. Specifically, the ability to differentiate between the unintentional liar and the intentional one or between the benevolent messenger of unpopular news and the dishonest forwarder of popular ones may prove to be hard. For example, the 'liar detection' system [Fernandes2004] that is based on the coherence of observations punishes the first entity that brings unpopular news, effectively removing incentives to act as a whistleblower.

## 2.9.3 Identity

The identity enabler allows Alice to consolidate evidence that has been collected at different moments of time – to correlate evidence about Bob. Within the model, identity is expected to be reasonably permanent – one of simplifications of the model. Bob is always Bob, only Bob and everybody knows that he is Bob. However, while discussing the relationship that is e.g. managed over the Internet (e.g. [Turkle1997]) we can see the erosion of this simplistic assumption as identities proliferate and confidence diminishes. While identity is central to the notion of confidence, assuring identity is both challenging and beneficial in such environments as pervasive computing [Seigneur2005] or even simple e-mail [Seigneur2004].

The relationship between confidence and identity is complex and is discussed elsewhere in this book. In short, identity is an enabler for Alice's confidence, Alice can be confident in an identity and finally the shared identity can reaffirm trust between Alice and Bob.

Bob's identity is subjectively perceived by Alice by the sum of evidence that she has associated with him [Giddens1991] – his identity is built on her memories of him (the same way that she built her own identity by self-reflecting on her own memories). Potentially this may lead to an identity split: assuming that evidence about Bob is of a mixed kind (some good and some bad), depending on the way Alice interprets the evidence, she may either end up with a mixed opinion about one Bob or with firm opinions about two Bobs – one good and one bad.

### 2.9.4 Accessibility

Apart from Alice's willingness to evaluate her confidence, she needs input to this process – she needs evidence. Without evidence her confidence will be reduced to faith [Arion1994], the belief that can exist without evidence [Marsh1994]. Our willingness to gather evidence is usually taken for granted, as there is a strong belief in the basic anxiety [Giddens1991] that is driving us towards experimentation and learning. However, traumatic experiences or potentially high complexity of evidence-gathering may inhibit people from engaging in such activity.

The model does not cater for such situations, but it allows for their interpretation by the construct of certainty. The less access Alice has to evidence, the less evidence she can gather and the less certain she is about any outcome of her reasoning.

## 2.10 Processing

The model itself does not mandate any particular method of processing evidence into the decision regarding confidence. Even though such processing may be (and often is) quite subjective, we can identify certain common traits that are applicable probably not only to the decision regarding confidence but to our decision-making process in general.

Giddens [Giddens1991] has noticed that the rules that we use to assess confidence fall into two categories, depending on our previous experience and depending on available complexity. Reflexive rules form an automatic behaviour that delivers satisfactory assessment rapidly and with low cognitive complexity. They are formed mostly on the basis of existing experiences (stored patterns that can be recalled when needed) and works almost as a set of ready-made behavioural templates.

Reasoning rules allow us to handle more complex cases at the expense of complexity and time. Here, our reasoning capability is invoked to rationally deal with available evidence, by exploring through reasoning. The process is quite slow, restricted in its maximum complexity and may require significant mental effort. As such, it cannot apply to known situations that require rapid decisions but works well for situations that are new or unexpected. Specifically, the appearance of evidence that contradicts existing knowledge may trigger the transition from reflexive to reasoning rules. Similarly, the repetitive appearance of similar evidence may lead through the learning process to the creation and use of the new reflexive rule.

## 2.11 Characteristics of Trust and Control

When considering modern society one cannot resist the feeling that control has the upper hand and trust is diminishing. However, instead of lamenting the destruction of trust it is worth considering why society so often chooses the control path, despite its obvious higher complexity.

The answer (at least partly) lies in the different perceived characteristics of control and trust, captured in Table 2.1. The concept of 'two trusts' provides a certain insight in it, but there is more detailed discussion that highlights differences between control and trust. The understanding of those differences may lead not only to the better understanding of the skewed balance between trust and control, but also may show methods to re-balance it.

Modern culture has re-defined the concept of risk [Giddens1991]: from risk associated with individual actions and encounters to pervasive risk that is reduced and at the same time increased by the existence of intermediating institutions. The shift in the allocation of risk results in the shift in the way confidence is evaluated – towards control. If there is not a

**Table 2.1**  Differences between trust and control

| Characteristic | Trust | Control |
|---|---|---|
| Activity | Passive | Active |
| Volatility | Volatile | Predictable |
| Internality | Internal | External |
| Contextuality | Holistic | Contextual |
| Temporality | Gradual | Instant |
| Measurability | Qualitative | Quantitative |

sufficient amount of trust-related interactions, then the demand for increased control is the only possible solution.

We start from an observation [Giddens1991] that modern life favours rational knowledge, doubt and critical reasoning – to the extent where everything can be questioned. This leaves less space for belief-based expectations (trust in this case) that cannot be critically questioned and rationally deconstructed. Control, on the other hand, can usually withstand rational analysis (unless we start questioning foundations of such control, i.e. trust in instruments), so that it seem to be more appropriate for modern life.

### 2.11.1 Activity

Trust caters for Alice's own inactivity as the assumed goodwill of the trustee should itself facilitate the positive course of actions regardless of her own activity. When Alice is relying on Bob's internal trustworthiness, the only thing that she has to do is inform Bob about her needs – as Bob cannot be expected to act upon her needs if he is not aware of them. This can be delivered reasonably simply and does not require any specific tools (even though it may require certain determination if Bob is not readily accessible).

Control is closely related to the preparedness and readiness to act, possibly rationally and in the pre-planned manner. By considering control Alice accepts that she may not only contemplate evidence but may also perform an actual action. Note that control is not only about action itself – Alice may never be required to perform any action regarding Bob. It is about her confidence that she (or someone else) is ready to act if needed – with the expected result.

### 2.11.2 Volatility

Trust has a high volatility. If Alice trusts, then she potentially exposes her vulnerability to Bob and becomes subject to his behaviour – she becomes dependent on Bob. Trust is essentially the exchange process where the uncertainty of the world is replaced by the uncertainty of the single person – the trustee. By trusting, we are entirely dependent on the behaviour of an individual that we cannot control. Such an individual becomes the single point of failure of the process – if he defaults, we have nothing to protect ourselves with.

Control, stressing differences, is a piecemeal decision with space for plenty of recourse and consequently with low volatility. One can build an elaborate system of control (within the limits of disposable complexity) so that even if one part of such system fails, other parts will hold and deliver at least a partial result. Uncertainty can be spread among several entities and there is no single point of failure. The system is still fallible, but it is a graceful degradation, not a total disaster.

### 2.11.3 Internality

Control requires instruments that are usually external to the relationship. Certainly Alice can control Bob directly (say, the way mother exercises control over her child), but this situation is unlikely, as it consumes too much of Alice's resources (what every mother would experience from time to time). She will rather seek instruments of control that can be summoned to help her if needed.

Alice must be ready to bear the burden of dealing with those instruments, as they must be included in her horizon. If, for example, she considers controlling Bob through the judicial system, she is facing the complexity of understanding and applying the law. While the task may seem daunting, the institutionalisation of our world means that Alice has a rich set of instruments to choose from, making control attractive.

In contrast, trust is internal to the relationship between Alice and Bob as much as trustworthiness in internal to Bob. Acting upon Bob with certain instruments may actually have a detrimental effect on his trustworthiness. Trust, not dependent on any instrument external to the relationship, develops by Alice's actions and observations within such a relationship.

### 2.11.4 Contextuality

Trust is considered to be the quality of an agent that permeates its behaviour across all the possible context. Trustworthy Bob is supposed to be trustworthy at work as well as at home, in the train, etc. Trusting Bob means knowing Bob across all those contexts – quite a task, specifically if Alice has no time for this or no interest in this. Even the most contextual element of trust – competence – may require a holistic approach to assess it correctly.

Control is narrowly contextual – Alice would like to control Bob here and now, with disregard to the future. Actually, as control is built on instruments that are external to Bob, Alice can rely on those instruments without even knowing Bob, but only knowing the environment he operates in [Giddens1991]. By gaining expertise in environmental instruments, Alice can swiftly build up a passing relationship not only with Bob, but with anyone who is controlled in a similar manner. Relying on control allows Alice to roam freely.

### 2.11.5 Temporality

Control is useful for short-term encounters, where Alice has no interest in the common future. It is not only that Alice cannot reliably coerce Bob into a long-term relationship, but it is also that the continuous control of Bob will eventually drain her resources. However, on the bright side, control is an instant mechanism: Alice does not have to engage in complex observations or interactions prior to the transaction. She can quickly build control on the basis of instruments, even though such control may not last.

In contrast, if Alice wants to build trust with Bob, she should possibly employ her empathy [Feng2004] and build a relationship over time. Even though such a relationship may benefit her in the longer time frame, it effectively locks her into Bob, as she has neither time nor resources to nurture alternative relationships. In a marketplace of disposable relationships that does not value loyalty and continuously entices customers with new offers, there is little place for the proper development of trust. As the time horizon of the relationship shrinks, the ability to nurture such relationship also decreases.

### 2.11.6 Measurability

Yet another reason why trust seems to be unpopular is the lack of a reliable way to quantify trust. Control has developed several methods to measure probability, risk or the financial outcome of control-based transactions, leading to greater predictability and accountability (conveniently ignoring trust in instruments of control). Trust is unfortunately devoid of such methods, as measuring intentions is neither straightforward nor reliable.

The concept of risk is well known in businesses and there is a large body of knowledge dedicated to methods related to the assessment of risk. Therefore, even though risk itself is not avoidable, it can be well predicted – and consequently it can be factored into the business model, e.g. in a form of higher charges or additional insurance. Over time, for a reasonably stable business, risk becomes a well-known, predictable factor, part of a business as usual.

Trust does not yield itself to calculations (at least not yet). Dealing with individuals and with their internal intentions is associated with high level of uncertainty, thus increasing the volatility and decreasing the predictability of the outcome. From the business planning perspective, greater known risk is a safer bet than possibly a lower but unknown one, so businesses make a rational decision to prefer control over trust.

### 2.11.7 Digital Technology

An important reason for the shift from trust to control-based relationships is the development of technology. Society has developed and is still developing significant amount of instruments of control at the high level of sophistication, at a relatively low price and relatively easy to use. Technology (specifically digital technology) has been particularly supportive in creating new tools of control. All those instruments are designed to decrease the need for trust. However, they are actually increasing the need for trust, but for another trust – trust in instruments.

Offsetting trust in people with trust in digital tools might have been a valid option, as the total complexity has initially decreased. However, as those tools rely on trust (in the system administrator, in the infallibility of digital technology), the long-term complexity may be quite high, which is evident e.g. in the growth of time that we spend protecting our computers from attacks. However, the path of control, once taken, cannot easily be undone, specifically when traditional relationships of trust have been abandoned as no longer relevant or modern enough. The only choice, it seems to be, is to proliferate tools of control up to the level where we collectively will not be able to control tools that control us. If this is not the desired future, some alternatives will be discussed throughout in this book.

## Bibliography

[Abdul-Rahman2005] Alfarez Abdul-Rahman: A Framework for Decentralised trust Reasoning. PhD thesis. Available at: http://www.cs.ucl.ac.uk/staff/F.AbdulRahman/docs/thesis-final.pdf. 2005.

[Arion1994] M. Arion, J. H. Numan, H. Pitariu and R. Jorna. Placing Trust in Human-Computer Interaction. *Proc. 7th European Conf. on Cognitive Ergonomics (ECCE 7)*: 353–365. 1994.

[Bacharach2001] Michael Bacharach and Diego Gambetta: Trust as Type Detection. In: Cristiano Castelfranchi and Yao-Hua Tan (eds): *Trust and Deception in Virtual Societies*, Dordrecht: Kluwer Academic Publishers. pp. 1–26. 2001.

[Barber1983] Bernard Barber: *The Logic and Limits of Trust*. New Jersey: Rutgers University Press. 1983.

[Castelfranchi2000] Cristiano Castelfranchi and Rino Falcone: Trust and Control: A Dialectic Link. *Applied Artificial Intelligence*, 14(8), September 1, 2000. Available at: http://www.istc.cnr.it/T3/download/Trust-and-control.pdf. 2000.

[Child2001] J. Child: Trust – The Fundamental Bond in Global Collaboration. *Organizational Dynamics*, 29(4): 221–239. 2001.

[Cummings1996] L. L. Cummings and Philip Bromiley: The Organizational Trust Inventory (OTI): Development and Validation. In: Roderick M. Kramer and Tom R. Tyler: *Trust in Organizations*. London: Sage Publications. 1996.

[Das1998] T. K. Das and B. S. Teng: Between Trust and Control: Developing Confidence in Partner Cooperation in Alliances. In: *Academy of Management J.*, 23(3): 491–512. 1998.

[Dennett1989] D.C. Dennett: *The Intentional Stance*. Cambridge, MA: MIT Press. 1989.

[Deutsch1962] Morton Deutsch: Cooperation and Trust: Some Theoretical Notes. *Nebraska Symposium on Motivation*. Lincoln, NB: University of Nebraska Press, pp. 275–318. 1962.

[Dunn2005] Jennifer R. Dunn and Maurice E. Schweitzer: Feeling and Believing: the Influence of Emotion on Trust. *J. of Personality and Social Psychology*, 88(5): 736–748. Available at: http://opim.wharton.upenn.edu/~schweitz/papers/Emotion_Trust.pdf. 2005.

[Farrell2004] Henry Farrell: Trust, Distrust, and Power. In: Russel Hardin (ed.) *Distrust*. New York: Russell Sage Foundation. 2004.

[Feng2004] Jinjuan Feng et al.: Empathy and Online Interpersonal Trust: A Fragile Relationship. *Behaviour and Information Technology*, 23(2), March–April: 97–106. 2004.

[Fernandes2004] Alberto Fernandes et al.: Pinocchio: Incentives for Honest Participation in Distributed Trust Management. In: C.D. Jensen et al. (eds) *iTrust 2004, LNCS 2995*, Springer-Verlag. pp. 63–77. 2004.

[Fitness2001] Julie Fitness: Betrayal, Rejection, Revenge, and Forgiveness: An Interpersonal Script Approach. In: M. Leary, (ed.) *Interpersonal Rejection*. New York: Oxford University Press. 2001.

[Frankema2005] Katinka Bijlsma-Frankema and Ana Cristina Costa: Understanding the Trust-Control Nexus. *Int. Sociology*. 20(30): 259–282. 2005.

[Gerck2002] Ed Gerck: Trust as Qualified Reliance on Information. In *The COOK Report on Internet*, X(10). January 2002, ISSN 1071–6327. Available at: http://nma.com/papers/it-trustpart1.pdf. 2002.

[Giddens1988] Anthony Giddens: *The Consequences of Modernity*. Cambridge: Polity Press. 1988.

[Giddens1991] Anthony Giddens: *Modernity and Self-identity: Self and Society in the Late Modern Age*. Cambridge: Polity Press. 1991.

[Goffee2005] Rob Goffee and Gareth Jones: Managing Authenticity: The Paradox of Great Leadership. *Harvard Business Review*, December: 87–94. 2005.

[Hardin2002] Russel Hardin: *Trust and Trustworthiness*. New York: Russell Sage Foundation. 2002.

[Hofstede2006] Gert Jan Hofstede et al.: Modelling Trade and Trust Across Cultures. In K. Stolen et al. (eds): *iTrust 2006, LNCS 3986*, pp. 120–134. 2006.

[Huang2006] Jingweni Huang and Mark S. Fox: An Ontology of Trust: Formal Semantics and Transitivity. In *Proc. of Eight Int. Conf. on Electronic Commerce ICEC2006*, Canada, pp. 259–270. 2006.

[Keser2003] C. Keser: Experimental games for the design of reputation management systems. *IBM Systems J.*, 42(3). 2003.

[Kramer2004] Roderick M. Kramer: Collective Paranoia: Distrust Between Social Groups. In Russel Hardin (ed.): *Distrust*. New York: Russell Sage Foundation. 2004.

[Lindskold1978] S. Lindskold: Trust Development, the GRIT Proposal, and the Effects of Conciliatory Acts on Conflict and Cooperation. *Psychol. Bull.* 85: 772–793. 1978.

[Liu2004] Jinshan Liu and Valerie Issarny: Enhanced Reputation Mechanism for Mobile Ad Hoc Networks. In C. D. Jensen et al. (eds): *iTrust 2004, LNCS 2995*. pp. 48–62. 2004.

[Lopes2006] Alexandre Lopes and Ana Cristina Bicharra Garcia: A Model for Generating More Precise Reputation Estimates in Case of Changing Behavior by the Rated Agent. In *Proc. of Fifth Int. Conf. on Autonomous Agents and Multiagent Systems AAMAS-06*. Workshop WS8 – *Ninth Int. Workshop on Trust in Agent Societies*. Hakodate, Japan. 2006.

[Luhmann1979] Niklas Luhmann: *Trust and Power*. Chichester: John Wiley & Sons, Ltd. 1979.

[Marsh1994] Stephen Marsh: Optimism and Pessimism in Trust. Technical Report CSM-117. In *Proc. IBERAMIA94/CNAISE'94*. Ed. J. Ramirez, New York: McGraw-Hill. 1994.

[McKnight2001] D. Harrison McKnight and Norman L. Chervany: Conceptualizing Trust: A Typology and E-Commerce Customer Relationships Model. In *Proc. of the 34th Hawaii Int. Conf. on System Sciences*. 2001.

[Mishra1996] Aneil K. Mishra: Organizational Responses to Crisis. The Centrality of Trust. In Roderick M. Kramer and Tom R. Tyler: *Trust in Organizations*. London: Sage Publications. 1996.

[Mollering2005] Guido Mollering: The Trust/Control Duality: An Integrative Perspective on Positive Expectations of Others. *Int. Sociology*, 20(3): 283–305. 2005.

[Mui2003] Lik Mui, Ari Halberstadt and Mojdeh Mohtashemi: Evaluating Reputation in Multi-agents Systems. In R. Falcone et al. (eds): *AAMAS 2002 Workshop on Trust, Reputation . . .* LNAI 2631, pp. 123–137. 2003.

[No-oteboom2005] Bart Noteboom: Framing, Attribution and Scripts in the Development of Trust. In *Proc. of Symposium on 'Risk, Trust and Civility'*, Victoria College, University of Toronto, 6–8 May. 2005.

[Pearson2002] Siani Pearson, et al: *Trusted Computing Platforms: TCPA Technology in Context*. Englewood Cliffs, NJ: Prentice-Hall. 2002.

[Prietula2001] Michael J. Prietula: Boundedly Rational and Emotional Agents Cooperation, Trust and Rumor. In C. Castelfranchi and Y.-H. Tan (eds): *Trust and Deception in Virtual Societies*. Dordrecht: Kluwer Academic Publishers. pp. 169–193. 2001.

[Riegelsberger2003] Jens Riegelsberger. Interpersonal Cues and Consumer Trust in E-commerce. CHI2003 Doctoral Consortium, Extended Abstracts CHI2003, 5–10 April, Ft. Lauderdale, FL. 2003.

[Riegelsberger2005] Jens Riegelsberger et al.: The Mechanics of Trust: A Framework of Research and Design. *Int. J. of Human-Computer Studies*, 62(3): 381–422. 2005.

[Seigneur2004] Jean-Marc Seigneur et al.: Combating Spam with TEA (Trustworthy Email Addresses). In *Proc. of 2nd Conf. on Privacy, Security and Trust*, Fredericton, New Brunswick, Canada, pp. 47–58. 2004.

[Seigneur2005] Jean-Marc Seigneur and Christian Damsgaard Jensen: The Role of Identity in Pervasive Computational Trust. In Security and Privacy in Pervasive Computing Workshop (Second Int. Pervasive Conf.), Vienna, Austria, 2004, Philip Robinson and Harald Vogt and Waleed Wagealla, apr, Kluwer, The Kluwer Int. Series in Engineering and Computer Science, Vol. 780. 2005.

[Tan2000] Yao-Hua Tan and Walter Thoen: Formal Aspects of a Generic Model of Trust for Electronic Commerce. In *Proc. of the 33rd Hawaii Int. Conf. on System Sciences*. 2000.

[Turkle1997] Sherry Turkle: *Life on the Screen: Identity in the Age of the Internet*. New York: Simon & Schuster. 1997.

[Tyler1996] Tom R. Tyler and Roderick M. Kramer: Whither Trust? In Roderick M. Kramer and Tom R. Tyler: *Trust in Organizations*. London: Sage Publications. 1996.

# 3

# Measuring Confidence

If you cannot measure it, you cannot improve it.

(Lord Kelvin)

## 3.1 Introduction

The development of modern science is marked by our ability to express certain phenomena in a numerical form – to measure. We will not discuss here the validity of this approach but simply accept the fact that without assigning numerical values, science apparently cannot progress in analysing any given phenomenon beyond the contemplation of its complexity.

Confidence (specifically trust) escapes clear measurement for several reasons, whether it is its unclear definition, its complexity, subjectivity or the fact that it is a mental process, unavailable to instruments. There is a strong argument [Castelfranchi2000] against the use of simplistic metrics to measure trust that validly points to the complexity of the process and the embeddedness of trust in so many instances of our activity that it may be impossible to capture (and measure) trust in a clear form. If we consider also that trust is one of the sub-constructs of every utterance [Habermas1986] (so that we cannot rely on language to be trust-neutral), the task of measuring trust, control or confidence seems daunting indeed.

However, even simplification of a natural phenomenon has a value as a tool that guides our understanding and sheds light on interesting aspects that would not be discussed otherwise. Assuming that confidence can be captured by a single number (or a small tuple of related numbers) is a simplification, but it may turn out to be useful as it opens up opportunities to investigate and model certain aspects of confidence such as cooperation, propagation, etc. We should only understand that what we measure may not be exactly trust, confidence or trustworthiness, but something (hopefully) very close to it.

Beginning with this warning, this chapter explores various ways in which confidence can be captured in its numerical equivalent and processed within the context of the model. We start from a discussion of risk and uncertainty. From there we will investigate some empirical methods that can be used to assess the extent of trust and trustworthiness (that seems to be harder to capture and more interesting than control itself). This is followed by a discussion about different metrics of confidence, not only from the perspective of their numerical range, but also in relation to their underlying assumptions. Those considerations lead to the discussion of different operations that can be performed on confidence.

## 3.2 Risk and Uncertainty

We can see from the model that the situation related to confidence is characterised by two elements: trust and control. We can also view it as being characterised by two different concepts: uncertainty and risk. There is an observation about the difference between risk and uncertainty ([Ormerod2005], following [Knight1921]) that the situation is characterised by risk if the outcome is not known but the probability function of the outcome is known. That is, while an individual transaction may end in success or failure, when repeated a sufficient number of times, we can see certain known distribution of outcomes (e.g. 10 per cent of transactions fail). Following this reasoning, the lack of confidence associated with situations that are characterised by risk can be offset by known methods (insurance, securitisation, diversification, etc.).

In contrast, situations characterised by uncertainty are those where the probability function is not known – or is not relevant. Such lack of knowledge can be attributed to several factors such as insufficient number of samples, lack of access to important information, inability to process information, irrelevance of personal experience, etc. Whatever the case, uncertainty cannot be offset by any instrument exactly because it is not known how such an instrument may operate. Ormerod argues that in an increasingly complex world the situation of uncertainty becomes prevalent, specifically at the individual level of experience.

The perception of a duality of risk and uncertainty is further reinforced by the identification of two types of 'uncertainty' (the terminology may be slightly confusing here) [Helton1997] (see also [Dimitrakos2003] for an alternative approach). The first type of uncertainty (aleatory uncertainty) is associated with the potential random behaviour of the system. Such uncertainty may randomly affect observations and can be best handled by probability theory. In contrast, the epistemic uncertainty (the second type of uncertainty) is associated with ignorance, bounded rationality and subjective inability to reason. Epistemic uncertainty cannot be reduced to probability (it is unlikely that it can be reduced at all) but may significantly influence the reasoning. For clarification, what we call here 'uncertainty' is the uncertainty of the latter kind, the epistemic kind while the concept of risk is closer to the former.

### 3.2.1 Relationship to the Model

Drawing the parallel between this reasoning (of risk and uncertainty) and the model of confidence, we can say (simplifying) that the control path is helpful in dealing with situations that can be characterised by risk while the trust path addresses situations that are related to uncertainty. It does not imply that we rigidly equate risk with the control path and certainty with the trust path. Situations seem to usually have elements of both, with uncertainty being more pronounced where there is little structure, experience or knowledge and risk is visible in structured, routine activities. However, we may consider risk a good method to express control (with consideration to the note below) and certainty a good method to express trust.

The control path is actually more complex: it is built on risk, on the assumption that the probability (or the probability distribution) is known. However, the fact that we use control instruments means that there should also be the element of confidence (ultimately trust) in such instruments. Therefore, the control path works on the basis of uncertain probability, i.e. on the construct that allows expression of probability of the assumption that the certainty has been achieved about the instrument.

## 3.2.2 Example

Let us explore a simple example. Let's assume that Alice would like to invest a certain amount of money in the fund that is managed by Bob. She has done her research on funds in general and Bob's fund in particular and she believes that she has a good understanding of what kind of performance she can expect – this is her risk-based probability distribution. However, she also faces uncertainty associated with her research – whether she has used reputable sources, whether Bob has cooked his books, or not, etc. Considering the control path (and heavily simplifying the story), Alice is uncertain about her risk distribution. She believes that she can safely predict Bob's performance only if her sources are reputable.

She also faces another uncertainty, this time one not associated with Bob's performance. At any moment Bob may execute an exit strategy, close his fund and disappear with all her money. This uncertainty has nothing to do with his past performance and cannot be estimated by probability distribution. It is only Bob who can decide whether he is going to default on his customers or not. Even knowing the probability distribution of the behaviour of the population of fund managers will not help Alice – Bob is the critical element (her single point of failure) and Alice cannot protect herself against his default. The situation will be different if Alice diversifies her investment into several funds, or takes out an insurance, moving also this part of her decision into the control path, but in that case she will probably be uncertain about someone or something else.

## 3.3 Empirical Metrics

Empirical methods try to capture the value of confidence by exploring the behaviour or introspection of people, to determine the perceived or expressed level of confidence. Those methods combine theoretical background (that determines what it is that they measure) with a defined set of questions and statistical processing of results. Alternatively [Elofson2001], intelligent agents can be used to elicit and then replicate human decision-making process

The outcome of empirical measurements is essential to verify a hypothesis, and it serves as the ultimate reference point in simulating human confidence in artificial environments. At the same time, they suffer from common problems such as the relevance to the underlying hypothesis, cultural dependence, language embeddedness, representative selection of participants, etc.

The methodology of empirical measurements has been developed to counteract and minimise those negative factors. Such a methodology usually concentrates on those elements of trustworthy behaviour that are (according to underlying theories) clear signs of trust. The willingness to cooperate and actual cooperation are commonly used as an evidence of and a measure of trust. The actual value (level of trust, level of trustworthiness) is assessed from the difference between the observed behaviour and the theoretical behaviour that should have been observed in the absence of any cooperation.

### 3.3.1 Surveys

Surveys attempt to capture the level of confidence (and are mostly concerned with trust) by means of both observations or introspection, i.e. without engaging in any experiments. Respondents usually provide answers to a set of questions or statements and responses are e.g. structured according to a Likert scale. Differentiating factors are the underlying theoretical background (what they measure) and contextual relevance (in what situations they are applicable).

There are several trust surveys with their associated scales, designed for different purposes. One of the earliest ones (yet still in use) is McCroskey's scales [McCroskey1966] that have been used to determine authoritativeness (competence) and character (trustworthiness) of speakers. The authoritativeness scale consists of 22 questions with responses selected from a Likert scale and contains statements designed to elicit the respondent's opinions and beliefs regarding different aspects of the speaker's competence, along six main semantic differential scales such as 'informed–uninformed'. The character scale consists of 20 statements, again used to gather opinions and beliefs, this time about the trustworthiness of the speaker, along another six semantic differential axes such as 'unselfish–selfish'. Original scales have been thoroughly verified against the set of profiles using a large group of respondents.

Of others, Rempel's trust scale [Rempel1985] and Rotter's scale [Rotter1971] are quite popular in determining the level of interpersonal trust in different settings. For a particular research area a more specific survey can be developed. For example, the interdisciplinary model of trust [McKnight2001], [McKnight2003] has been verified using a survey while [Corritore2005] uses a survey to establish the relationship between design elements of the website and perceived trustworthiness of it.

The Organizational Trust Inventory (OTI) [Cummings1996] is an example of an exhaustive, theory-driven survey that can be used to determine the level of trust within the organisation. The underlying theory defines trust as a collection of three dimensions of beliefs, each dimension assessed on three components: affective state, cognition and intended behaviour. The level of trust can be estimated as the difference between the actual organisational behaviour and a theoretical behaviour that is characterised by an absence of trust. The survey contains 81 statements and responses can be related back to original nine factors. There is also a shorter version of the survey that contains only 12 items that is similarly reliable but can be administered and decoded in much shorter time.

An alternative to a survey is a one-side bet that allows for the quick estimation of a person's subjective probability of a transaction. Assuming that the potential gain is fixed, the amount that a person bets can be used to estimate his confidence. If a game is open-ended but a person has limited resources, the amount that is being bet can be used to indicate the level of confidence in the situation.

## 3.3.2 Games of trust

Another method to empirically measure confidence (again, we are here mostly concerned with trust) is to engage participants in experiments and to treat the outcome of such experiments as an estimate of confidence. Several economic games and game-like scenarios have been tried, with certain preferences for those that allow the estimation of confidence in monetary terms (see [Keser2003] for an interesting overview).

Games of trust are designed in such a way that their Nash equilibrium differs from the Pareto optimum so that no player alone can maximise his own utility by altering his selfish strategy without cooperation, while cooperating partners can benefit. However, these games are not strictly cooperative, as the communication between players is restricted. Therefore any appearance of cooperation can usually be attributed to trust (in some games there is an element of control involved). Trust can therefore be estimated on the basis of monetary gain on cooperation and then expressed e.g. as a fraction of the difference between the gain from the behaviour driven by maximisation of individual's utility and behaviour where the cooperation maximises the common utility.

There are several advantages of using games to estimate the level of trust or confidence. The familiar yet fictional environment makes participants engaged but less concerned about their real-life status. Controlled conditions of the experiment allow several variations (anonymity, iterations, etc.) while the controlled pace of the game may inhibit overly deliberate responses for the benefit of more intuitive-based ones. Finally, money acts both as a familiar but emotional stimulus and a simple numeric estimate of the outcome.

## Basic Game of Trust

The original 'game of trust' has been described in [Berg1995] as an abstracted investment game. The game is played in pairs and rules of the game are known to both players. Alice, the first player (known also as a buyer or principal) is endowed with $10 and may invest a certain amount of money in Bob who is another player (seller, agent). Bob has also been endowed with $10 but he also receives from the 'bank' three times the amount invested by Alice. He has the opportunity to transfer all or some of his money to Alice (but may decide not to transfer anything).

It has been observed that if both players follow their economic best interests, Alice should never invest and Bob will never be able to re-pay anything, so that both of them ends up gaining $10 each. Monetary flow, its volume and character are attributable entirely to the existence of trust (as there is no control instrument available). Under optimum conditions to trust, Alice should invest all her money in Bob, and he should return her $20, keeping another $20 for himself. The amount of money invested by Alice can be used as a measure of her trust while the amount of money returned by Bob on top of Alice's investment is a measure of his trustworthiness.

Interestingly, what is commonly observed challenges purely selfish justifications. Even under the cloak of anonymity (i.e. where there is no expectation of further interactions), buyers invested in the average $5.16 and sellers returned $4.66 [Berg1995]. The payoff in indefinitely repetitive, non-anonymous games (where the reciprocity can be expected as the relationship develops) is generally higher, even though results vary depending on details of the arrangement. Note that the payoff in definite repetitive games can be theoretically very similar to non-repetitive games as players can iteratively deduce their preferred strategy from the fact that the other party is likely to default in the last step (see e.g. [Ullmann-Margalit2001]), a testimony to the strength of the continuity factor in trust building.

## Alternative Versions

Rules of the game can be reversed into what can be called a game of distrust [Bohnet2005]. In such game, Alice initially has nothing and Bob has $40. Alice may withdraw money from Bob (up to $30) but she will receive only one-third of it (the 'bank' will hold the rest). In the next step Bob can return any of his remaining money to Alice, and this time Alice will receive all this money. Again, if Alice does not trust Bob then her best move is to withdraw $30 (of which she will receive $10) and leave Bob with remaining $10. However, if she withdraws nothing (trusting Bob) then Bob can return her $20 and still keep $20 for himself.

Signalling in the game of trust combines the act of speech and transactional exchange, unlike in real life where the communication usually precedes and facilitates the action. [Ariau2006] proposes an additional declaratory step in the game of trust with incomplete information (i.e. the payoff matrix is not known) to allow both parties to determine the extent of commitment between players. This should allow them to learn to trust each other across repetitions of the game and eventually facilitate the convergence of both players to Pareto-optimal Nash equilibrium.

Changing the initial distribution of endowments (and associated rules) alters the perception and behaviour of players. Assuming that two basic strategies: the economically rational one (no trust) and the optimal one (full trust) should end here in the same financial result, altering the perception causes Alice to trust (invest) twice as much and receive about a half of what she has received in a trust game, as Bob behaves in a less trustworthy manner.

## Prisoner's Dilemma

Other interesting games are e.g. binary-choice trust games [Camerer1988], the gift-exchange game [Fehr1993] and various other forms of social games. Specifically the Prisoner's Dilemma (formalised by W. Tucker in [Poundstone1992]) is popularly used to link trust with economic utility and demonstrate the rationality behind reciprocity. The Prisoner's Dilemma is defined as a game between two prisoners (hence the name of the game) where each prisoner must make in isolation an independent decision whether he will cooperate or betray (defect) the other one. The payoff function is set in such a way that each prisoner always gains certain benefit if he defects, may gain higher benefit if both prisoners cooperate but may incur significant losses if he cooperates and the other chooses to defect. Rational approach should lead both players to defect so that the decision to cooperate is interpreted as a sign of trust.

The average level of trust can be determined by playing the game in random pairs within the large group. Playing the game repetitively between the same pair allows development of the relationship between participants, specifically if combined with the ability to punish those participants who did not cooperate (which brings elements of control into the scope of the game). Several strategies have been created for players to develop the relationship towards cooperation. Specifically, the strategy of 'tit-for-tat with forgiveness' [O'Riordan2001] allows the quick development of a relationship but permits recovery from incidental crisis, that may be caused e.g. by communication noise or misunderstanding.

## Multi-player Games

Regarding games with several participants, trade-based games are very popular. Different forms of close market simulations allow participants to exchange goods for money. For example [Bolton2003], the game may require a buyer to initiate the transaction by sending money to a seller. The seller then can either complete the transaction by shipping goods or default (and keep the money). Such a game also can distinguish between trust (that can be measured as a number of initiated transactions), trustworthiness (a number of transactions completed) and efficiency (the percentage of transactions completed as a fraction of all possible transactions).

If the game is combined with the reputation mechanism (introducing control in addition to trust), it allows for a group-level view on the behaviour of its participants. Buyers can provide feedback to the reputation system and can use the reputation to influence their purchase decisions. Higher reputation usually leads to higher profit so that it is either the final profit or the reputation scoring that can be used as a measure of confidence. Such games can actually be played in a real environment (e.g. [Resnick2006]) leading to interesting findings.

Simulated market-like games may suffer from the same symptoms as the real market sometimes does: facing the prospect of limited continuity, participants may build a reputation only to benefit from destroying it in the last move (e.g. [Bolton2003]).

## 3.4  Value of Confidence

The model assumes the existence of two basic operators: the 'addition' $\oplus$ that sums confidence from several sources and the 'multiplication' $\otimes$ that discounts one confidence by another. Those two operations can be also called 'fusion' and 'discount', respectively.

The model itself does not provide any guidelines about the exact meaning of those operators. There is a certain latent expectation that those operators may be relevant to confidence that is understood as subjective probabilities, with values from 0 to 1. However, the reality is more complicated: the pragmatics of those operators may be different, and the difference is visible when it comes to the way risk and uncertainty are handled. The representation of risk usually uses one of two solutions, both associated with the concept of probability: either probability distribution or subjective probabilities. Representing uncertainty has developed into solutions that integrate uncertainty into risk and those that keep it orthogonal. No solution is inherently better than others, but they have different implication and application areas.

Even if we simply want to express confidence as a number, the amount of different schemes and ranges that are currently in use is almost equal to the number of available models (hence it is quite large). There are systems that assume only binary values [Adams2002], that use fixed scale [Zimmermann1994], where confidence ranges from −100 to +100 (while excluding zero) [Grandison2003], from 0 to 1 [Mui2002] or from [−1 to +1) [Marsh1994]; where confidence is discrete or continuous, one-dimensional or has many dimensions [Gujral2006]. For a further overview, see [Abdul-Rahman2005]. There is also disagreement about the semantics of the value (specifically on the negative side of the range). There are also models that use ordered set of values without attempting to convert them to any particular numerical range (e.g. [Nielsen2004]).

### 3.4.1 Expressing Risk

If Alice really knows Bob (and the environment he operates in) then she can construct the complete model of his behaviour. She may observe him for a very long time, understand his reasoning so that even if he behaves erratically, she may discover a certain *probability distribution* of his behaviour. Such distribution can always tell her the complete (even though statistical only) truth about Bob. This is the first (and probably the most beneficial) way of expressing risk.

However, Alice may be less interested in understanding Bob and more interested in her expectations regarding Bob's future activity – whether he will help her or not. What we will end up with is Alice's level of expectation regarding particular events in the future: to what extent she thinks such an event is likely to happen. We will therefore have a *subjective probability* of the event. Of course, if Alice knows Bob's probability distribution, she can calculate the expected probability, but the concept of subjective probability also caters for the fact that Alice is unlikely to know everything about Bob and that Alice may not be able to evaluate everything that she knows.

### Probability Distribution

The most convenient from the research perspective (yet the most unlikely from the real life perspective) is the situation where the probability distribution of possible actions of particular entities is known in advance and is known not to change. If this is the case, agents (unlike humans) can entirely rely on the control path in evaluating their expectations regarding confidence in other agents on the basis of rational calculation from all available data. In a much more realistic version the distribution is not known in advance but its estimated by the known model on the basis of the outcome of past interactions, e.g. by using empirical Bayesian estimators and Beta distribution.

Depending on the level of details, the probability distribution may even include events that are quite unlikely to happen. For example, the distribution may state that Bob supports Alice 68 per cent of the time, not support her 29 per cent, does not respond to communication at 2 per cent of cases and is likely to withdraw from the experiment with the probability of 1 per cent. The more detailed the probability distribution is, the better for Alice. However, the really detailed distribution requires an indefinitely long observation time and an indefinite number of interactions – something that Alice is unlikely to wait for.

The use of probability distribution is particularly popular in the area of autonomous agents (e.g. [Luck2005]) where independent agents interact (e.g. trading their wares) in an artificially created world. Being currently mostly of academic interest (even though practical implications are quite interesting), creators of agents and their world can often assume that all probability distributions are known and that agents have enough number-crunching capabilities to deal with complex equations. Similarly, they can assume that within this limited world agents behave according to known models of probability distribution.

Probability and statistics assume that the future can be (statistically of course) predicted from the past. As such, they make two silent claims: about the stability of the world and about the non-changeability of an agent (Bob). The first claim is common to all schemes presented here and can be accepted without further consideration. The second claim may lead to incorrect predictions if Bob is able to change his intentions (hence his behaviour). Note that simple statistical methods are not concerned with Bob's intentions. In fact, they are not concerned with Bob at all, as they describe observed statistical regularities. There are potentially several methods to overcome this limitation: from adding the 'uncertainty', 'ageing' and 'forgetting' factors to using statistical models that are able to accommodate intentions and trend (and changes to it) rather than stable values, e.g. the hidden Markov model or a dynamic Bayesian network.

The use of probability distribution does not mean that there is no space for uncertainty. The distribution talks only about risk in terms of probabilities of known types of outcome. There are still uncertainties regarding unknown types of outcome or the validity of observations. How uncertainty is catered for will be discussed later, for now we assume that in the case of probability distribution confidence (more exactly: risk embedded in confidence) is expressed as a following function (3.1):

$$risk^{subject}_{Alice \rightarrow Bob} \equiv P^{subject}_{Alice \rightarrow Bob} \tag{3.1}$$

where:

$P^{subject}_{Alice \rightarrow Bob}$  probability distribution of Bob's behaviour regarding subject, as seen by Alice

## Subjective Probability

The concept of *subjective probability* [Gambetta2000] originates from the fact that Alice has very limited processing capability so that she must keep her reasoning simple, driven by bounded rationality [Simon1957]. Similarly, Alice's ability to collect evidence about Bob is quite limited – she may not have access to all evidence, she may not have time to collect them or to wait for them. Therefore Alice is unlikely to develop and explore the probability distribution of Bob's behaviour, but she is more likely to develop a simple expectation regarding the likelihood of some future events.

There are two significant elements that distinguish probability distribution from subjective probability. First, subjective probability discusses the level of expected probability regarding a particular event (Bob's beneficial behaviour), not the probability distribution – so that it only answers the question how likely it is for the event to happen (according to Alice's beliefs). Second, subjective probability is subjective, i.e. it is derived from Alice's beliefs and incomplete information that is available to her, not from complete and objective observations.

For subjective probabilities, the risk factor of confidence is the expected value of probability, not the complete probability distribution, so that risk is defined as the following value (not a function) (3.2):

$$risk^{subject}_{Alice \rightarrow Bob} \equiv p^{subject}_{Alice \rightarrow Bob} \tag{3.2}$$

where:

$p_{Alice \to Bob}^{subject}$   Alice's subjective probability of Bob's beneficial behaviour regarding *subject*

The construct of subjective probability agrees better with an intuition regarding expectations of future events: people tend to think in terms of subjective probability of events, not in terms of complete probability distribution. Therefore the use of subjective probabilities is quite common in research. Subjective probabilities are used in several empirical measurements and – to certain extent – in reputation-based systems and in agent networks.

## 3.4.2 Sources of Uncertainty

There are several sources of uncertainty within the model. Variability of enablers contributes to uncertainty rather than to the level of confidence. In addition, time also alters the perception of uncertainty.

The lack of similarity introduces uncertainty regarding the semantics of evidence and intentions. If Carol is significantly different from Alice, then Alice is uncertain whether they both share the same interpretation of evidences – there is a 'semantic gap' between what Carol communicates and what Alice understands. For example, if Carol prefers fast food while Alice is more into gourmet dining, then Carol's assessment that the particular restaurant is good has no relevance to Alice – actually this is probably the restaurant that Alice should avoid. Carol's undemanding palate (that is no match to Alice's sophistication) may make her deliver an assessment that may be misleading. Collaborative filtering (e.g. [Golbeck2004] or [Benkler2006]) to a certain extent addresses and alleviates this aspect of incompatibility of similarity by grouping agents of similar preferences at the expense of the amount of available evidence, effectively offsetting one type of uncertainty with another.

Further, if Bob is not similar to Alice, then Alice cannot reliably assume that Bob is an intentional agent and consequently cannot attribute intentions to him. Here the potential problem is more damaging than in the case of dissimilarity of evidences – if Bob is not an intentional agent then trusting him is not warranted, as he cannot hold any trustworthy intentions. In an interesting psychological twist, Alice is willing to attribute intentionality to non-humans [Dennet1989] and to deny intentionality to apparent humans, if they are believed not to be similar to her.

The level of honesty of an agent that is reporting certain evidence is an obvious source of uncertainty. Even if Alice assumes that there is no misunderstanding regarding the semantics of evidence, Dave may himself be misinformed, incompetent or may behave in a malicious manner. Note that not all dishonesty is malicious – sometimes Dave may be willing to provide honest information but he himself has been misled. From Alice's perspective unfortunately all those cases look similar. If Alice does not know how Dave collected and processed evidence, what his motivations are or his level of competence, then she is uncertain about his reports. Interestingly, Alice may even be dishonest with herself, tweaking first-hand evidence – including evidence of other's honesty.

Alice cannot even be certain that some agents exist or that their identity has not changed in time. If Carol stops providing evidence, Alice can assume that Carol unfortunately has ceased to exist or that she no longer has time to visit new restaurants. If Dave has always been a great fan of Chinese take-away and suddenly he competently keeps reporting his experiences from Michelin-starred restaurants, is this the same Dave who suddenly has struck it rich or has his identity been stolen?

The number of available evidence (the availability) also influences Alice's uncertainty. Even though no amount of evidence can make Alice certain about the future, the more evidence she has, the better (subjectively at least) she is able to predict the future. Finally, decay of the value of information in time leads to uncertainty as well. Alice can reasonably expect that other agents may change their behaviour in time – not too fast (as it leads to uncertainty of their identities), but gradually. Therefore old evidence is less reliable in assessing future behaviour and Alice should treat it with caution. Further, Alice's memory may hold a less clear record of such evidence, while details and context may fade away.

There are several sources of uncertainty, but for simplicity in all future considerations we will combine them together into one value that can express our uncertainty. Actually, for the purpose of computations it may be more beneficial to talk about certainty, the extent to which the situation is certain. Note that this simplification may need improvement if we start considering malicious agents. For example, if Dave is perceived as benevolently dishonest (e.g. he tries to be honest but has been misled by others) then uncertainty associated with his evidence can be set high so that Alice is less likely to listen to him. However, if Dave is malicious, then Alice may be better off if she listens to him and then inverts whatever he says. This, if modelled, may call for 'negative certainty'. We will save those considerations for later.

## 3.5 Combining Risk and Uncertainty

Assuming that we can measure (estimate, guess, predict, etc.) independently risk and uncertainty, such uncertainty must be combined with risk in order to create the complete metrics that can be used to express confidence. There are two methods to achieve this goal. First, uncertainty can directly influence the value of confidence, e.g. by decreasing it, so that there is only one value that is processed. Such an approach is illustrated below by the construct of discounting.

In the second approach, uncertainty may remain orthogonal to risk so that all operations are performed on both: risk and uncertainty at the same time. Of the several methods that can be potentially used in this area, two are illustrated below. The first (decoration) is based directly on the Dempster–Shafer theory of evidence while the other (logic of uncertain probabilities) extends this theory towards the integration of probability distribution.

Another alternative, not discussed here, is to use fuzzy logic. Surprisingly, the fuzzy approach to confidence and trust is not as popular as might have been expected. Fuzzy systems ([Falcone2003], [Castelfranchi2003]) seem to be beneficial in linking natural language expressions with values that can be used for evaluation and processing them for meaningful results. The 'trust zone' [Manchala2000] can be determined on the basis of the most characteristic variables (such as cost, history, indemnity, etc.). The lack of certainty about future actions [Griffiths2006] can also be captured by fuzzy logic. The application of fuzzy logic to confidence and trust has been studied in the context of peer-to-peer networks [Damiani2003] to improve peer rating. Also, for grid computing [Song2004], it has been demonstrated that fuzzy logic allows security issues to be solved in a reliable and efficient manner. However, those works concentrated more on replacing probability-based models with ones that are based on fuzzy logic, less on exploiting the advantages of fuzzy logic in capturing uncertainty.

### 3.5.1 Discounting

Discounting is conceptually the simplest method of combining risk and certainty into confidence. Discounting applies a simple function to subjective probability (that is a measure of risk) and certainty, to produce confidence (3.3). Assuming the range from zero to one for both probability and certainty, the function can be as simple as a multiplication:

$$conf_{Alice \rightarrow Bob}^{subject} = f(p_{Alice \rightarrow Bob}^{subject}, cert_{Alice \rightarrow Bob}^{subject}) \qquad (3.3)$$

The process of discounting is opaque as it effectively hides the level of certainty from any further processing. However, uncertainty does not disappear just because we do not talk about it. Discounting, while evaluating the confidence, must substitute information that is missing due to uncertainty. As no other information is available, discounting replaces it with a stance, a belief about the default behaviour.

The problem associated with discounting can clearly be seen when Alice does not know anything about Bob – so that she proceeds with no evidence. Should she, in the absence of any evidence (thus at the highest possible uncertainty) decide to be confident, not confident or assume a certain value in-between? The stance Alice assumes is arbitrary as it reflects her understanding of what the neutral value and default behaviour are. Alice can e.g. assume zero as her neutral value (i.e. if she is not certain then she is not confident), 0.5 (if she is not certain then she is confident on average) or even one (granting others the benefit of doubt).

Differences among stances can be illustrated with a simple example. Let's assume the network shown in Figure 3.1. Alice receives evidence about Bob from two sources: Carol and Dave. She is uncertain about the honesty of both, and she is using her confidence in their honesty as a measure of her certainty (we ignore all other potential sources of uncertainty) while she is using what they tell her as her measure of her subjective probability.

If Alice takes the pessimistic stance (i.e. not knowing implies no confidence), she may apply the following formula (3.4) to evaluate her confidence, that will lead to the outcome of 0.25.

$$conf_{Alice \rightarrow Bob}^{subject} = \frac{\sum\limits_{x_1..x_n} conf_{Alice \rightarrow x_i}^{truth} \cdot conf_{x \rightarrow Bob}^{subject}}{n} \qquad (3.4)$$

This formula follows the Dempster and Shafer 'discount and combine' method [Shafer1976] but without recording the uncertainty as such, i.e. we resort here only to calculating belief in the positive action of Bob and do not discuss Alice's belief in the negative actions of Bob nor her uncertainty about the outcome. What we find is the estimate of the lower bound of probability that Bob will support Alice (hence the pessimistic stance).

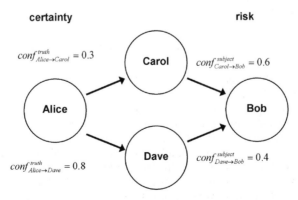

**Figure 3.1** Simple network with discounting

However, Alice may assume the optimistic stance and fill doubt with confidence rather than with the lack of it, leading to the result of 0.7. If Alice decides to proportionally replace doubt with confidence (the weighted stance), she ends up with 0.45. She may also assume the negative stance and think that because Carol is confident about Bob but Alice is not confident about Carol's honesty, then Alice should be suspicious about Bob, subtracting her confidence rather than adding it. The majority stance may assume that in a difference of opinions Alice should follow the majority and she should actually decrease her trust in the dissident minority. The number of potential stances is quite large, as the selection of a particular stance depends on the purpose of the overall evaluation.

The basic pessimistic stance is quite popular in research related to autonomous agents. Similar versions of this formula can be found e.g. in [Golbeck2004] and in [Marsh1994]. Note that the formula used here does not allow Alice to be more confident than the mostly honest neighbour, quickly leading to the loss in the maximum value of confidence that Alice can achieve.

As stated before, Alice's own uncertainty regarding such confidence is lost in this calculation, as it becomes integrated in the value of the confidence itself. It is therefore not possible to tell how Alice came to the particular conclusion. If Alice is presenting her confidence in Bob as her opinion to someone else, this information is lost: Alice is treated as an authoritative and opaque source of information about Bob.

Assuming that the stance is understood and accepted, discounting is quite powerful as a metrics, as it combines the relative simplicity of calculations, good intuitive understanding and a direct relationship to the probability theory.

## 3.5.2 Decoration

If we are dealing with subjective probabilities, it is possible to 'decorate' such probability with the indicator of certainty and to perform operations on both elements of confidence at the same time. The confidence is therefore seen as a pair, where the probability part is associated with the perception of risk while certainty is captured separately (3.5):

$$conf_{Alice \to Bob}^{subject} \equiv (p_{Alice \to Bob}^{subject}, cert_{Alice \to Bob}^{subject}) \tag{3.5}$$

Dempster–Shafer's theory of evidence ([Shafer1976], see [Sentz2002] for an interesting overview) combines the subjective probability with uncertainty by defining confidence as a pair of two probabilities. Simplifying, from the analysis of mass associated with available evidence within the fixed area of discernment (here called the 'frame of discourse'), confidence is defined as follows (3.6):

$$conf_{Alice \to Bob}^{subject} \equiv (belief_{Alice \to Bob}^{subject}, plausibility_{Alice \to Bob}^{subject}) \tag{3.6}$$

The lower bound is called 'belief' and reflects the probability of expectation that a given even will happen (e.g. Bob will support Alice). The upper bound is called 'plausibility' and represents the one's complement of the belief that the given event will not happen. The area between belief and plausibility represents the extent of uncertainty.

For illustration only, let's consider the network that is identical to the previous example (Fig. 3.2). In this case, however, two opinions that are available to Alice are represented already as the (Belief, Plausibility) pair, on the basis of beliefs regarding honesty and opinions delivered by Carol and Dave. Opinion that came from Carol retained its high uncertainty (there was little confidence in Carol's honesty) while opinion from Dave reflects certainty about his belief.

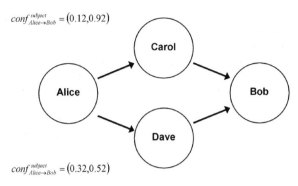

$$conf^{subject}_{Alice \to Bob} = (0.12, 0.92)$$

$$conf^{subject}_{Alice \to Bob} = (0.32, 0.52)$$

**Figure 3.2** The network with (*Belief, Plausibility*) pairs

Note that this example is used only to illustrate different methods of expressing uncertainty, not to compare them. Different ways of expressing uncertainty may have different philosophical and logical foundations and may not be easily comparable. Specifically, this representation of uncertainty assumes independence of evidence, i.e. Carol and Dave must not share their experience about Bob.

Similar to discounting, there are several ways by which Alice can combine those two opinions (confidences). In order to better understand differences between those rules it is worth thinking of Alice's frame of discourse as a product of two sets of three statements each, with probability associated with each statement delivered respectively by Carol and Dave. The square of nine areas outlined in Figure 3.3 represents probability resulting from different combinations of opinions, e.g. the lower left rectangle represents agreement between Carol and Dave that Bob will help while the upper left represents a conflict of opinions: Carol believes that Bob will not help while Dave believes he will help.

There is a general agreement that the size of the lower left rectangle contributes to Alice's belief and the upper right rectangle determines disbelief – the distance from certainty to plausibility. However, the remaining seven rectangles can be interpreted differently – as contributing to belief, disbelief or uncertainty. In addition, Alice may ignore certain rectangles while normalising areas.

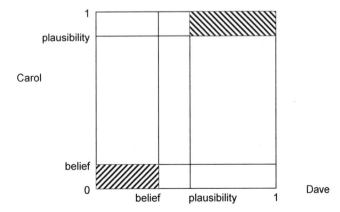

**Figure 3.3** The product frame of discourse

Originally Shafer's rules of combination excluded areas of disagreement and then proportionally distributed the weight of agreed areas, which is conceptually close to the weighted stance. By applying this method we can receive confidence of approx. (0.162, 0.838) that can be transformed back to an expected probability of 0.5 and an uncertainty of 0.676, reasonably close to the results that have been received for the weighted stance.

Applying different rules (Yager's modified Dempster's rule, Inagaki's unified combination role, Zhang's centre combination rule, Dubois and Prade's disjunctive consensus rule, etc.) leads to different results, depending whether uncertainty is maximised, evenly distributed, minimised and whether conflicts of opinions are excluded or included in calculations.

### 3.5.3 Uncertain Probability

We can consider the metrics where the uncertainty is permanently embedded in every probability distribution from the very start, thus always being an element of a consideration (3.7). This may potentially prevent the loss of information about uncertainty, but it requires a different logic, as simple decoration is no longer sufficient.

$$conf^{subject}_{Alice \to Bob} \equiv (P^{subject}_{Alice \to Bob}, cert^{subject}_{Alice \to Bob}) \tag{3.7}$$

The logic for uncertain probabilities was introduced by Josang ([Josang2001], [Josang2004]), on the basis of the Dempster–Shafer theory of evidence (e.g. [Lucas1991]). This elegant concept combines probability distribution with uncertainty, so that each opinion (that can be interpreted as a statement about confidence) can be viewed as a distribution of probability distributions where each distribution is qualified by associated uncertainty. The intuition behind the logic for uncertain probabilities is simple, even though the details may look quite complicated. The foundation of the logic is the observation that the opinion (an evidence or a confidence) can be represented as a four-tuple (3.8):

$$\omega_x \equiv (b(x), d(x), u(x), a(x)) \tag{3.8}$$

where:

$\omega_x$     opinion about $x$
$b(x)$     belief about $x$ (i.e. belief that $x$ will happen)
$d(x)$     disbelief about $x$ (i.e. belief that $x$ will not happen)
$u(x)$     uncertainty about $x$
$a(x)$     relative atomicity of $x$

Values of belief, disbelief and uncertainty are interconnected in the following equation, making this logic similar to the concept of opinion as a pair of (*Belief, Plausibility*), which is intuitively similar to the decorated subjective probability. The main difference is in fact that uncertain probabilities operate on a given family of probability distributions (beta distribution), not on a single expected value. There is a relationship between three out of four elements of a four-tuple (3.9):

$$b(x) + d(x) + u(x) = 1 \tag{3.9}$$

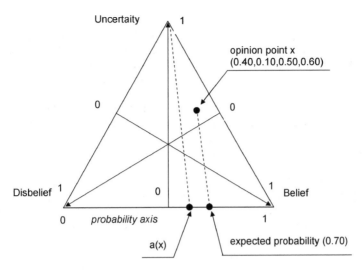

**Figure 3.4** Opinion triangle and opinion point

Opinion (confidence) can be graphically represented as points within the opinion triangle (Figure 3.4.), with three interconnecting axis representing belief, disbelief and uncertainty while atomicity is represented by value on a horizontal axis. The probability expectation associated with a given opinion can be derived as a projection of the opinion point onto the horizontal axis parallel to the line defined by the atomicity.

The logic for uncertain probabilities is an interesting example of a confidence metric where uncertainty is inherently embedded in the calculation process and is visible at the output. It is not the only one, as it is even possible to use a quadruplet (trust, distrust, unknown, ignorance) to express the value of confidence [Ding2003], for as long as the appropriate operations are defined. In order to apply the logic for uncertain probabilities to our simple example, we must make certain assumptions as to how data represented there has been collected. Assuming that each member has formulated his or her opinion on the basis of an equal number of independent interactions (e.g. 10 interactions), we can define our opinions as follows (Figure 3.5).

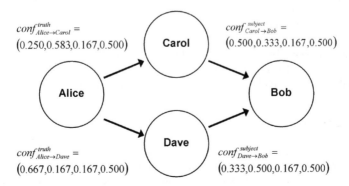

$$conf^{truth}_{Alice \to Carol} = (0.250, 0.583, 0.167, 0.500)$$

$$conf^{subject}_{Carol \to Bob} = (0.500, 0.333, 0.167, 0.500)$$

$$conf^{truth}_{Alice \to Dave} = (0.667, 0.167, 0.167, 0.500)$$

$$conf^{subject}_{Dave \to Bob} = (0.333, 0.500, 0.167, 0.500)$$

**Figure 3.5** The network of opinions

Note that the situation that Alice is facing is conceptually different from the situation that she experiences e.g. with subjective probabilities. Alice's confidence about Carol's and Bob's honesty still should discount what she learns from them, but here we are facing another uncertainty: the one that comes from the limited availability of evidence. Such uncertainty has not been of Alice's concern in previous examples, but it is an inherent element of our considerations here.

We can model all the confidences Alice is dealing with as opinions that have certain inherent uncertainty due to a limited amount of evidence and then apply discounting and consensus operators. We can intuitively expect that Alice's overall uncertainty will be higher (as we have introduced more uncertainty into the example) and that it may be reflected in a lower expected probability.

Alice's opinion about Bob can be now calculated as follows, on the basis of opinions about honesty and opinions about confidences. In order to make a comparable example, the discount operation will favour uncertainty (which roughly refers to the pessimistic stance) and the consensus will assume dependent opinions (as both Carol and Dave report on Bob). The formula is presented in (3.10):

$$conf_{Alice \to Bob}^{subject} = \left( conf_{Alice \to Carol}^{truth} \otimes con_{Carol \to Bob}^{subject} \right)$$
$$\oplus \left( conf_{Alice \to Dave}^{truth} \otimes conf_{Dave \to Bob}^{subject} \right) \tag{3.10}$$

where operators $\otimes$ and $\oplus$ are defined as discounting and consensus, respectively.

By substituting numerical values and calculating the outcome, we get (0.187, 0.243, 0.569, 0.0) and the expected probability is equal to 0.187. There is a difference between the result produced by both methods: discounting subjective probabilities returned 0.25. Assuming the larger number of proofs (thus reducing additional uncertainty that we have introduced into the model), this method may deliver a slightly higher expected probability of around 0.289, this time above what has been provided by discounting. Differences can be attributed to different underlying assumptions about distribution and the way uncertainty is handled. Specifically, one can apply the opposite belief favouring discount and end up with the result of approx. 0.45, close to the results provided by the weighted stance. Apparently, the logic for uncertain probabilities allows more detailed analysis of the relationship but the rough outcome is reasonably similar.

The main value of the logic for uncertain probabilities may lie not in more accurate results, but in the ability to embed uncertainty into every opinion so that it can be carried and processed over the long chain of connections in a consistent manner, allowing Alice to deliver statements that combine both her expectations (confidence) and the associated certainty (which can also be interpreted as her trust in her own statement), such as 'I quite believe that Bob will rather help me' – where 'quite' refers to her certainty and 'rather' – to her confidence. Further, the logic provides representation and explanation for certain paradoxes (e.g. the Ellsberg paradox [Ellsberg1961]) that may not be catered for by subjective probabilities alone. From the perspective of the model it is additionally important that the logic seems to reflect operations that are suggested by the model in a satisfactory manners.

## 3.6 Conclusion

There is an inherent problem with assigning quantitative values to qualitative phenomena. Concerning beliefs (such as confidence) Alice is quite ready to state that she is more confident in Carol than in Bob. However, whether her confidence in Carol is 0.6 and in Bob is 0.5, or 0.65 and 0.59 respectively is probably impossible for her to tell or for us to decide. While surveys, questionnaires or games are essential to link theories to real life, they are not perfect. Assigning a value to confidence remains a tricky issue.

Of the different methods to represent confidence, the logic for uncertain probabilities relates well to the model and ideas discussed in this book. However, the logic for uncertain probabilities may not reflect well what we experience in a usual course of events. Alice seldom delivers very insightful introspection and it is unlikely that she will be able to assess both her confidence and her certainty about confidence. Therefore the logic for uncertain probabilities may be (at least temporarily) confined to the area of automatic reasoning and autonomous agents, before more sophisticated experiment methods become available.

The model described in this book intentionally uses subjective probabilities to evaluate confidence, trust and control. Subjective probability is used for several purposes: as a simplifying concept, as an explanatory tool, as a measurement of the strength of beliefs and as a potential link with the observable behaviour.

It is a simplifying concept because this model (and this book) is about Alice and it is taking the human perspective on the evaluation of confidence. Confidence here is a method to deal with complexity, so that we can reasonably expect that the processing of confidence will not add extra complexity by itself, i.e. that the way confidence is evaluated will provide little overhead comparing to the original situation. If the metrics itself is too complex (from the human point of view), then there should be no benefit in using trust or control.

However, subjective probability is used here as an explanatory tool in a sense similar to that proposed in [Gambetta2000]: to conveniently describe the presupposed internal state of mind, the system of beliefs, intentions and reasoning that is associated with confidence. Subjective probability is used as a common denominator to explain things that range from the subjective value of evidence to the perceived minimum confidence to trust in someone's statement.

Selecting the simple metrics allows us to concentrate on the model itself rather than going into details of evaluation. The simplicity of computation associated with subjective probabilities (there are only few formulas that are derived from basic statistics) allows us to spend more time on things that are more important: the concept, structure of the model, classes of evidences, implications for different application areas, etc.

If Alice had more processing power (e.g. should Alice use the trustworthy computer) or if we speak about agent Alice rather than human Alice or if Alice lived in a smaller world, the situation might have been different. Preferences will be likely for uncertain probabilities as a metrics, evaluation of probability distribution and the preference of decision-making process until all potential avenues are explored. As already stated, uncertain probabilities are fit for this task: they allow expression of evaluation along the control path and they can easily be adopted to cover the trust path as well.

# Bibliography

[Abdul-Rahman2005] Alfarez Abdul-Rahman: A Framework for Decentralised Trust Reasoning. PhD thesis. Available at: http://www.cs.ucl.ac.uk/staff/F.AbdulRahman/docs/thesis-final.pdf. 2005.

[Adams2002] Carlisle Adams and Steve Lloyd: *Understanding PKI: Concepts, Standards, and Deployment Considerations*. Sams Publishing. 2nd edition. 2002.

[Ariau2006] Stephane Airiau and Sandip Sen: Learning to Commit in Repeated Games. In *Proc. of the Fifth Int. Joint Conf. on Autonomous Agents and Multiagent Systems (AAMAS06)*. 2006.

[Benkler2006] Yochai Benkler: *The Wealth of Networks: How Social Production Transforms Markets and Freedom*. New Haven, CT: Yale University Press. 2006.

[Berg1995] J. Berg, J. Dickhaut and K. McCabe, Trust, Reciprocity, and Social History, *Games and Economic Behavior*, 10: 122–142. 1995.

[Bohnet2005] Iris Bohnet and Stephan Meier: Deciding to Distrust. KSG Working Paper No. RWP05-049. Available at: http://ssrn.com/abstract=839225. 2005.

[Bolton2004] Gary E. Bolton, Elena Katok and Axel Ockenfels: How Effective Are Electronic Reputation Mechanisms? An Experimental Investigation. *Management Science*, 50(11): 1587–1602. 2004.

[Camerer1988] Colin Camerer and Keith Weigelt: Experimental Tests of a Sequential Equilibrium Reputation Model. *Econometrica*, 56(1): 1–36. 1988.

[Castelfranchi2000] Cristiano Castelfranchi and Rino Falcone: Trust Is Much More than Subjective Probability: Mental Components and Sources of Trust. In *Proc. of the 33rd Hawaii Int. Conf. on System Sciences (HICSS2000)*. Vol. 6. Available at: http://www.istc.cnr.it/T3/download/Trust-more-than-probability.pdf. 2006.

[Castelfranchi2003] Cristiano Castelfranchi, Rino Falcone and Giovanni Pezzulo: Trust in Information Sources as a Source for Trust: A Fuzzy Approach. In *Proc. of AAMAS'03*, July 14–18, 2003, Melbourne, Australia. 2003.

[Corritore2005] Cynthia L. Corritore et al: Measuring Online Trust of Websites: Credibility, Perceived Ease of Use, and Risk. In *Proc. of Eleventh Americas Conf. on Information Systems*, Omaha, NE, USA, pp. 2419–2427. 2005.

[Cummings1996] L. L. Cummings and Philip Bromiley: The Organizational Trust Inventory (OTI): Development and Validation. In Roderick M. Kramer and Tom R. Tyler: *Trust in Organizations*. London: Sage Publications. 1996.

[Damiani2003] E. Damiani et al: Fuzzy Logic Techniques for Reputation Management in Anonymous Peer-to-peer Systems. In *Proc. of the Third Int. Conf. in Fuzzy Logic and Technology*, Zittau, Germany. 2003.

[Dennett1989] D.C. Dennett: *The Intentional Stance*. Cambridge, MA: MIT Press. 1989.

[Dimitrakos2003] Theo Dimitrakos: A Service-Oriented Trust Management Framework. In R. Falcone et al. (eds): *AAMAS 2002 Workshop on Trust, Reputation . . .* LNAI 2631, Berlin: Springer. pp. 53–72. 2003.

[Ding2003] Li Ding, Lina Zhou and Timothy Finin: Trust Based Knowledge Outsourcing for Semantic Web Agents. In *2003 IEEE / WIC Int. Conf. on Web Intelligence, (WI 2003)*, 13–17 October 2003, Halifax, Canada. 2003.

[Ellsberg1961] Daniel Ellsberg: Risk, Ambiguity, and the Savage Axiom. *Quarterly J. of Economics*, 75: 643–669. 1961.

[Elofson2001] Greg Elofson: Developing Trust with Intelligent Agents: An Exploratory Study. In Cristiano Castelfranchi and Yao-Hua Tan (eds): *Trust and Deception in Virtual Societies*, pp. 125–138. Dordrecht: Kluwer Academic Publishers. 2001.

[Falcone2003] Rino Falcone, Giovanni Pezzulo and Cristiano Castelfranchi: A Fuzzy Approach to a Belief-Based Trust Computation. In R. Falcone et al. (eds): *AAMAS 2002 Ws Trust, Reputation*, LNAI 2631 pp. 73–86. 2003.

[Fehr1993] Ernst Fehr, Georg Kirchsteiger and Arno Riedl: Does Fairness Prevent Market Clearing? An Experimental Investigation. *Quarterly J. of Economics*, 108(May): 437–60. 1993.

[Gambetta2000] Diego Gambetta: Can We Trust Trust? In Diego Gambetta (ed.) *Trust: Making and Breaking Cooperative Relations*, electronic edition, Department of Sociology, University of Oxford, Chapter 13, pp. 213–237. Available at: http://www.sociology.ox.ac.uk/papers/gambetta213-237. 2000.

[Golbeck2004] Jennifer Golbeck, James Hendler: Reputation Network Analysis for Email Filtering. *Proc. of the First Conf. on Email and Anti-Spam*, 30–31 July, Mountain View, California. 2004.

[Golbeck2004] Jennifer Golbeck and Bijan Parsia: Trusting Claims from Trusted Sources: Trust Network Based Filtering of Aggregated Claims. *Int. J. of Metadata, Semantics, and Ontologies*, 1 (1). Available at: http://trust.mindswap.org/papers/ijmso.pdf. 2005.

[Grandison2003] Tyrone Grandison: Trust Management for Internet Applications. PhD thesis, University of London. 2003.

[Griffiths2006] Nathan Griffiths: A Fuzzy Approach to Reasoning with Trust, Distrust and Insufficient Trust. In M. Klusch, M. Rovatsos, and T. Payne (eds): *CIA 2006*, LNCS 4149, pp. 360–374. 2006.

[Gujral2006] Nishit Gujral, David DeAngelis, Karen K. Fullam and K. Suzanne Barber: Modelling Multi-Dimensional Trust. In *Proc. of Fifth Int. Conf. on Autonomous Agents and Multiagent Systems AAMAS-06*. Hakodate, Japan. 2006.

[Habermas1986] Jürgen Habermas: *The Theory of Communicative Action. Vol. 1. Reason and the Rationalization of Society*. Cambridge: Polity Press. 1986.

[Helton1997] J. C. Helton: Uncertainty and Sensitivity Analysis in the Presence of Stochastic and Subjective Uncertainty. *J. of Statistical Computation and Simulation*, 57: 3–76. 1997.

[Josang2001] Audun Josang: A Logic for Uncertain Probabilities. *Int. J. of Uncertainty, Fuzziness and Knowledge-Based Systems*, 9: 279–311, June. 2001.

[Josang2004] Audun Josang, David McAnally: Multiplication and Comultiplication of Beliefs. *Int. J. of Approximate Reasoning*, 38/1: 19–55. 2004.

[Keser2003] C. Keser: Experimental Games for the Design of Reputation Management Systems. *IBM Systems J.*, 42(3). 2003.

[Knight1921] Frank H. Knight: *Risk, Uncertainty, and Profit: Hart, Schaffner & Marx*. Boston: Houghton Mifflin Company. 1921.

[Lucas1991] P. Lucas and L. Van Der Gaal: *Principles of Expert Systems*. Reading, MA: Addison-Wesley Publishing. 1991.

[Luck2005] Michael Luck et al: Agent Technology: Computing as Interaction. A Roadmap for Agent-based Computing. *AgentLink*. 2005.

[Manchala2000] Daniel W. Manchala: E-Commerce Trust Metrics and Models. *IEEE Internet Computing*, March–April: 36–44. 2000.

[Marsh1994] Stephen P. Marsh: Formalising Trust as a Computational Concept. PhD thesis, University of Stirling. Available at: http://www.nr.no/~abie/Papers/TR133.pdf. 1994.

[McCroskey1966] James C. McCroskey: Scales for the Measurement of Ethos. *Speech Monographs*, 33: 65–72. Available at: http://www.jamescmccroskey.com/publications/22.htm. 1966.

[McKnight2001] D. Harrison McKnight and Norman L. Chervany: Conceptualizing Trust: A Typology and E-Commerce Customer Relationships Model. *Proc. of the 34th Hawaii Int. Conf. on System Sciences*. 2001.

[McKnight2003] D. Harrison McKnight, Chuck Kacmar and Vivek Choudhury: Whoops... Did I Use the Wrong Concept to Predict E-Commerce Trust? Modelling the Risk-Related Effects of Trust Versus Distrust Concepts. In *Proc. of the 36th Hawaii Int. Conf. on System Sciences (HICSS'03)*. Available at: http://csdl2.computer.org/comp/Proc./hicss/2003/1874/07/187470182b.pdf. 2003.

[Mui2002] Lik Mui et al.: A Computational Model of Trust and Reputation. In *35th Hawaii Int. Conf. on System Science (HICSS)*. 2002.

[Nielsen2004] Mogens Nielsen and Karl Krukow: On the Formal Modelling of Trust in Reputation-Based Systems. In Juhani Karhumäki, et al. (eds): *Theory Is Forever: Essays Dedicated to Arto Salomaa on the Occasion of His 70th Birthday*. Lecture Notes in Computer Science 3113. Berlin: Springer. 2004.

[O'Riordan2001] Colm O'Riordan: Forgiveness in the Iterated Prisoner's Dilemma and Cooperation in Multi-Agent Systems. In Mike Luck and Mark d'Inverno (eds): *UKMAS*. 2001.

[Ormerod2005] Paul Ormerod: *Why Most Things Fail: And How to Avoid It*. London: Faber and Faber. 2006.

[Poundstone1992] William Poundstone: *Prisoner's Dilemma*. New York: Doubleday. 1992.

[Rempel1985] John K. Rempel, John G. Holmes and Mark P. Zanna: Trust in Close Relationships. *J. of Personality and Social Psychology*, 49(1): 95–112. 1985.

[Resnick2006] Paul Resnick: The Value of Reputation on eBay: A Controlled Experiment. *Experimental Economics*, 9(2): 79–101. Available at: http://www.si.umich.edu/~presnick/papers/postcards/PostcardsFinalPrePub.pdf. 2006.

[Rotter1971] John B. Rotter: Generalized Expectancies for Interpersonal Trust. *American Psychologist*, 26(5): 443–52. 1971.

[Sentz2002] Kari Sentz and Scott Ferson: Combination of Evidence in Dempster-Shafer Theory. *SAND 2002–0835*. Binghamton University. 2002.

[Shafer1976] G. Schafer: *A Mathematical Theory of Evidence*. Princeton, NJ: Princeton University Press. 1976.

[Simon1957] Herbert Simon: A Behavioral Model of Rational Choice. In *Models of Man*. Chichester: John Wiley. 1957.

[Song2004] Shansang Song, Kai Hwang and Mikin Macwan: Fuzzy Trust Integration for Security Enforcement in Grid Computing. In *Proc. of IFIP Int. Symposium on Network and Parallel Computing (NPC-2004)*. LNCS 3222. pp. 9–21. 2004.

[Ullmann-Margalit2001] Edna Ullmann-Margalit: Trust out of Distrust. Available at: www.law.nyu.edu/clppt/program2001/readings/ullman_margalit/Trust%20out%20of%20Distrust.pdf. 2001.

[Zimmermann1994] Phil Zimmermann (ed.): *PGP User's Guide*. Cambridge, MA: MIT Press. 1994.

# 4

# Comparative Analysis of the Model of Confidence

Compatibility means deliberately repeating other people's mistakes.

(David Wheeler)

## 4.1 Introduction

No model exist in a vacuum, but it usually complements or contradicts others while possibly delivering fresh views and addressing missing phenomena. The complexity-based model of confidence is no exception – it was designed in a rich scope of research that has already produced several models of confidence (usually called 'models of trust'). Of those, there are several models that already address the human process of the assessment of trust in the context of digital media. Specifically, e-commerce has generated significant interest as it naturally lies on the intersection of technology and social sciences.

From the perspective of analysis presented in the chapter, the complexity-based model of confidence postulates that disposable complexity can be exchanged for certainty of a level of confidence, by combining (fusing) confidence regarding four types of entities: world, domains, agents and transactions. Confidence regarding each entity can be assessed on the basis of evidence of trust and control. Simplifying the recursive nature of the model (as control relies on confidence in instruments that can be described through a recursive application of the model), the structure of Alice's beliefs in confidence postulated by the model can be shown in Figure 4.1.

The assessment of confidence is driven by availability of evidences. There are three classes of evidences that build trust: competence, continuity and motivation. Similarly, there are three classes of evidences for control: knowledge, influence and assurance.

*Trust, Complexity and Control: Confidence in a Convergent World*   Piotr Cofta
© 2007 John Wiley & Sons, Ltd

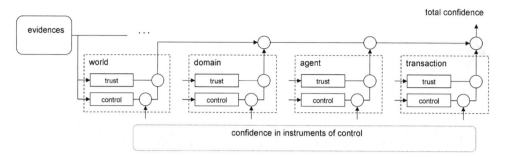

**Figure 4.1**   The simplified structure proposed by the model

On the basis of [Cofta2006], this chapter compares the complexity-based model of trust with some established models that address the similar area and problems. It is demonstrated that:

- The model contains all the non-disputable elements of established models and will not be in conflict with other elements of those models.
- The model provides a good explanation of certain phenomena that were insufficiently explained by previous models.
- The model provides a foundation to better explain the relationship between elements of the model.

## 4.2 Models

There are several models that introduce a structure and explain the process of the assessment of confidence, trust creation, trust-based decisions, etc. (as those models go under different names). Quite a few of those models relate to our complexity-based model, even though they may address different phases, aspects or components of it. The analysis presented here has concentrated on models that share a similar approach to the model presented here. Specifically:

- Selected models addressed the process of the assessment of trust (confidence).
- Models originated from sociological or psychological analysis but they are applicable either to the technology or to the technology-mediated relationship between people.
- There is the potential to use those models for the purpose of the quantitative analysis of confidence (trust).

Further,

- Models are well established within the research community, popular and have withstood the research critique.
- It is beneficial if models originate in or have been applied to the same application area that is important for the creation of this model, i.e. e-commerce.

Note that it is not required that models should share the same underlying concept of the importance of complexity or that they should share the same theoretical underpinnings. Indeed, there is value in comparing our model with behavioural ones, that has been created on the basis of empirical observations.

The following three models were used for the comparison:

- McKnight's interdisciplinary model of trust [McKnight1996] with later modifications [McKnight2003]. This popular model describes the assessment of trust in e-commerce with special focus on two types of entities: domain and agents. In order to demonstrate a similar yet competing approach, Li's TTFM model [Li2004] is also analysed.
- Egger's MoTEC model of trust in e-commerce [Egger2003]. This model explores the creation of trust from the user's perspective, thus concentrating on the trust in the entity.
- Tan's generic model of trust for e-commerce ([Tan2000], [Ganzaroli1999]) that focuses on the transactional context, explicitly combining elements of trust and control.

The comparative analysis concentrates on the structural similarity between models, i.e. similarity of basic constructs and their relationships that are used to explain how confidence is attained (or assessed). Other structural similarities or differences are also noted, if present, e.g. in the form of a visible separation of trust and control or in the recursive usage of the model. The existence of enablers is usually not discussed (with some exceptions) as, in general, models do not directly address enablers but they assume their existence.

The relevance of our complexity-based model has been also verified by using it to structure and analyse the confidence-building process of a successful e-commerce website. This analysis is presented later in this book.

## 4.3 Comparative Analysis

The comparative analysis presented here is structured along the four basic aspects of the relationship between models:

1. *Scope.* This aspect analyses which contexts from our complexity-based model are covered by the discussed model and verifies whether the context of the discussed model can be entirely expressed in the terms of our model.
2. *Approach.* Basic assumptions that are fundamental to the model are discussed to see whether our model contains all the important assumptions of that discussed and whether missing assumptions can be explained.
3. *Components.* Using the set of constructs (specifically classes of evidence) from our model, this aspect verifies that all the important components of the discussed model are covered by our model.
4. *Relationships.* If the model postulates certain relationships, the discussion explains how such relationships relate to ones included in our model.

### 4.3.1 Interdisciplinary Model of Trust

This well-known model was developed by McKnight [McKnight1996] (with later changes in [McKnight2003]). It was successfully applied to e-commerce to predict the extent of trust associated with a particular agent (represented by the website). The structure of constructs defined by this model has been also applied to predict the impact of convergence on e-commerce [Cofta2005].

### Scope

Comparing this model with our complexity-based model, it is necessary to begin by noting that McKnight's model concentrates on domains and agents addressing the remaining entities ('world' and 'transaction') to lesser extent. The model defines 10 constructs that contribute to the creation of trust (this was later extended to 13 constructs) and defines several relationships between them. The model was complemented by extensive qualitative research that demonstrated relationships between different constructs proposed by the model.

## Approach

The model structures Alice's conceptual process that results in trusting behaviour to link conceptual and operational constructs of confidence (trust). The purpose of the model has been to combine the interdisciplinary typology related to trust in a single model, to facilitate the potential measurement of trust, with the potential application to measure trust in e-commerce. The model in fact combines psychology, sociology and social psychology and is built on the basis of inter-relationship between different constructs of trust used within those disciplines.

The model does not claim that Alice is gaining confidence or that Alice is assessing confidence, but it postulates that her beliefs regarding confidence can be described as a set of constructs, and that there are links between such constructs that allow practical analysis of otherwise inaccessible reasoning by Alice. Such constructs are created from conceptual categories derived from theoretical foundations, while their particular sub-constructs are expected to be measurable by sociological scales. Relationships between constructs are directional and indicate the influence of one belief on others, e.g. the disposition to trust influences all categories while trusting intentions are influenced by all other categories.

Comparing with our complexity-based model, it must be noted that McKnight's model is quite close in its approach with the exception that it does not attempt to explain the driver behind the confidence (or trust), nor potentially limiting factors. In the case of our model, complexity is used to explain both.

## Components

There is a significant similarity between components of McKnight's model and our model of confidence. Three out of four of McKnight's main conceptual categories (disposition to trust, institution-based trust, and trusting beliefs) roughly refer to three out of four types of entities of our model: world, domain and agent. Trusting intentions, the fourth conceptual category, can be seen as an equivalent of our transactional assessment of confidence.

Note that McKnight has been using constructs in the sense of elements of Alice's system of beliefs, not referring to any particular entities. Our model is postulating that Alice holds her own assessment of confidence regarding abstract entities and that such an assessment can be combined in the transactional one. Despite certain differences, both models have a very similar approach: our assessment of confidence is a belief and as such is in fact an element of Alice's system of beliefs. In a similar vein, we can assume that McKnight's sub-constructs should be related to our categories of evidence within different contexts. As every category and every path (trust or control) can be also represented as a belief, the expected relationship seems to be justifiable.

Table 4.1 summarises the relationship between McKnight's extended list of constructs and categories of evidence as defined by our model. In the leftmost column, Table 4.1 lists all McKnight's constructs with constructs from the extended model indicated in the 'Ext' column. For each construct the relevant category of evidence is identified by providing the entity it relates to, whether it is a trust or control-related category and the name of the actual category.

Of the 13 constructs listed, 11 clearly map into different respective categories of our model. In the case of McKnight's disposition to trust (that encompasses constructs relevant to the 'world' context), McKnight has used consolidated sub-constructs, one for trust and one for control. This is in agreement with the overall purpose of his model that should concentrate on the domains and agents. Similarly, there is no construct that addresses the transactional

**Table 4.1**  Comparison with McKnight's model

| Construct | Ext | Entity | T/C | Category |
|---|---|---|---|---|
| *Disposition to trust* | | | | |
| Faith in humanity | | world | trust | general |
| Faith in professionals | x | domain | trust | competence |
| Trusting stance | | world | control | general |
| *Institution-based trust* | | | | |
| Structural assurance | | domain | control | reassurance |
| | | domain | control | influence |
| Situation normality | | domain | trust | continuity |
| | | domain | control | knowledge |
| *Trusting beliefs* | | | | |
| Competence | | agent | trust | competence |
| Benevolence | | agent | trust | motivation |
| Integrity | | agent | trust | motivation |
| Predictability | | agent | control | knowledge |
| *Trusting intentions* | | | | |
| Willingness to depend | | | | |
| Predictability of depending | | domain | trust | continuity |
| | | domain | control | knowledge |
| Willingness to explore | x | | | |
| Perceived quality | | agent | trust | competence |

trust. Predictably, some constructs refer to more than one category of evidence, but it is worth noting that only four categories are not addressed by McKnight's model.

It is interesting to have a quick look at those categories that are not addressed by McKnight's model. In general, those omissions can be attributed to the relatively early stage of the e-commerce when the model was originally created. At such an early stage not all expectations had been clearly formulated so that the original model was not able to capture them. Following is a brief discussion of those missing constructs:

- *Domain – motivation*. The expectation that e-commerce (which was the original domain of the model) is motivated to act for the benefit of its customers is essential to the adoption of e-commerce. However, as the model developed in the early stages of e-commerce, customers have not yet created a perception of e-commerce sharing certain common properties, contrary to the common perception of such domains like traditional commerce, banking services, etc. Such a perception may currently exist (even though it may not be entirely beneficial for e-commerce) and such an expectation is valid within the model.
- *Agent – continuity*. It is surprising that the original model does not cater for the expectation of continuity. It is known [Bolton2003] that the perceived lack of continuity may lead to significant disruptions in the trust relationship. However, as continuity in the context of the domain is covered twice within the original model, such as omission within the context of the agent may have less influence on the viability of the model.
- *Agent – influence*. The ability to influence the transaction ('being in control') is frequently listed as an enabling factor for trust creation, leading to an increased demand regarding the transparency of transactions. However, the similar perception was not yet associated with agents when the model was proposed. The development of e-commerce created such a perception, differentiating between those vendors that inform and allow control of the transaction and those that do not, making this expectation justified.

- *Agent – assurance*. The omission of this construct can be attributed again to the relatively early stage of the e-commerce market at the time of the creation of the model. In such a market there was little reassurance provided by the particular agent, so that the main focus was on domain-wide reassurance provided e.g. by legal systems. With the development of large e-commerce sites such as Amazon or eBay, the expectation of reassurance provided by the agent becomes more important.

Additional discussion is needed of two constructs that are not directly referred to in the proposed model: willingness to depend and willingness to explore. Both constructs seem to be related in fact to the level of complexity the person is willing to bear under given circumstances. The willingness to depend represents the unwillingness to deal with complexity (i.e. the high willingness to depend can be translated into the low acceptable complexity). Our model explains that the willingness to depend, by reducing acceptable complexity, will drive the person towards trust (which usually bears the lower penalty regarding complexity). Therefore the willingness to depend should result in a higher level of trust, even though such trust may not be always accepted voluntarily, but rather out of necessity.

The willingness to explore is closely related to one of the enablers – availability, even though it is positioned differently. The role of the enablers is to deliver evidence that can be used by Alice to assess the level of confidence. Willingness to explore is positioned as a perception that comes from generally positive beliefs in the world and in a domain, making them a prerequisite of receiving any evidences regarding an individual agent. As the relationship between different enablers is not discussed within our model, McKnight's approach can be accepted here as an additional explanation for the model.

It is worth noting that McKnight's model does not address other enablers such as similarity, honesty or identity. This also can be attributed to the fact that electronic commerce, his primary area of interest, was not well developed when the model was developed, so that there were no clear constructs (specifically no observable sub-constructs) that related to the remaining enablers. Identity and similarity had not yet been recognised as a potential problem and it was not possible to clearly separate honesty from other enablers in one-to-one relationships.

## Relationships

A separate explanation is needed when it comes to comparing the structure of relationships. Our model assumes that Alice assesses confidence within a horizon of a transaction and that her transactional confidence is a 'fusion' (similar to a sum, but including respective uncertainties) of fractional confidences in entities included in such a horizon. The implication of this assumption is that high confidence in one category can substitute for the lower confidence in another category or for the lack of experience in a given category. However, within the scope of a transaction, there is no relationship between confidences in different categories, i.e. if Alice is confident in a world that she is not necessarily more confident in Bob just because of her high confidence in a world in general.

McKnight's model suggests that there is a relationship of influence and dependence between categories, i.e. that a higher level of confidence in one category positively influences a higher level of confidence in another category. For example, if Alice is confident in a world in general, she is likely to be more confident in Bob.

This difference can be explained by the mechanism that has been postulated by our model: stance. While there are several ways Alice can perform operations on confidence (fusion, discounting), she exhibits a particular personal trait in handling those operations is a particular way. Such a 'systematic error' is perceived as Alice's stance and is particularly visible in the absence of solid evidence: Alice may be e.g. pessimistic and assume low confidence or she may be optimistic and assume high confidence.

It is disputable whether the stance is a personal trait affected by Alice's past and current experience, or whether her stance in a particular context is driven by her experience already aggregated in a more abstract context. If the former is true, then Alice exhibits the identical trait in all contexts, so that apparently a higher level of confidence in one context is positively correlated with a higher level of confidence in another, even though the reason for such correlation lies beyond both contexts. If Alice's aggregated experience sets her stance within a particular context, then the relationship observed by McKnight is even more clearly visible. There is no clear answer to this problem, but considering more recent results [McKnight2006], the latter may hold true: Alice's stance may depend on her level of confidence in a more abstract entity.

## 4.3.2 Technology Trust Formation Model

Li's Technology Trust Formation Model (TTFM) [Li2004] builds on foundations of both Theory of Removed Action (TRA) and Theory of Planned Behaviour (TPB). While some of its building blocks are very similar to McKnight's model, its authors believe that TTFM is more powerful in explaining the creation of trust.

The model directly acknowledges the complementary character of trust and control that is reflected within the model by 'trusting attitudes' and 'perceived behavioural control'. The third identified component, 'subjective norm' relates to the notion of reputation and can potentially also be classified as a construct of control.

The TTFM has been designed to predict trust formation process up to trusting intentions, but it does not include trusting behaviour. Therefore naturally behavioural elements are missing from the model. Even though the model to certain extent criticises McKnight's inter-disciplinary model of trust, it also heavily draws on basic constructs defined by McKnight. Therefore, instead of providing the compete discussion of the model, the analysis concentrates only on those elements that differ from McKnight's model. Table 4.2 lists in its upper part

**Table 4.2**  Comparison with Li's TTFM model

| Construct | Entity | T/C | Category |
|---|---|---|---|
| *Directly related to McKnight's model* | | | |
| *Personality trusting base (disposition to trust)* | | | |
|    Faith in humanity | world | trust | general |
|    Trusting stance | world | control | general |
| *Institutional trusted base (institution-based trust)* | | | |
|    Structural assurance | domain | control | assurance |
| | domain | control | influence |
|    Situation normality | domain | trust | continuity |
| | domain | control | knowledge |
| *New elements* | | | |
| *Cognitive trusting base* | | | |
|    Reputation/second-hand knowledge | agent | control | knowledge |
|    Stereotyping | agent | trust | continuity |
|    Perceived control in situation | transaction | control | general |
| *Calculative trusting base* | | | |
|    Positive/negative outcomes | transaction | control | knowledge |
|    Cost/benefit from violations | transaction | control | assurance |

those constructs that have a direct relationship to McKnight's components while constructs that are new to TTFM are listed in its lower part. The discussion will be restricted only to those new elements.

It seems that all constructs (both old and new) of the TTFM are covered by our complexity-based model of confidence. It is worth noting, however, that TTFM has a certain novelty by stressing different forms of reputation (see Table 4.2, cognitive trusting base), something that was less pronounced in McKnight's model. Also the introduction of norms (stereotyping) delivers a strong notion of continuity with respect to agents, something that is also discussed later in this book, together with the future of trust management systems.

Similarly, the transactional element is very visible here, even though it is restricted to the control path (see Table 4.2, calculative trusting base). Constructs that address control in general as well as its particular elements (knowledge, assurance) are present. This can be attributed to the inclusion of a calculative trusting base into the model, where the 'calculativeness' of trust in fact represents the control aspect.

However, it is also worth noting what is missing from TTFM. Transactional trust (not control) is not addressed at all while competence and motivation are not addressed in any context. Control over agents is not strongly addressed (but it is partly addressed by knowledge and assurance). In general, the model stresses the control side of confidence, specifically the role of knowledge and calculations, with certain deficiencies when it comes to trust. As the model has less constructs than McKnight's, its relative explanatory power might be more limited. Also, in light of e.g. [Lacohee2006], the omission of 'soft' elements of confidence may be a deficiency of this model.

### 4.3.3 MoTEC Model of Trust in e-Commerce

Egger's MoTEC model of trust in e-commerce [Egger2003] addresses the 'trust challenge' where the lack of trust can be seen as an inhibitor of the adoption of computer-mediated services (e.g. e-commerce). Specifically, MoTEC addresses the initial trust in e-commerce, the level of confidence that the user should have towards the newly visited website. The model itself is the result of several iterations of study in consumer trust creation in e-commerce, on the basis of a rich and diverse factual base. The model was created on the basis of existing research and then refined through empirical experiments.

### Scope

The model takes the very personal and individualistic approach to the assessment of trust in e-commerce, concentrating on the asymmetric role of the consumer. Consequently, Egger's model treats world and domain contexts in quite general terms, as the model is focused on the agent and transaction.

The model inherits from the human–computer interaction the user-centric approach that constrains it to more 'soft' side of the relationship (related to trust in terms of the complexity-based model), slightly disadvantaging the 'hard' (control-based) side.

Note that contrary to the complexity-based model, Egger's model does not require trust to match trustworthiness. Interestingly, the problem of trustworthiness is not discussed throughout the model, which leaves the perception that the model can be used to create the perception of trust where trust is not due, i.e. by designing a more trust-inducing website the merchant may actually deceive the customer and abuse the unjustified trust. This problem is discussed in more detail later in the book in conjunction with the proposition of Trust-Enhancing Technologies.

## Approach

The model inherits heavily from research in human–computer interaction, correctly arguing that in the computer-mediated relationship, the digital tool has a significant impact on the overall trust relationship. The model postulates the 'toolbox' or 'checklist' approach, listing several properties of the human–computer interaction that should make a human trust the computer (website), specifically in the initial phase of interaction when the relationship has not yet been developed.

The model starts by capturing the most important characteristics of a user's trustworthy behaviour. By seeking justification in user cognitive processes, the model defines four dimension of trust creation, as follows:

- pre-interaction filters
- interface properties
- informational content
- relationship management

Within each dimension, the model provides a list of components and sub-components that together contribute to the creation and maintenance of trust.

## Components

Table 4.3 summarises the relationship between components that are present in Egger's MoTEC model and constructs from our complexity-based model of trust. For the purpose of this comparison, constructs from our model are limited only to classes of evidences, within four contexts. MoTEC components are listed in order of dimensions (as presented originally in the model), with certain grouping of sub-components to retain the similar level of details throughout the whole comparison. Names of dimensions and groups are in italic; names of components are in plain text.

For every component from MoTEC, the similar category of evidence is provided, by defining the context, trust or control domain and the actual category of evidence. Note that the relationship is not always one-to-one, with several MoTEC's components addressing more thanone category.

In this case, the relationship between components of Egger's model and categories of evidences from the complexity-based model is less clear than in the case of McKnight's model. However, it is necessary to understand that Egger's model encapsulates elements of user interaction and expresses them in terms specific to components of the interaction process. In contrast, the complexity-based model or McKnight's model defines constructs in more generic terms (beliefs they are related to rather than their representation) so that it can be reasonably expected that several MoTEC's categories should match the same construct.

As can be expected, Egger's model is very detailed when discussing the agent and transaction (which are the main focus of the model). World and domain contexts are addressed in a more sketchy way. It is worth noting that there is no category addressing control of domain, equivalent e.g. to McKnight's institution-based trust. This can be partly attributed to the general attitude of the model that tends to ignore non-digital communication means, effectively leaving social control instruments out of the picture.

Regarding agents, all categories of evidence regarding both trust and control are reflected in the model, often several times, showing different ways that can be used to convey evidence of particular expectations. Note that the original titles of categories from Egger's model can sometimes be misleading regarding its actual content. It has been necessary to refer to the

**Table 4.3**  Comparison with Egger's MoTEC model

| Category | Entity | T/C | Category |
|---|---|---|---|
| *Pre-interaction filters* | | | |
| *User psychology* | | | |
| General propensity to trust | world | trust | general |
| Trust in IT and the Internet | domain | trust | general |
| Attitude towards e-commerce | domain | trust | general |
| *Pre-purchase knowledge* | | | |
| Reputation of the industry | domain | trust | general |
| Reputation of the company | agent | trust | general |
| Transference | agent | control | knowledge |
| | agent | trust | continuity |
| *Interface properties* | | | |
| Branding | agent | trust | competence |
| Usability | agent | trust | competence |
| | agent | trust | continuity |
| *Informational content* | | | |
| *Competence* | | | |
| Company | agent | trust | competence |
| | | control | reassurance |
| Products and services | agent | trust | competence |
| | | control | influence |
| *Risk* | | | |
| Security | agent | trust | competence |
| | | control | knowledge |
| | | control | reassurance |
| Privacy | agent | control | knowledge |
| *Relationship management* | | | |
| *Pre-purchase interaction* | | | |
| Means of contact | agent | trust | competence |
| Responsiveness | agent | trust | competence |
| | | | motivation |
| Quality of help | agent | trust | competence |
| Personal touch | agent | trust | motivation |
| *Post-purchase interactions* | | | |
| Order processing | transaction | control | influence |
| Fulfilment | transaction | control | knowledge |
| After-sales | transaction | control | reassurance |

complete description of the component (supported by a sample questionnaire) to determine the most appropriate matching.

Regarding transaction context, Egger's model includes all the constructs regarding control while it does not address trust-related classes of evidence. This may be due to the fact that digital communication is usually not conducive to trust building within the context of individual transaction or to the fact that the model assumes that the decision to purchase is a rational one (e.g. on the basis of price analysis), leaving no room for trust-based purchase. Note, however, that e.g. eBay experience [Resnick2006] demonstrates that there is a premium to be earned on trust, including transactional trust, so that certain components of transactional trust should be included in the model.

## Relationships

MoTEC does not postulate any particular relationship between its components, beyond the identification of four dimensions of trust. Within each dimension, components of the interaction that should lead to higher trust are not structured, but it is expected that by fulfilling more of those components, trust will eventually increase.

### 4.3.4 Tan's Transactional Trust Model

Tan's model ([Tan2000] and also [Ganzaroli1999]) has been chosen mostly because of its clear identification of the duality of trust (i.e. trust and control), that is reflected also in the complexity-based model of trust. Tan's model itself has been developed for the purpose of e-commerce so that it closely matches the scope of the complexity-based model.

The original model (discussed here) takes into account only actions of the individual agent. Later extensions have added social context to the model. Due to the significant similarity to other models, the social aspect of this model is not discussed here.

## Scope

Tan's model is a relatively simple one and focuses on the context of transaction. It identifies two sources of trust: trust in the other party and trust in control mechanisms. Further, for the purpose of transaction, the model proposes the existence of the minimum trust threshold related to the perceived risk and gain.

Other types of entities (world, domain and agents) are addressed only briefly in a discourse attached to the model and are not fully integrated into the model itself. Consequently, they are not included in the analysis presented here. The brief analysis demonstrates that components that exist in those contexts are very similar to McKnight's approach and are covered by the complexity-based model.

## Approach

The model analyses the perspective of the single agent (potentially the user) involved in e-commerce transactions, with the expectation that the model can be used to determine if and why the agent should involve in the transaction. The initial concept of a lone agent operating in a social vacuum is being gradually extended to cover social and institutional contexts of trust.

The value of the model is mostly in introducing the duality of trust by addressing the concept of party trust (directly in the other party) and control trust (where the other party is constrained by control mechanism but trust is needed to accept control). Both concepts are covered by the proposed model, however, the existence of two different 'trusts' within the same model can make the comparison confusing.

## Components

Tan's model of transactional trust lists several components that together contribute to the transactional trust. Those components are listed in Table 4.4 and are compared to categories of evidences in our complexity-based model of confidence.

The immediate observation is that Tan's model combines trust in an agent with control of the transaction, thus linking two different contexts. There is relatively little in the model regarding control of agents.

Control of agents requires instruments of control that can work in the context beyond the agent itself, e.g. instruments that are provided by the law. The single agent, removed from the social context, has usually no access to such instruments or has no trust in those

**Table 4.4**   Relationship to Tan's model

| Construct | Entity | T/C | Category |
|---|---|---|---|
| *Trust in the other party* | | | |
|   *Objective component* | | | |
|     Social signs | domain | trust | general |
|   *Subjective component* | | | |
|     'Good vibrations' | agent | trust | general |
|     Personal experience | agent | control | knowledge |
| | | trust | motivation |
|     Communality | agent | control | knowledge |
| | | | |
| *Trust in control mechanisms* | | | |
|   *Objective component* | | | |
|     Control assessment | transaction | control | influence |
| | | | reassurance |
|   *Subjective component* | | | |
|     Understanding of controls | transaction | control | knowledge |
|     Communality | transaction | control | knowledge |
| | | | |
| Transactional trust | transaction | trust | general |

instruments. The agent can exercise certain control over agents, specifically if such an agent has comparable market power. Therefore it is the restricted scope that has removed control of agents from the model.

Further, the transactional trust is the outcome of the model, not one of its components. Such trust (which can be called 'confidence' in terms of the complexity-based model) is created within the model. Consequently, such trust cannot be one of the input components of the model. This approach is shared by several models within the scope of e-commerce.

However, it is understood that the transactional trust as the input component is possible and that it should be separated from the trust in the agent. Such trust captures our reaction on evidence that are delivered within the context of the transaction and that may contradict existing knowledge about the agent. Transactional trust is actually one of the methods to recover from the temporary distrust. The existence of transactional trust is clearly demonstrated when we are willing to deal with the untrustworthy agent on the basis of our perception of its behaviour within the context of one transaction.

## Relationships

Tan's model does not provide any particular relationship between its components, but only postulates that all components contribute to the transactional trust.

## 4.4  Conclusion

This chapter compares our complexity-based model of trust to four selected models of trust creation, all in the area of e-commerce. This comparative analysis demonstrates that all the important constructs of those models can be mapped into respective elements of our complexity-based model. Even though the mapping is not always one-to-one (which is the result of different approaches of models), there is very high similarity and convergence.

Elements that cannot be directly mapped into classes of evidence in the complexity-based model, can clearly be identified with other components of this model.

Further, the complexity-based model can be used to identify components that are missing from each model, thus providing indication about potential improvement areas for those models. Specifically, the ability to systematically analyse models using the unifying set of criteria has clearly demonstrated that differences in the intended scope and approach result in omissions that may be important for those models. The complexity-based model of confidence can be particularly useful in explaining phenomena not fully addressed by other models.

## Bibliography

[Bolton2004] Gary E. Bolton, Elena Katok and Axel Ockenfels: How Effective are Electronic Reputation Mechanisms? An Experimental Investigation. *Management Science*, 50(11): 1587–1602. 2004.

[Cofta2005] Piotr Cofta: Impact of Convergence on Trust in e-commerce. In *Proc. of Networking and Electronic Commerce Research Conf. NAEC2005*. 2005.

[Cofta2006] Piotr Cofta: Comparative Analysis of the Complexity-based Model of Trust. In *Proc. of Fifth Int. Conf. on Autonomous Agents and Multiagent Systems AAMAS-06*; Rino Falcone et al. (eds): WS8: Ninth Int. Workshop on Trust in Agent Societies (TRUST), Hakodate, Japan. 2006.

[Egger2003] Florian N. Egger: From Interactions to Transactions: Designing the Trust Experience for Business-to-Consumer Electronic Commerce. PhD thesis, Eindhoven University of Technology (The Netherlands). ISBN 90-386-1778-X. 2003.

[Ganzaroli1999] Andrea Ganzaroli, Yao-Hua Tan, Walter Thoen: The Social and Institutional Context of Trust in Electronic Commerce. In *Proc. of 2nd Workshop on Deception, Fraud and Trust in Agent Societies*, Seattle. 1999.

[Lacohee2006] Hazel Lacohee, Stephen Crane, and Andy Phippen: *Trustguide*. Available at: http://www.trustguide.org/Trustguide%20-%20Final%20Report.pdf. 2006.

[Li2004] Xin Li, Joe S. Valacich and Traci J. Hess: Predicting User Trust in Information Systems: A Comparison of Competing Trust Models. In *Proc. of the 37th Hawaii Int. Conf. on System Sciences HICSS2004*. 2004.

[McKnight1996] D. Harrison McKnight and Norman L. Chervany: The Meanings of Trust. In University of Minnesota. Available at: http://www.misrc.umn.edu/wpaper/wp96-04.htm. 1996.

[McKnight2003] D. Harrison McKnight, Chuck Kacmar and Vivek Choudhury: Whoops... Did I Use the Wrong Concept to Predict E-Commerce Trust? Modelling the Risk-Related Effects of Trust versus Distrust Concepts. In *Proc. of the 36th Hawaii Int. Conf. on System Sciences (HICSS'03)*. Available at: http://csdl2.computer.org/comp/Proc./hicss/2003/1874/07/187470182b.pdf. 2003.

[McKnight2006] D. Harrison McKnight and Vivek Choudhury: Distrust and Trust in B2C E-Commerce: Do They Differ? In *Proc. of Eight Int. Conf. on Electronic Commerce ICEC'06*, Fredericton, Canada. pp. 482–491. 2006.

[Resnick2006] Paul Resnick: The Value of Reputation on eBay: A Controlled Experiment. *Experimental Economics*, 9(2): 79–101. Available at: http://www.si.umich.edu/~presnick/papers/postcards/PostcardsFinalPrePub.pdf. 2006.

[Tan2000] Yao-Hua Tan and Walter Thoen: Formal Aspects of a Generic Model of Trust for Electronic Commerce. In *Proc. of the 33rd Hawaii Int. Conf. on System Sciences*. 2000.

# 5

# The Dynamics of Confidence

People who have given us their complete confidence believe that they have a right to ours. The inference is false, a gift confers no rights.

(Friedrich Nietzsche)

## 5.1 Introduction

This book concentrates mainly on Alice's transactional ability to assess her confidence in a transaction and particularly in Bob. This is something that has a short-term rather than relational value – the assessment is valid only at a specific time and only in a specific situation and does not extend beyond it. This is a moment in time, not the flow of it. Quite often the reality is that Alice (and her assessment) are embedded in a context of a relationship. The existence of such a relationship does not change the assessment – Alice still should be able to check to what extent she is confident about Bob. However, the existence (or even the expectation) of the relationship may change what Alice does, with potential long-term implications.

There is an interesting paradox: trust is the best long-term rational strategy: it is inexpensive, it potentially applies to a variety of cases and it may last for a long time. Therefore we should expect trust to flourish and we should see everybody being involved in a trust-based relationship. However, at the same time, the process of building trust cannot be fully explained on a purely rational basis. The same logic that dictates that trust is the best option tells us that there is no way to achieve it. If we restrict ourselves, e.g. to the game theory, we can see benefits of trusting but we do not see the path that may lead from ignorance or distrust towards trust.

It is apparent that the development of the trust-based relationship requires time and it does not seem that there is a quick fix that can substitute for the relative duration of a relationship ([Tyler1996b], also [Stewart2006] when the time needed to build a trust-based relationship is quoted as from 6 to 18 months), even among professionals in building relationships. With the exception of 'swift trust' (that does not develop trust), the development of a trust-based relationship is a significant investment of time and resources.

This does not imply that confidence (not trust) cannot be achieved in a shorter period of time or even instantly. As confidence can be reached either through trust or through control

*Trust, Complexity and Control: Confidence in a Convergent World*   Piotr Cofta
© 2007 John Wiley & Sons, Ltd

(or combination of both), there is always an avenue for confidence in short-term relationships: to employ control rather than trust. The ability to replace trust with control (and vice versa) also shows a possible way to develop a relationship: start with control, proceed towards trust.

Trust as the best long-term strategy is widely confirmed (e.g. [Webb1996]). However, this also means that trust is not the best strategy in all relationships that are not long-term, regardless of how long the term should be to qualify as long rather than short. It seems reasonable to build lifetime relationships on trust, but everything shorter may raise a question–even if it lasts 25 years or so. Experiments [Bolton2004] show that for relationships that are not permanent (that are missing the aspect of continuity) the most efficient strategy is to build trust and then take advantage of it in the last move – something that does not inspire much confidence.

The development of a relationship has usually been studied from two different perspectives. The first is the interaction between people (specifically within a context of a society or an organisation). The second is the interaction between agents (specifically in the context of signalling protocols in an interactive game of trust). Both perspectives are valid and deliver complementary observations, but they leave the picture incomplete. We miss the development of the mutual relationship between people and technology and some bits of the development of trust between devices. Both are addressed in different chapters of this book.

This chapter serves mostly as a background for discussion, even though it also demonstrates how the model of confidence can be used to explain different stages of the relationship. We start from the development of a confidence-based relationship, discussed from the two perspectives mentioned above. From there we discuss the phenomenon of 'swift trust', that seems to contradict the assumption of a slow and laborious process of trust building. Next, we explore avenues that lead from control to trust – how to achieve the most important first step towards trust-based relationship. Finally, we will look at the process of damage and restoration and possible dead ends of the development of the relationship.

## 5.2 Development of Confidence-Based Relationship

If we look at the relationship rather than at the transactional assessment of confidence, we can see that the behaviour of both parties may change. Instead of relying on control, Alice may be willing to 'invest' her trust in the relationship while Bob may be willing to reciprocate. Both understand that in the long run, trust is more beneficial than control. Models of the interpersonal dynamic relationship (e.g. [Luna-Reyes2004]) confirm this assumption.

Solomon and Flores [Solomon2001] observe that the relationship develops towards trust by crediting trust above and beyond what can be justified by trustworthiness, by what they call 'practising trust'. From the perspective of the model we can see that if Alice is interested in developing the relationship, she may intentionally differentiate between her confidence (mostly trust) and her confident behaviour, thus signalling the belief that is actually not present. Note that this does not undermine the importance of the ability to assess confidence: expressing undeserved confidence for the sake of relationship development is a rational choice but e.g. always expressing undeserved confidence is not.

A trust-based relationship does not develop immediately. Lewicki [Lewicki1996] proposes that the development of a relationship towards trust should grow in a stage-wise manner. Initially, the relationship is based on calculus (rational control) to progress to the stage of knowledge-based and finally to the identity-based trust. Some relationships may remain at one of early stages and only some make it to the final one. The transition to the new stage may require a change in a mindset – the qualitative change driven by the amount of time (and evidence).

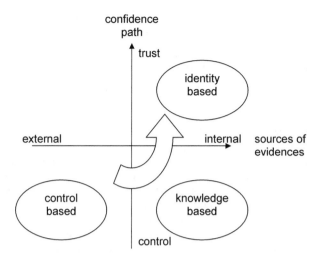

**Figure 5.1** The development of relationship

Comparing this stage-wise development with the model we can see that actually the relationship develops along two axis (Figure 5.1). First, it changes from being externally supported to become internally driven. Second, it moves from being control-driven to trust-driven. Transitions, as proposed by Lewicki, require changes in the frame of discourse, as different types of evidence are admitted into the relationship. Transition from the relationship that is mostly control-based to the relationship that is mostly trust-based seems to be done in increments, driven by the desire to minimise complexity but restricted by the size of the 'trust gap' that the person is willing to invest.

Note that interestingly the alternative path (that leads through external sources of trust, such as e.g. trusted intermediaries or Trusted Third Party) is not recognised as the valid path to develop a trust-based relationship. In fact, trusted intermediaries prevent trust formation by acting as an opaque proxy. If Carol is a trusted intermediary that stands between Alice and Bob, then Alice has no means to look 'beyond' Carol, to learn about Bob's properties and to eventually develop trust. This resonates well with the observed lack of transitivity of trust discussed elsewhere.

### 5.2.1 Cognitive Trust (Control, External)

The first stage of the relationship relies heavily on instruments of control that are external to the relationship to deliver the required level of confidence. Specifically, evidence of assurance and influence can be very obvious while knowledge is accepted from outside of the relationship (e.g. from mutual friends). For example, the company may seek strict contractual terms or possibly advance payment while a personal relationship will be restricted to casual meetings in public places. At this stage evidence of trust, even if present, may be ignored as the development focuses on enablers.

The ability to gather evidence is the most important enabler at this stage – without evidence the relationship cannot progress. We may expect that this stage will be devoted to developing confidence in identity and establishing certain understandings of similarity. However, the enabler that is the prerequisite to the transition to the next stage is confidence about the honesty of the other party. As the next stage will be based on control that exists within the relationship (so that evidence will be delivered by parties that are already within the relationship), honesty is the key element [Dasgupta2000].

### 5.2.2 Knowledge-based Trust (Control, Internal)

Assuming that both parties are confident about their honesty (as well as other enablers), they may eventually move to the next stage of the development where they will build on control that is internal to the relationship. Specifically, the relationship will be based on the knowledge, the understanding of what is 'normal' and 'expected'.

Parties will still gather evidence, but this time they will seek patterns and rules that guide the behaviour of the other party. They will react positively to evidence that reinforces their perception of the other party and negatively on deviations from expected behaviour. At the same time, however, they will develop increased tolerance towards what may be called 'normal variations' in behaviour: slightly over- or under-delivery, late arrivals, etc. Those variations will not affect their general expectations.

The main concern during this stage is to understand what the motives, competences and beliefs of the other party are – to discover evidence to trust. The process of deducing, predicting, discovering and verifying those evidence may be complicated and time-consuming, but it is necessary if the relationship is going to progress to the next stage.

### 5.2.3 Identity-based Trust (Trust, Internal)

The highest level of relationship is built on the common identity of both parties., i.e. it is achieved when they share the same identity: they are driven by mutually supportive motivation and they share similar expectations about their continuity. Further, they recognise and value their competences. An identity-based relationship is closely related to the construct of reputation and honour (e.g. [Sierra2006]), where the integrity of personal identity guarantees the relationship and can be supported by social institutions [Valsiner2005].

Self-identity is built though the self-reflecting process of internalising experiences [Giddens1991]. By sharing experiences, both parties may eventually become parts of each other's self-identity – part of each other's life. Such a shared identity (i.e. common parts of self-identities) forms a very strong psychological bond that may lead to the formation of a very strong relationship of trust.

An identity-based relationship is self-satisfying and supports all forms of enablers. The shared identity serves as a proxy for the identity enabler: Alice knows Bob (and presumably has known him for some time) so that her knowledge itself provides sufficient guarantee of Bob's identity. Their shared identity is the best proof of who Bob is. This shared identity, being a collection of shared memories, satisfies the accessibility. Indeed, as the shared identity cannot develop without shared experiences, such experiences are the same that are needed to develop the relationship of trust. Finally, by internalising the shared identity as a part of her self-identity, Alice has a strong perception of similarity (almost sameness) with Bob – she can feel that she knows Bob well.

## 5.3 Relationship as a Signalling Protocol

It is beneficial to think of a relationship as of a form of a signalling protocol where interaction and exchange of goods or services as such are relevant mostly as signals of intentions within the established protocol. Such signals can be interpreted by Alice within the context of her chosen strategy and can alter her perception of Bob's strategy. The actual cost of interaction (including potential losses on exchange) becomes the cost of sending and receiving information and may be considered less important than the benefit of having the interaction and relationship. The value is therefore shifted from maximising the outcome of the exchange into building the relationship itself and partners tend to maximise the perceived value of relationship rather than their gains. In some cases (e.g. in the game of trust) those two goals seem to be compatible, but often partners have to make a choice.

The overall goal of a relationship is to communicate a message to the partner and to understand to what extent the partner's message is compatible with ours. The message is usually simple: Alice would like to tell Bob that she is willing to cooperate (assuming that she is interested in cooperation), but she will not allow herself to be exploited. She also wants to understand whether the message that comes from Bob (Bob's intentions, Bob's strategy, Bob's internal algorithm, Bob's trustworthiness, etc.) is similar, if not identical. If the messages converge, then Alice can proceed towards cooperation; if not – then she should probably take a more protective approach.

The process of signalling, even within established protocols, is subject to several potential problems – from the simple case of noise in communication channel to more subtle misunderstanding of signals. Sending the message and receiving response incur costs for both parties, and such costs can be justified only if the expected benefit outweighs it. Finally, both parties may freely change protocols and strategies throughout the relationship, rendering past signalling useless. All that suggests that the optimum strategy is the incremental one, where data are gathered before making decisions and small commitments precede large ones.

From the perspective of signalling and protocols, the relationship can also be characterised by three stages (see Figure 5.2.), drawn along similar axis of externality–internality and trust–control. The relationship starts from the stage of observation, then moves to the stage of interaction to finally reach the stage of cooperation. Each stage has a different goal and employs a different set of protocols. Confusion about the actual stage may actually damage the relationship.

### 5.3.1 Observation (Control, External)

This stage is concerned with a low-cost data gathering. Observation, whether done by Alice or by others for Alice, incurs a very low cost, so that it may be preferred to interactions. Interactions and signalling, if any, are of low value and are heavily guarded by social protocols. Signals, whether from observations or from interactions, are interpreted within a

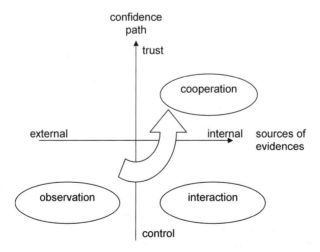

**Figure 5.2** Three stages of signalling

socially recognised rules and strategies, with little recognition of Bob's individuality. Bob is expected to behave in a socially appropriate manner. This stage is guided by externalised control: institutions of control are needed to both guard against losses and to provide an interpretation context for signalling. Conformance to social norms and rules, availability for observations, exchange of small gifts – all this facilitates the development of the relationship.

### 5.3.2 Interaction (Control, Internal)

Even though the previous stage may have contained some interactions, it is the second stage that is based on personalised interactions. Alice, by interacting with Bob, is sending the key message that she wants to cooperate but does not want to be exploited – the principle of her strategy. At the same time she listens to Bob to establish what his strategy is. The main difference between observations and interactions is that here Alice relies on Bob himself to deliver signals. Social support is less important, the relationship is internalised and personalised. Bob is no longer seen through the lenses of social protocol, but through intricacies of his own strategy. Alice is less concerned with Bob's compliance to social rules because she is more interested in discovering what exactly are the rules that guide Bob. The cost of interaction gradually increases, but is still governed by the economy of offsetting its cost with expected future gains combined with the prevention of significant losses.

### 5.3.3 Cooperation (Trust, Internal)

There is a significant change between the two previous stages and the cooperation stage. Both previous stages relied on control and contained expenditures associated with sending and receiving signals. The cooperation stage starts when both parties believe that they understand the strategy (belief, intentions, norms) of the other party – and approve it. Once such an understanding has been reached (which is the equivalent of establishing trust between parties), both parties can switch to cooperation protocols and cooperation strategy. Such protocols assume the goodwill strategy of the other party and prefer cooperation despite occasional problems. Protocols become lightweight and the cost of commitment is no longer an issue. Both parties employ certain safeguards or monitoring features to be able to react to changes in strategies (possibly out of prudence), but such features are not designed to offset or contain losses.

## 5.4 Swift Trust

Swift trust seems somehow a contradiction to what has already been said about the formation of confidence: it seems to be built on trust but it appears rapidly in semi-formal non-routine goal-oriented groups with limited lifespan and low expected recurrence, where control mechanisms are not readily available – apparently in situations where confidence (least trust) should not be present (e.g. [Meyerson1996]).

Swift trust should not be mistaken for what can be considered a 'benefit of a stranger', the rapid trust-like relationship with the passing stranger (e.g. while travelling together). The assumption about temporary relationship with a stranger means that there is little expectation and little vulnerability involved so that there is a little need for confidence (particularly for trust). If Alice meets a stranger that she is unlikely to meet again, she does not expect much from him (except maybe passing courtesy) and she does not risk much (except maybe passing inconvenience) – specifically she does not risk her reputation.

Swift trust should not also be mistaken for what can be called an 'instant relationship', where two strangers form a strong relationship immediately upon meeting each other. Not entering the discussion about the emotional side of it (love at first sight, etc.), we can contend that this phenomenon has a lot to do with similarity, one of the enablers of confidence. Similarity is associated mostly with determining by Alice whether Bob is able to have intentions (so that trustworthiness can be attributed to such party). However, similarity works also in determining whether Bob is similar to someone who has already been trusted or is similar to Alice herself. In the first case Alice is able to expect a similar trustworthiness by extrapolating from her experience. In the latter case Alice can extrapolate her trust in herself and – drawing on a similarity – can expect trustworthiness from Bob. The fact that Alice is more willing to trust someone who is similar to her has been noted [Feng2004] and indeed this assumption is the foundation of automated recommendation systems such as in collaborative filtering (e.g. [Resnick1994]).

Swift trust has been studied in relation to group formation [Lewicki1996], among people who cooperate for a time span that is relatively short (and limited) but in situations where vulnerability is relatively high and a certain outcome is expected - therefore where barriers to confidence are higher than in a 'passing stranger' situation. Situations where swift trust operates are outside what should be considered 'normal': people are selected on the basis of their professional merits, groups are created for the purpose of the tasks with disregard of long-term careers, they operate under the high pressure of timely delivery and are disbanded once the goal is achieved with little expectation of further involvement.

Close inspection of swift trust reveals that it is actually a superimposition of three different phenomena that are present at the same time, all conforming to the model presented here. Therefore, swift trust should not be considered a 'proper trust', but a local manifestation of them. These are: manifestation of existing relationships, imposed confident behaviour and domain-based confidence.

### 5.4.1 Manifestation of Existing Relationships

Assuming that the selection follows professional criteria, groups where swift trust emerges employ a relatively small group of professionals. Therefore even though the group itself is short-lived, the professional relationship between group members both precedes and outlives the group. Even though it may be unlikely that they will meet again in a similar configuration, their professional reputation is known and may be affected by their performance within a group. Therefore the relationship that exists within the group is a local manifestation of relationships that exist elsewhere. The swift trust allows this relationship to be exposed within the group, but it does not add much to the creation of such a relationship – it existed before the group was formed.

It is specifically obvious in the role of a group leader. It is expected that the leader of the group will bring his or her reputation to the group, to create and cement the group. Specifically, such a leader may override existing relationships e.g. in case of emergency [Gomez2006]. Again, the group is using the relationship that has existed before the group was created. If the leader was not known to members in advance (or his or her reputation has been tarnished), then within the group little can be done to rectify it.

### 5.4.2 Imposed Confident Behaviour

Groups with swift trust operate in situations where mutual vulnerability is high – members rely on each other to reach the goal. High vulnerability causes high levels of stress so that this situation may not be bearable for members who can develop certain defence mechanisms. One of the mechanisms that can be used to cope with the situation is to expose confident behaviour, not because of actual confidence, but because such behaviour is socially acceptable within the group. However, as such confident behaviour is artificially imposed rather than voluntarily accepted, it may disappear at the first possible moment – whether this is the termination of a group or even earlier, leading potentially to backstabbing and intrigues.

Such behaviour may also turn into the more stable relationship of confidence if it is reciprocated and learned, specifically by new members who have less experience of the dynamics of such groups.

### 5.4.3 Domain-based Confidence

As people are brought into the group on the basis of their professional merits, the initial confidence that is extended towards them may be built on the basis of domain, not individuals. Such domain-based confidence is related to the confidence in one's profession. As a professional, an individual is perceived as a bearer of professional ethics and competences, so that the confidence extended to such person is built on the basis of past experiences with people of a given profession, general reputation of a profession, legal requirements, etc.

Such confidence has nothing to do with the group itself – the group (again) is using what has been created outside of the group and is not contributing to it. Certainly, positive or negative experience with a given professional may alter the perception of confidence in the profession, but this is a part of the development of a regular relationship, nothing specifically attributable to the group.

## 5.5 Emergence of Trust

One of the most puzzling questions about trust is that there does not seem to be any rational avenue that leads from initial uncertainty to trust. The emergence of trust does not seem to be justified. Not only the general expectation regarding human nature, but also theories of rational behaviour and everyday social practice demonstrate that it is generally more beneficial to settle for the smaller but certain gain (that comes with the lack of trust) than to reach out, trust and risk being harmed. As has been noted as a paradox [Ullmann-Margalit2001], we trust others in their persistence to rationally pursue their best interest over ours, and because of this we cannot trust them. Still, trust can be universally observed, so that there should be way to build it.

The fundamental observation is that trust requires two parties who are willing to maintain the dynamic relationship (e.g. [Solomon2001]). Trust is not built because Alice one day decided to trust Bob, but it is built because Alice demonstrated to Bob that she trusts him and Bob has reciprocated with his trustworthy behaviour. Alice decided to trust because Bob seems to be trustworthy and Bob is trustworthy exactly because Alice seems to trust him. For as long as both parties accept and support trust, trust become the reality. Therefore, trust can self-sustain itself, once it is given a chance. The first time Alice decides to trust Bob (or at least decides to behave as if she trusts Bob) is a decisive moment. However, why should she do it? Why should she move out from the safe, control-guarded relationship towards the one where she is vulnerable?

The first answer acknowledges that humans are not perfectly rational in their behaviour, but they are driven by emotions [Dunn2005] as much as by the intellect [Prietula2001]. Therefore Alice may prefer to suspend her reasoning and trust Bob because of her emotions – she may love him, she may admire him, she may be willing to suffer for him – there are several reasons. This explanation is at the same time plausible and unsatisfactory. It is plausible because it refers to our emotional part that can be perceived as a real and intense everyday experience and is capable of explaining several observable cases of trust development. However, it is unsatisfactory because it is almost unthinkable that in order to realise the rationally perceived gain, we must go through the phase that cannot be rationally justified. If there is no rational path towards trust, then neither organisations (companies, etc.) nor computers will ever hope to go beyond control.

Local random fluctuation, error in judgement, imperfection in planning: interestingly they are potential sources of the decision to trust. If Alice is waiting for Bob to demonstrate his trustworthiness, she may positively misinterpret his behaviour and respond with trust, thus initiating the relationship of trust. Bob may not be aware of such a fact or he may not do it willingly, but such behaviour may be the result of his error or imperfection. Stories from the trenches [Solomon2001] demonstrate that if both parties seek trust, then individual incidents can build the relationship of trust even between apparent enemies. However, even though people usually seek evidence of trust, this source of decisions to trust is also not satisfactory: random fluctuations are unable to explain the popularity of trust. If we need error to build trust, then it should be quite rare, thriving only at the edges of an otherwise controlled world.

We have already mentioned that Alice may treat trust as a kind of investment: she invests her trust in Bob (thus risking more and working more) in expectation that Bob will reciprocate so that in the end she will work less and risk less. This explanation coincides well with the generally perceived longevity of trust-based relationship: Alice will not invest in a relationship that is not supposed to last, as in such a case she will not be able to enjoy the benefits of trust. Further, Alice may follow rational, economic judgement and assign only limited resources, and create a 'probation period' for Bob to develop the relationship. If Bob does not reciprocate within a set time, Alice will write off her investment and continue with Bob on the basis of control. The explanation is plausible from the perspective of rational behaviour, but then it may be flawed from psychological sense. During the probation period Alice is not trusting Bob, she only pretends to do so. Bob may feel hurt, understanding that Alice is trying him without investing much and may reciprocate with the behaviour that looks trustworthy even if he is at the same time plotting his revenge

Yet another explanation can be found in lowering Alice's rational barriers to trust [Ullmann-Margalit2001]. Assuming that Alice (and Bob) are seeking opportunities to trust, the question may be not in finding ways to overcome a barrier but in lowering such barrier. If Alice can rationally justify even temporary trust in Bob, then she may be willing to extend such trust. However, in order to justify it, she must face the situation that is not the traditional 'hard' game of trust (where losing hurts and costs) but a game where trusting and losing does not impose significant consequences. This kind of 'soft' game ensures that Alice's losses will be minimised or compensated, i.e. that she can safely experiment with trust. Interestingly, it is not necessarily that 'soft' is the only game in town, it is sufficient that it is the game that seems to be possible. The premise of such game facilitates the belief in its existence and this facilitates the perception that it is being played – nothing more is needed to initiate the process of trust.

Yet another answer can be drawn from game theory, using a modified game of trust. While strangers are usually untrustworthy and untrusting, the emergence of a cluster of trusting persons can facilitate the process of building trust [Ho1999]. This is partly supported [Hawthorn2000] by the observation that while stable social trust cannot be created by any spontaneous action by society itself (and specifically trust among strangers is neither justifiable nor achievable), the existence of a cultivated cluster of trustworthy agents may eventually lead to the creation of wide trust.

Whatever the reason and the source of the first decision to trust, one thing is certain: a trust-based relationship requires reciprocal behaviour, and this requires expectations regarding the future. Alice can make a decision to trust Bob and she can assess the extent of Bob's trustworthiness, but without Bob's future participation in the relationship, such a decision and assessment will be transitional at best. If Alice is not convinced that she is going to meet Bob somewhere in the future and that she may need him again, then short-term opportunistic behaviour may easily prevail [Bolton2004], and trust will not develop. Such expectations of reciprocity are usually communicated by norms, as a part of evidence of continuity

[Riegelsberger2003]. This provides yet another interesting observation: Alice's ability to assess Bob's trustworthiness contains the assessment of Bob's potential to reciprocate: her decision to trust depends on her expectation whether the relationship may develop towards trust.

## 5.6 Breach of Trust

There are only a few states that are more loaded with emotions than the breach of trust, betrayal of a partner, treachery, desertion. Dante assigned traitors to the inner circle of inferno and most societies treat them with disrespect. Personal experience of betrayal is highly traumatic and may affect one's ability to develop a meaningful relationship in the future. It is hard to be rational about breach of trust.

However, clarification is needed. First, the breach of trust applies only to situations where trust has been involved, i.e. in the final, identity-based stage of cooperation. Control-based stages do not involve betrayal, as there is no trust to be breached (see also [Lewicki1996] for differences between violations at different stages). Certainly, contracts can be left unfulfilled and people may not live up to our expectations. However, for as long as we do not trust them, we are not facing the pain of betrayal. If goods arrive late, there is compensation, if a computer is faulty, there is a warranty, if a stranger commits a crime, there is the police. Instruments of control keep the other party at bay, containing the damage. If Bob underperforms, Alice may adjust her expectations with regard to him (e.g. by lowering his reputation rating) and may eventually decide not to cooperate or to revert from knowledge-based stage to the control-based one, but that's all. The loss of control-based confidence is less expensive then the loss of trust.

Next, there is a distinction between the breach of trust and the misplaced trust. Misplaced trust is sometimes called mistrust [Marsh2005], even though mistrust may have other similar meanings – e.g. the generalised default on trust [Abdul-Rahman2005]. The difference is in the guilty party and in correctness of one's assessment. If, say, Alice decided to trust Bob and Bob did not live up to her expectations, whose fault is it?: his or hers? If Bob knew the peculiarities of Alice's situation and has accepted the fact that she is counting on him, then the fault is his and Alice rightfully feels betrayed. If, however, Bob was unaware, unable or unwilling to accept the responsibility, then the error is on Alice's side – she misplaced her trust in Bob.

From the perspective of interaction protocols and signalling, breach of trust and the misplacement of trust are signs of two different problems. In the former case it is Bob who potentially has changed his strategy (and Alice did not notice it). In the latter case it is Alice who has misinterpreted Bob's strategy while it potentially remained unchanged. In either case Alice should adjust her relationship with Bob. However, the adjustment is different for the breach of trust and for misplaced trust. If trust has been breached, then all that Alice knows about Bob has been invalidated. Effectively, it resets their relationship back to the early, control-based stage (if any relationship remains at all). All Alice's investment in observation, signalling and interaction is effectively lost, increasing further the immediate loss that has already been incurred due to Bob's betrayal. The loss is so significant that the purpose of the whole relationship is in question.

In the case of misplaced trust, all Alice's past investments (except possibly for the most current one) are justified and everything that she knows about Bob is still valid. The only thing that she has to do is to re-think, re-interpret the situation – something that can be potentially done at low cost. The relationship can be saved – specifically that it has been Alice's fault.

## 5.7 Restoration and Forgiveness

Once Bob has betrayed Alice, she is facing an important dilemma: what's next? She may find the breach non-reparable, terminate the relationship and close the past. She may demote the relationship to its lower level: from trust to control. However, she may also be willing to restore the relationship. Alice may have good reasons to try to overcome betrayal: Bob may be still in the learning phase and betrayal is a part of this process, she may feel obliged to support Bob, the breach may be of lesser importance, she may have very limited choices and she has to continue with Bob, etc. Whatever her choice, Alice is facing the problem of forgiving Bob.

The important question of forgiveness delivers several interesting results when explored. First, forgiveness should not be confused with forgetting. If Alice forgets, then she later draws her conclusions on a smaller set of information, so that she becomes uncertain. If she forgives, then the act of forgiveness, as well as the reason for it, are still remembered and can be used for any further evaluations. An act of forgiving can also be gradually forgotten (as it is yet another evidence), but itself forgiving is a valid element of the relationship.

Forgiveness is believed to play an important role in the restoration of trust [Solomon2001]. However, it is not the same relationship that has been before – forgiving does not erase the past, but builds towards the future. It is further noticed [Camp2001] that forgiveness is necessary to maintain the integrity of social groups over the period of time and that the willingness to forgive depends on the nature of betrayal, an observation important when it comes to trusting technical infrastructure. The relationship between forgiveness and confession [Fitness2001] demonstrates the link between acts of regret and forgiveness within a context of social scripts. Note, however, that forgiveness does not always improve cooperation, specifically in a volatile environment where it is relatively hard to identify a non-cooperating person [Prietula2000].

The concept of forgiveness has been formalised in the context of regret and remediation [Marsh2006]. It has also been studied in the context of cooperation in multi-agent systems and iterated Prisoner's Dilemma (e.g. [O'Riordan2001]), in the context of the popular strategy of tit-for-tat, claiming significant advantages of the strategy that includes elements of forgiveness. The interaction in a noisy environment (e.g. [Au2006]) benefits from different forms of forgiveness where the ability to restore the relationship can overcome problems with possible false signalling of intentions.

### 5.7.1 Forgiveness in a Game of Trust

The tit-for-tat strategy (e.g. [Axelrod1984]) is one of the best known for modelling cooperation under imperfect information and has been sometimes claimed to be the best strategy for situations that are structured as the Prisoner's Dilemma. Further, the tit-for-tat strategy has been confirmed to be quite popular [Smith2002]). The strategy itself is quite simple: Alice starts her relationship with Bob with her willingness to cooperate (that indicates trust). She continues to cooperate for as long as Bob reciprocates, but if he defects, she responds with defection in the next round. Depending on the variant of strategy, she may permanently defect (the spiteful version), she may punish Bob for as long as he defects (classical tit-for-tat), she may increase the non-cooperation depending on history, introduce randomness, weight overall history of relationship, etc.

However, our question is not how to make Alice win, but to check whether forgiving makes a difference, for her and for the society she lives in. We would like to advise Alice whether and if she should forgive Bob or not, and what may be consequences. Therefore we do not concentrate on the best strategy, but on the impact that forgiving has on existing strategy. For that purpose we will consider the modified variant of tit-for-tat, known as 'generous tit-for-tat'

(GTFT), the simple alteration where a player can forgive instead of retaliate. Assuming that studying games can tell us about ourselves, we may learn something.

As we are going to play the game of trust, it is important to first understand the relationship between this game and the model. Earlier in the book it was mentioned that the game of trust, if played by people, can be used to estimate the level of trust. In the absence of any communication mechanism between players (except for the game itself) there is no place for instruments of control, so that cooperation seems to be built on trust alone. Settings of a game usually exclude social control, enforce anonymity and remove external pressure of any kind. It can be argued that certain forms of self-control still exist, but for practical reasons they can be ignored.

Simulating the run of a game in a multi-agent society, entirely programmed in the computer, raises the question of relevance, as we do not learn about people but about the behaviour of software. However, there are enough similarities to justify the benefit of such experiment. Within the confines of the gaming protocol, both parties try to adopt the winning strategy. However, in order to win, both parties must develop the strategy that enables the maximum level of cooperation, attainable only through a digital equivalent of mutual trust – cooperation while vulnerable and not informed. What we experience is the game between different understandings of how such relationship should develop. For as long we do not confuse societies of agents with societies of people, we may benefit from observing similarities between them.

It is beneficial to think of the iterated game of trust in the context of signalling rather than in the context of purely economical exchange. Interestingly, the only way for Alice to discover Bob's strategy is to interact (so that the game is devoid of the 'observation' stage and starts immediately at the 'interaction' one). Throughout the game both parties are more concerned with discovering other's strategies than with actual exchange. If the strategy warrants trust, then they tend to switch to the 'cooperation' stage, if not, they try to reinforce their own message. Interestingly tit-for-tat is a very good signalling strategy for both stages: it starts from an invitation to cooperation, it reinforces cooperating actions of the other party but it inhibits exploitation.

## 5.7.2 Simulation

Let's consider an experiment: a world of agents that play the iterative GTFT. Agents interact in pairs and in each turn, an agent may decide whether it wants to cooperate or defect. The payoff matrix is set with usual constraints [Axelrod1984] and reflects the 'hard' version of the game [Ullmann-Margalit2001], i.e. both partners benefit if both cooperate, but the one that is willing to cooperate experiences losses if its partner decides to defect.

GTFT introduces the forgiving factor: each agent, while seeing that its partner has recently defected, can, with certain probability, decide to forgive its partner and try to re-enter the relationship. Forgiving is an opportunity for both parties to overcome potential communication problems, conflict of strategies, continuous retaliation, random lapses in judgement, temporary conflicts, etc. As such, the forgiveness should benefit the society of agents, but it may decrease gains realised by an agent that decides to forgive.

The simulation (not presented here) allows us to make a couple of interesting observations. First, there are two very different cases when we speak about forgiveness, even though both may lead to the same temporary observable results. If agents randomly defect (i.e. there is no correlation between defection at a given moment in time and the future actions of an agent), forgiveness is a powerful yet sometimes expensive tool to maintain social relationships. However, if some agents persistently defect then forgiveness is of no importance and only generates added cost to those who try to cooperate. There is a significant difference between e.g. a society of 100 agents where each one may refuse to cooperate 10 per cent of the time and the society of 100 agents where 10 agents always refuse to cooperate.

For random defectors, the society benefits from forgiveness. It is apparent that forgiveness contains the continuous deterioration of the number of cooperations. The higher the forgiveness rate, the better the society accommodates the increasing defection rate and the higher is the overall social benefit. From the perspective of a society there is a benefit of forgiving even if the level of defection is very high, as forgiving still brings marginal social benefit (even if for the same rate of forgiving the gain diminishes as defection grows).

The perspective of the individual agent is different, as the payoff attributed to cooperation diminishes very rapidly with the increased defection rate. Quite soon an agent that forgives incurs losses and an agent that defects and is being forgiven realises the gain (which may be perceived as unjustified). The social payoff is increasingly built on the rip-off of those agents that are willing to cooperate. Therefore individual agents, if they have access to information about the increase of the overall level of defection, may want to alter their policy of forgiving quite early, to hedge against expected losses.

The recommended strategy for Alice regarding forgiveness depends on several aspects of the society. If defectors seem to be persistent, then it is best not to forgive them, but to sever ties with them and not to cooperate. Otherwise, the level and the dynamics of a change in defection are important. If the level of defectors is high, then agents have no choice but to default to non-cooperation as their primary strategy. If the level of defecting agents seems to be low and stable or is diminishing, then forgiving will bring benefits for both agents and the society. If the level if increasing, the agents may be willing to pre-empt potential losses with an early change in their policies to non-cooperation (so that non-cooperation will become a self-fulfilling prophecy).

## 5.8 Dead Ends

Not all relationships end with trust, and not all survive. Some of them induce in participants what can be called 'dead ends' – the self-reinforcing circles from which there is no apparent escape. Two cases of dead ends are particularly interesting: the negative lock-in of social paranoia and the positive lock-in of hyper-optimism. Of those, the negative lock-in is more dangerous. While hyper-optimism may be unjustified, it is still better to be overly optimistic (and occasionally fail) than to be pessimistic and not to try at all [Marsh1994].

Both are the results of a misunderstanding that comes from the misinterpretation of evidence, quite often combined with the attempt to resolve different frames of discourse simultaneously. Both lead to the same observable behaviour: Alice refuses to accept new evidence about Bob or interprets such evidence in a discriminating way.

From the perspective of signalling, there is no 'objective' evidence of trust or control. While in a socially-defined (external) context certain evidence can be interpreted with reasonable certainty, the internal context causes the interpretation to be private, specific to the relationship – thus subject to misinterpretation. In a true Byzantine way, every evidence can be interpreted either towards confidence and trust or against it. Specifically evidence of trust is susceptible to such manipulations. Once Alice starts to re-interpret evidence to conform to her assessment (rather than to assess it in a way that is as objective as possible), she gradually slides into the self-reinforcing stage of paranoia: the reality is redefined to conform with existing assessment. Evidence is re-interpreted, honesty of other agents is questioned if not conforming to the assessment, etc.

Let's consider a simple example. Alice believes at certain moment that Bob is trustworthy. She receives evidence from Carol that Bob has betrayed her, thus questioning his trustworthiness. Alice can either alter her assessment about Bob or she can start believing that Carol is dishonest. In situations of high uncertainty both avenues seem to be equally likely, so that it is quite possible for Alice to make a wrong decision. The thoroughly rational Alice

should suspend her judgement and seek further evidence, but this approach is quite rare.
Let's say that Alice decided to retain her trust in Bob so that she started believing that Carol
is dishonest. If Carol delivers fresh evidence about Bob's untrustworthiness, if will only reinforce Alice in her beliefs. If Dave tell Alice that Carol is honest, she may not believe him.
Eventually, she will not believe anyone except her own faulty assessment and Bob.

## 5.9 Conclusion

The development of a relationship flows from control towards trust and from external to
internal. Even though the road may be long and not all relationships may reach their final
stages, the desire to move towards trust seems to prevails in societies as well as among
agents. Developing trust requires an understanding of the beliefs, strategies and intentions of
the other party. While the observation and interaction can deliver some assessment of Bob's
performance and Alice's risk, the transition towards trust is possible only if Alice understands
and accepts Bob.

Relationship that is built on reciprocated trust (rather than on control) may have several
benefits. It is an efficient relationship – little effort is needed to achieve significant outcome;
it enables creativity and opportunities and finally it benefits our mental health. It even seems
that it is better to err on the side of trust than on the side of control (or distrust) – e.g. that
it is usually better to forgive and be optimistic. However, trust-based relationship is not a
effort-free one – it requires maintenance.

It is hard to provide general guidance regarding the best strategy for Alice. Some relationships are simply not designed to be elevated to the level of trust and Alice should be
aware of it. Short-term relationships, incidental cooperation, lack of expectation regarding
reciprocity are the most obvious signs of relationships that should be left at its control stage.
Even though the running cost of such relationship is higher, Alice is facing less risk and less
uncertainty.

If the relationship seems to be designed for long, Alice is better off if she considers moving
towards trust. If all what she wants is to maximise her own profits, then the best for her is
to stay on the edge of trust – between control-based and trust-based stages. Essentially she
should play the tit-for-tat strategy in one of its many flavours: extend an offer to trust but
withdraw at the first sign of non-cooperation, then try again. She may be even malicious,
balancing on the side of control rather than trust: cooperate only as long as it is required for
Bob to reciprocate, then default, then start cooperating again. What exactly the strategy is,
depends on the communication method that she can use as well as her overall attitude and
her perception of humankind. Only if it is the relationship itself that is of higher value to her
than occasional loss of profit, should Alice enter the trust phase and potentially stay there.
She risks disappointments, but she may expect high rewards as well.

## Bibliography

[Abdul-Rahman2005] Alfarez Abdul-Rahman: A Framework for Decentralised trust
    Reasoning. PhD thesis. Available at: http://www.cs.ucl.ac.uk/staff/F.AbdulRahman/
    docs/ thesis-final.pdf. 2005.
[Au2006] Tsz-Chiu Au and Dana Nau: Maintaining Cooperation in Noisy Environments.
    Available at: http://www.cs.umd.edu/~chiu/papers/Au06nectar.pdf. 2006.
[Axelrod1984] Axelrod R. M. *The Evolution of Cooperation*. New York: Basic Books, 1984.
[Bolton2004] Gary E. Bolton, Elena Katok and Axel Ockenfels: How Effective are Elec-
    tronic Reputation Mechanisms? An Experimental Investigation. *Management Science*, 50(11):
    1587–1602. 2004.

[Camp2001] L. Jean Camp, Helen Nissenbaum and Cathleen McGrath: Trust: A Collision of Paradigms. In *Financial Cryptography: 5th Int. Conf., FC 2001*, Grand Cayman, British West Indies, 19–22 February. LNCS 2339. 2001.

[Dasgupta2000] P. Dasgupta: Trust as a Commodity. In Diego Gambetta (ed.): *Trust: Making and Breaking Cooperative Relations*, electronic edition, Department of Sociology, University of Oxford, Chapter 4, pp. 49–72. Available at: http://www.sociology.ox.ac.uk/papers/dasgupta49-72.pdf. 2000.

[Dunn2005] Jennifer R. Dunn and Maurice E. Schweitzer: Feeling and Believing: The Influence of Emotion on Trust. *J. of Personality and Social Psychology*, 88(5): 736–748. Available at: http://opim.wharton.upenn.edu/~schweitz/papers/Emotion_Trust.pdf. 2005.

[Feng2004] Jinjuan Feng et al.: Empathy and Online Interpersonal Trust: A Fragile Relationship. *Behaviour and Information Technology*, 23(2): 97–106. 2004.

[Fitness2001] Julie Fitness: Betrayal, Rejection, Revenge, and Forgiveness: An Interpersonal Script Approach. In M. Leary, (ed.): *Interpersonal Rejection*. New York: Oxford University Press. 2001.

[Giddens1991] Anthony Giddens: *Modernity and Self-identity: Self and Society in the Late Modern Age*. Cambridge: Polity Press. 1991.

[Gomez2006] Laurent Gomez and Ulrich Jansen: Trust Establishment in Emergency Case. In K. Stolen et al. (eds): *iTrust 2006*, LNCS 3986, pp. 443–448. 2006.

[Hawthorn2000] Geoffrey Hawthorn: Three Ironies in Trust. In Diego Gambetta (ed.): *Trust: Making and Breaking Cooperative Relations*, electronic edition, Department of Sociology, University of Oxford, Chapter 7, pp. 111–126. Available: http://www.sociology.ox.ac.uk/papers/hawthorn111–126.pdf. 2000.

[Ho1999] Teck-Hua Ho and Keith Weigelt: Trust Building Among Strangers: An Experimental Investigation of Mutual Trust and Trustbuilding. WP 99-01. A Working Paper of the Reginald H. Jones Center, Philadelphia, PA: The Wharton School, University of Pennsylvania. 1999.

[Lewicki1996] Roy J. Lewicki and Barnard Benedict Bunker: Developing and Maintaining Trust in Work Relationships. In Roderick M. Kramer and Tom R. Tyler: *Trust in Organizations*. London: Sage Publications. 1996.

[Luna-Reyes2004] Luis F. Luna-Reyes, Anthony M. Cresswell and George P. Richardson: Knowledge and the Development of Interpersonal Trust: A Dynamic Model. In *Proc. of the 37th Hawaii Int. Conf. on System Sciences. HICSS*. 2004.

[Marsh1994] Stephen Marsh: Optimism and Pessimism in Trust. Technical Report CSM-117. In *Proc. IBERAMIA94/CNAISE'94*. Ed. J. Ramirez, New York: McGraw-Hill. 1994.

[Marsh2005] Stephen Marsh and Mark R. Dibben: Trust, Untrust, Distrust and Mistrust – An Exploration of the Dark(er) Side. In P. Herrmann (ed.): *iTrust2005*, LNCS 3477, pp. 17–33. 2005.

[Marsh2006] Stephen Marsh: Trust, Forgiveness, Regret: An Internal Remediation Triangle. In *Proc. of Reinventing Trust, Collaboration and Compliance in Social Systems. A Workshop for Novel Insights and Solutions for Social Systems Design*. Available: http://hornbeam.cs.ucl.ac.uk/philip/reinvent06/presentations/CHI06-SteveMarsh.pdf

[Meyerson1996] Debra Meyerson, Karl E. Weick and Roderick M. Kramer: Swift Trust and Temporary Groups. In Roderick M. Kramer and Tom R. Tyler: *Trust in Organizations: Frontiers of Theory and Research*. London: Sage Publications. 1996.

[O'Riordan2001] Colm O'Riordan: Forgiveness in the Iterated Prisoner's Dilemma and Cooperation in Multi-Agent Systems. In Mike Luck and Mark d'Inverno (eds): *UKMAS* Oxford. 2001.

[Prietula2000] Michael J. Prietula: Advice, Trust, and Gossip Among Artificial Agents. In A. Lomi and E. Larsen (eds): *Simulating Organizational Societies: Theories, Models and Ideas*. Cambridge, MA: MIT Press. 2000.

[Prietula2001] Michael J. Prietula: Boundedly Rational and Emotional Agents Cooperation, Trust and Rumor. In C. Castelfranchi and Y.-H. Tan (eds): *Trust and Deception in Virtual Societies*. Dordrecht: Kluwer Academic Publishers. pp. 169–193. 2001.

[Resnick1994] Paul Resnick et al.: GroupLens: An Open Architecture for Collaborative Filtering of Netnews. In *Proc. of the 1994 ACM Conf. on Computer Supported Cooperative Work*. Chapel Hill, North Carolina, pp. 175–186. http://delivery.acm.org/10.1145/200000/192905/p175-resnick.pdf?key1=192905&key2=8948234411&coll=portal&dl=ACM&CFID=68878040&CFTOKEN=74261221.1994.

[Riegelsberger2003] Jens Riegelsberger. Interpersonal Cues and Consumer Trust in E-commerce. CHI2003 Doctoral Consortium, Extended Abstracts CHI2003, 5-10 April, Ft. Lauderdale, FL. 2003.

[Sierra2006] Carles Sierra and John Debenham: Trust and Honour in Information-based Agency. In *Proc. of AAMAS'06 May 8–12 2006*, Hakodate, Japan. 2006.

[Smith2002] Jonas Heide Smith: The Architectures of Trust. Supporting Cooperation in the Computer-Supported Community. MA thesis. The University of Copenhagen. 2002.

[Solomon2001] Robert C. Solomon and Fernando Flores: *Building Trust in Business, Politics, Relationships and Life*. Oxford: Oxford University Press. 2001.

[Stewart2006] Thomas A. Stewart and David Champion: Leading Change from the Top Line: Interview with Fred Hassan. *Harvard Business Review*, 84(718): 90–97. 2006.

[Tyler1996a] Tom R. Tyler and Roderick M. Kramer: Whither Trust? In Roderick M. Kramer and Tom R. Tyler: *Trust in Organizations*. London: Sage Publications. 1996.

[Tyler1996b] Tom R. Tyler and Peter Degoey: Trust in Organizational Authorities: The Influence of Motive Attributions on Willingness to Accept Decisions. In Roderick M. Kramer and Tom R. Tyler: *Trust in Organizations*. London: Sage Publications. 1996.

[Ullmann-Margalit2001] Edna Ullmann-Margalit: Trust Out of Distrust. Available at: www.law.nyu.edu/clppt/program2001/readings/ullman _margalit/Trust%20out%20of%20Distrust.pdf. 2001.

[Valsiner2005] Jaan Valsiner: Civility of Basic Distrust: A Cultural-psychological View on Persons-in-society. Paper presented at Symposium Risk, Trust, and Civility, Toronto, Victoria College. May 6–8. 2005.

[Webb1996] Eugene J. Webb: Trust and Crisis. In Roderick M. Kramer and Tom R. Tyler: *Trust in Organizations*. London: Sage Publications. 1996.

# 6

# The Dark Side

Light is the left hand of darkness.

(Ursula K. Le Guinn)

## 6.1 Introduction

A better understanding of social and psychological aspects of confidence and trust may solve several current problems related to cooperation in and through large information systems. However, we would be naïve to think that in this process we can filter out the less desirable sides of our nature and implement only what we consider wanted and approved. Every phenomenon has its darker side, and trust (as well as confidence) is not devoid of it.

Therefore, no discussion on trust can ignore its 'dark' counterparty: distrust. Similarly, the discussion about control (the dialectic partner of trust in the formation of confidence) would be incomplete without discussing dis-control. Of those two: distrust and dis-control, distrust attracts more attention. It is not surprising, considering that trust is the only irreducible element of the confidence so that distrust can have potentially greater damaging effects than dis-control.

Distrust is probably the only construct that is even more complex and less understood than trust. Considering that trust has 17 different meanings [McKnight1996], and that there are several different ways to negate it, the number of potentially conflicting concepts of distrust can be really high. However, distrust must not be ignored just because it is hard to define. Even though distrust bears often the negative connotation of the undesired, dark and morally doubtful, distrust is indeed a valid partner of trust.

This chapter, on the basis of [Cofta2006], discusses distrust and dis-control (together with the associated concepts of mix-trust and un-trust) from the perspective of the complexity-based model of confidence presented in this book. The usage of this model does not overly restrict the discussion, as several thoughts represented here can easily be applied to other models. The model therefore serves mostly by delivering a useful structure and leading the discussion, not as a restricting exoskeleton.

*Trust, Complexity and Control: Confidence in a Convergent World*   Piotr Cofta
© 2007 John Wiley & Sons, Ltd

## 6.2 Related Works

Considering the morally doubtful provenance of distrust (as well as its volatile nature), it is not surprising that distrust has attracted less research interest than trust, specifically when trust is associated with such positive concepts as cooperation, the development of economy, general prosperity and improvement in social life. If we add to this the fact, that distrust bears significant negative emotional connotations, is often associated with painful memories and damaging events, we can see that this important research area may be left unexplored and unwanted. In fact, Webb [Webb1996] notices that there are almost three times more papers discussing loyalty and trust than betrayal and distrust.

Certain important properties of distrust have already been identified. Luhmann in discussion of distrust [Luhmann1979] points to the fact that distrust, similarly to trust, is a tool to restrict complexity, and that distrust is the functional equivalent of trust. Paradoxically, distrust may be even stronger in complexity reduction than trust: by distrusting Bob, Alice eliminates all potential futures that involve him, potentially reducing futures by a significant extent. In comparison, if she trusts Bob, she can eliminate only futures where he is not helping her.

Russell Hardin's review [Hardin2004a] brings a wealth of sociological discussion about distrust, touching on several important points regarding the mechanism of distrust as well as its moral justification, impact on the society, politics, etc. The ethics of distrust is particularly important due to the negative connotations associated with it, that may prevent rational analysis. The discussion by Markoczy [Markoczy2003] and Larson [Larson2004] deals with the common perception of moral values of trust and distrust, demonstrating that certain behaviour that is commonly attributed to distrust is not only justifiable but even socially desired. Not everything that looks like distrust is distrust and not everything that is distrust is undesired.

Several authors have noticed that there is not only a gap between what can be considered full trust and full distrust, but also that information available is never complete or certain. Abdul-Rahman, in his monograph [Abdul-Rahman2005] introduces concepts of mistrust and ignorance. Ullmann-Margalit [Ullmann-Margalit2004] discusses the continuum of trust, identifying the important area where neither trust nor distrust prevail. There is an understanding that certain situations cannot be characterised as either trust or distrust or that they can be characterised by both trust and distrust at the same time [McKnight2006].

Marsh's work on distrust [Marsh2005] attempts to close the important gap between the social and computational perception of distrust (as well as trust) by 'calling to arms' the research community and by offering certain valuable considerations and formalisations (e.g. reinforcing the concept of mistrust) in this area.

Several models that deal with the creation of trust also cover distrust – even though sometimes the attention is concentrated on trust. McKnight's model of trust in electronic commerce [McKnight2003] has introduced distrust into the existing model in reaction to visible discrepancies, followed by the more in-depth analysis of the phenomenon of distrust [McKnight2006]. The complexity-based model that is discussed here was designed to accommodate distrust, even though the discussion of distrust was not included in the original description.

Practical considerations regarding complex information systems (e.g. cooperating agents) have introduced the concept of distrust as a mathematical complement of trust, usually without thorough justification of its semantics. For example, the EigenTrust [Kamvar2003] model introduces distrust as a reaction to negative experiences while Gans [Gans2001] studies the propagation of distrust in the agent network. Similarly, several formal models such as

Grandison's Sultan [Grandison2003] introduce the notion of distrust indirectly, by allowing negative values of trust.

## 6.3 The Starting Point

Let's look at the concept of distrust. It seems to be reasonably simple: distrust (following the logic of the language) is 'not-trust', the negation of trust, something more than simply the lack of trust. If Alice is distrusting Bob, then her beliefs are different than those of not trusting him (or not having beliefs at all). Note, however, that this particular comment applies to the English language. The author's mother tongue, for example, does not differentiate between the lack of trust and distrust. It is possible that other languages may have the vocabulary that provides more or less shades of trust and may cater for different relationships between trust and distrust.

Defining distrust as more than just a lack of trust points to the fact that distrust should be not only the lack of trust but in fact it should be the negation of it, the potentially equivalent parallel construct residing 'on the dark side'. However, while defining trust (simplifying) as the expectation about other's beneficial behaviour, we can see that such a statement can be negated in several ways. We can talk about the lack of expectation, about an expectation of harmful behaviour or about lack of expectation of harmful behaviour (two negations in one sentence) – each statement defining something very different. Referring to [Shafer1976] in [Josang2004], we can see that our frame of discourse has increased fourfold, and probably we should reduce it.

The lack of expectation (or opinion) can be attributed rather to ignorance than to distrust, regardless whether it is about beneficial or harmful behaviour. If there is no evidence to formulate expectations (or if Alice cannot create such an expectation on the basis of available evidence), then Alice is ignorant, not distrusting and Alice is in a situation of high uncertainty. Being ignorant does not mean trusting either, as an ignorant person cannot formulate any expectations regarding future behaviour.

From another perspective, if we define trust as a subjective estimate of another's trustworthiness (the way we introduced it early in this book), we can see that the 'not-trust' may mean that either Alice's estimate is negated (so that Bob may be trustworthy while she perceives him differently) or that her estimate is right and we face the phenomenon of 'negative trustworthiness', i.e. we face the reasonable expectation of harmful behaviour.

Even though in theory both options are possible (i.e. Alice's decision-making process may be negated or Bob may really be out to get her), the first option is likely to eliminate itself in a natural (but drastic) process. Alice errs here on the side of caution, resulting in 'false negative', rejecting Bob despite his good intentions. Alice's inability to come up with the correct estimate will eventually marginalise her in society and will also marginalise our interest in her. We can call her behaviour 'misplaced distrust' (or paranoia, depending on the level of her error), in parallel with the concept of misplaced trust (where Alice trusts Bob to find out that he is not trustworthy), but this is not distrust as such, as Alice's assessment of Bob's intentions is incorrect.

Both lines of thoughts converge here and leave us with the last (and most correct) form of negative trust: where distrust is the expectation of harmful behaviour (expectation of negative trustworthiness). Distrusting therefore means that Alice reasonably believes that Bob is willing to harm her (instead of helping her).

We can now start considering distrust as a separate construct, not simply the lack of trust. This is supported e.g. by [McKnight2006] where distrust has been identified as a separate structure, with its own specific sub-constructs. If combined with trust, distrust has more predictive power then trust [McKnight2006], specifically in situations of higher risk. However, this can be attributed more to the fact that in situations of risk people tend to hold back and re-consider their position while not allowing positive emotions to influence the decision. This is also supported by observations [Lumsden2006] that pessimistic personalities demonstrate behaviour that can be called rational, exploring and verifying available evidence.

The fact that distrust is a separate construct does not say much about the mechanism of the origin of distrust. Some authors (e.g. Giddens [Giddens1988], in relation to globalisation) suggest that distrust should be the default option, but others [Marsh1994a] point to benefits of assuming a more optimistic stance. If distrust is the default behaviour, then it can be achieved purely through will, with no evidence and no reasoning. However, if we assume that the assessment of distrust is somehow similar to the assessment of trust, we can think of distrust being in fact governed by the similar, evidence-driven, model (see also [McKnight2006] for a similar reasoning).

To be distrusted by Alice (we will assume for now), Bob must provide evidence that such distrust is justified. Following the model, Bob must demonstrate harmful motivation, competence to do harm and the environment that supports harm. This is the starting point.

## 6.4 The Ethics of Distrust

Distrust has a bad publicity. There is a popular understanding that distrust is undesired as it negatively impacts on economical development [Fukuyama1996] and generally negatively influences human relationship, politics and everyday life. A world without distrust seems to be a much better place. However, there are some works (e.g. [Kern1998]) suggesting that is not always the case.

The negativity associated with distrust is partly true and justified. Indeed, the existence of distrust prevents Alice and Bob from cooperating or imposes very high cost on such cooperation, as detailed control must be implemented. Further, as distrust cannot be easily converted into trust (there is a long and complex recovery path), potential damage done by distrust may outweigh the benefits of trust. This, combined with the volatility of relationships that can easily slip into distrust, makes distrust dangerous and undesired.

The moral and emotional valuation of distrust may obscure the analysis conducted here, so it is worthwhile to spend some time clarifying the potential value of distrust – or at least removing some of the misconceptions about its negativity. Let's first remove from the scope of this discussion the behaviour that may look like distrust while it is not. This will allow us to concentrate on distrust in its proper form.

### 6.4.1 Distrust-like Behaviour

Prudent and vigilant behaviour may seem to indicate distrust. Controlling and verifying others can easily be interpreted as a signal of distrusting them and can thus be reciprocated. Some research claims that the vigilance, as a form of distrust, breeds distrust in a kind of vicious circle. However, in fact neither prudence nor vigilance (within reasonable limits) is distrust (see [Markoczy2003] for an interesting discussion). They are necessary for the model to function properly so that they are beneficial from the perspective of confidence and trust.

The brief look at the model of trust shows that it is built on the availability of evidence. Without evidence it is impossible to establish trust – or distrust – leaving people ignorant. However, even if evidence is proactively sought and gathered in the early stage of a relationship, people gradually become complacent with what they already know. Such self-complacency closes the opportunity to acquire new evidence, thus locking them in one of two bi-polar paranoid states: trust or distrust.

Vigilant and prudent behaviour can be interpreted as simply not closing further opportunities to collect evidence. Indeed, by 'trusting and verifying' one can receive important evidence that the trust vested in someone has been abused (mistrusted). Therefore, neither vigilant nor prudent behaviour can be qualified as distrust, but as a reasonable evidence-seeking activity.

For example, it is not a sign of distrust to accountants to conduct the audit of the company finances, but it is prudent and vigilant behaviour (which is also the fiduciary responsibility of the management). Certainly, if the audit uncovers some wrongdoings, then it may lead to distrust – but it is not the failure of the audit.

One may consider where the line must be drawn between acceptable vigilance and unacceptable intrusion into one's life. In societies, this has been defined e.g. by privacy protection law while the practicality of evidence processing sets the upper limit of what can be achieved. The line can be drawn somewhere in-between, where the vigilance becomes the nuisance rather than just the evidence-enabling activity and where the complexity reduction gained from trusting is increasingly eroded by the additional workload associated with vigilance. Unfortunately, technology is continuously pushing this limit towards intrusion, as the cost of detailed vigilance is decreasing.

It is worth noting that both vigilance and prudence are usually associated with the transition from trust to distrust. However, in principle there is nothing preventing them from working both ways: the prudent person is not only the one that verifies those that are trusted, but also the one that verifies those that are not. In the latter case, evidence of trust may emerge that can gradually increase the trustworthiness of the person. Such 'positively aimed prudence', even though socially rare, is essential to the process of the restoration of trust. Forgiveness (discussed separately in this book) can be considered an example of such a process.

## 6.4.2 Distrust in Social Contexts

Now it is time to deal with the often negative connotation regarding 'proper' distrust (i.e. distrust as the expectation of hostile behaviour) that is treated as the undesired social behaviour. Let's consider two examples that are usually mentioned in this context: trust in the government and trust in commerce.

There has been never-ending discussion (e.g. [Barber1983]) about the crisis of trust in government, suggesting that governments become generally distrusted and this negatively affects the political and economical prospect of nations. As has been validly pointed out [Hardin2004a], governments (at least in modern democracies) are not supposed to be trusted. Certain levels of distrust towards social institutions are essential to prevent their degeneration [Valsiner2005].

Trust was potentially desired in serfdom, but the modern concept of the division of power is built on the foundation that needs no trust, but control. Thus government should be kept in check, watched, controlled and corrected if needed – everything but not trusted. Hence, distrust is the social virtue of the enlightened citizen, not trust. Certainly, trust in government simplifies life (as trust usually does), both for the government and for citizens, but this is not the prerequisite of a successful country.

Similarly, trust in economic relationships is presented as a capital [Fukuyama1996] that can be gained (and presumably spent). Such social capital lubricates the machinery of trade, thus lowering the cost of trade and improving the welfare. Trust therefore is considered 'good' while distrust is 'bad'. Again, the moral valuation is mixed with the pragmatic one: trust can indeed be more optimal from the economy standpoint and trusting may yield better economic results. Trust definitely simplifies protocols and generally decreases complexity.

However, this does not imply that trust is the desired virtue of every merchant. Prudence may be more desired than indiscriminate trust, where such prudence will lead to the appropriate understanding of confidence and business risk. It is secondary whether this will lead to trust or distrust.

The consideration above does not imply that trust is not important in all social contexts. In some areas such as health and social care [Smith2005], trust cannot be replaced by control to deliver sufficient confidence, so that in the absence of trust (and specifically in the presence of distrust), the whole area may not perform to our expectations – something that we may experience too often.

### 6.4.3 Benefits of Distrust

Finally, distrust delivers some benefits. First, distrust acts as a natural dividing mechanism [Kern1998]. As distrust excludes cooperation, barriers created by distrust can be effectively used to divide people (or computers) into groups with no communication between groups – into 'us' and 'them'. This, in turn, improves and facilitates cooperation within each group, leads to group solidarity, rapid formation of common identity and ultimately the development of trust within the group. Considering the overall efficiency, dividing may actually deliver better results than unlimited cooperation and trust. Combined with certain forms of communication between groups (through mutually trusted members [Yan2003]), distrust allows a structure to be built that is efficient, scalable and controllable – the foundation of a modern organisation.

Distrust can also work as an emotional compensation mechanism. In such a mechanism, if Alice want to exclude Bob from her world (for reasons other than his distrustful behaviour), she intentionally presents him as distrusted, possibly even re-interpreting available evidence. Through this mechanism she rationalises and justifies her decision to reject and exclude him, and transfers the emotional blame onto him. As long as Alice is not reporting such perceived distrust to others, the mechanism works well for her and does no harm to him.

## 6.5 Distrust and the Model of Confidence

Looking at the complexity-based model of confidence one can see that trust is deeply embedded in the model as one of its elements. However, for the sake of simplicity, we should restrict ourselves here only to discuss trust and distrust, not the whole confidence evaluation process. Therefore, the discussion will concentrate on one element of the model only, somehow extracted from its context. For this reason, the context should be briefly re-stated below.

Trust (and presumably distrust) is one of two elements in the confidence assessment process, with control (and potentially dis-control) being another one. Trust influences the model in several ways. First, the extent of trust (the 'level of trust') directly contributes to the extent of confidence attributed to the entity. Second, the extent of trust in honesty (together with control) influences the weight of evidence, so that evidence from honest sources influences the outcome to the greatest extent. Finally, trust is an irreducible prerequisite for control, as instruments of control should be eventually trusted (directly or indirectly).

The fact that the extent of trust influences the value of evidence leads to the phenomenon of self-locking trust where trusting someone may lead to overvaluing evidence from such person, that effectively 'locks in' the relationship on the trust side, leading to the unjustified trust. One can expect a similar mechanism on the side of distrust where the person considered distrusted will be locked in in such position so that every bit of evidence coming from such person will be discounted.

## 6.6 The Assessment of Distrust

Assessment of trust is built on evidence and such evidence comes in three different classes: continuity, competence and motivation [Barber1983]. Continuity supports the perception that the 'order of the world' will remain stable so that no undesired discontinuity will get in the way of Bob acting to Alice's benefit. Evidence of competence demonstrates that Bob is able (technically, resource-wise, etc.) to actually act to Alice's benefit. Finally, evidence of motivation demonstrates that Bob is willing to act to Alice's benefit, e.g. due to the fact that her best interest is encapsulated in his best interest [Hardin2004a].

Those three classes of evidence form three dimensions of trust, along three main axes: one for continuity, one for competence and one for motivation. We can for the moment assume that evidence of one class does not significantly contribute to other classes (even though Alice may attribute different weights to them). Indeed, Bob may be willing and powerless or he may be competent but unconcerned about the future.

This concept can be illustrated in Figure 6.1 where three axes form the three-dimensional 'cube of trust'. Assuming (for the purpose of this discourse) that the relative 'strength' of evidence in all three dimensions can be normalised to values from 0 (no evidence) to 1 (strong evidences), the point (1,1,1) can be interpreted as the absolute trust while remaining points

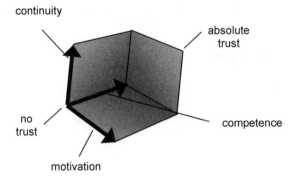

**Figure 6.1**  The trust cube

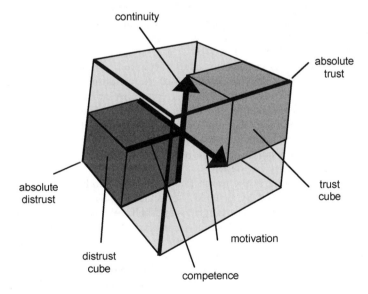

**Figure 6.2**  Trust and distrust

of the cube can have different 'shades' of trust assigned to them, down to the lack of trust (but not to distrust) at the point (0,0,0).

Assuming that distrust is the negation of trust, one can see that the evidence of a particular kind can be negated in two different ways – similar to the way we have negated trust into distrust. First, the existence of the evidence can be negated so that there will be no evidence of the particular kind. Second, the content of the evidence can be negated so that the evidence will demonstrate the contrary to what it originally supported.

Repeating the reasoning that has led us to the definition of distrust, if evidence is missing, then trust is supposed to decrease, to the extent where it disappears entirely – but this is the 'positive' lack of trust, where trust is latent, inhibited only by the lack of evidence. If, however, the evidence is negated in its content, the trust is reversed so that another relationship emerges. Following Marsh [Marsh2005], the lack of trust that is caused by insufficient evidence will be called un-trust while the relationship created by negative evidences will be called distrust.

We can extend Figure 6.1 by drawing all three axes into the negative area (Figure 6.2) and identify the point $(-1, -1, -1)$ as the point of absolute distrust, i.e. the point where there is strong supporting evidence to the contrary of trust. Symmetrically to the 'trust cube', the 'distrust cube' can be drafted with points representing different levels and shades of distrust.

## 6.7 Evidence of Distrust

Let's look at evidence that is required to support distrust. Three classes of trust-bearing evidence are: continuity, competence and motivation. Those classes should yield themselves to negation that will produce three classes of evidence of distrust. Note that the absence of evidence is not sufficient – it is the negation that creates distrust.

### 6.7.1 Continuity

What is the negation of continuity? Interestingly, discontinuity is not enough – discontinuity only means that the current continuity will no longer be valid somewhere in the future, possibly before Bob can satisfy Alice's trust in him. Discontinuity is therefore only an equivalent of the lack of evidence supporting continuity. Certainly discontinuity does not support trust (e.g. [Bolton2003]), but it does not necessary generate distrust. It is yet another case of un-trust that has been mentioned before.

The 'negative continuity' is yet another continuity, but a different one. Evidence should demonstrate that Bob is bound by (or believes in) the continuity that differs from Alice's, so that either he may survive the damage to her continuity or his continuity can terminate abruptly without any visible relationship to her. This may come (among other) from Bob being free to terminate the relationship at no cost, being controlled by forces unknown to Alice, being bound by moral obligations that are not shared by her, etc. The adherence to his alternative continuity will enable him to influence their common future but it may be of a kind of influence that she may not desire.

For example, let Bob belong to the group that is perceived to be of different moral order than Alice's (the case discussed by Lewicki [Lewicki2003]). This may be, for example, an organized crime group that follows certain rules and believes that those rules will bind them well into the future. Assuming that Alice is an ordinary law-abiding citizen, such a group demonstrates different (withholding all moral judgements) continuity. Bob may be loyal to his gang and may be willing to support his leaders. Alice may be loyal to her family and she may be willing to support her country. Apparently, Bob's future has little to do with Alice's. Thus he may be distrusted on the basis that his behaviour in the future is governed by rules that contradict hers.

## 6.7.2 Competence

The lack of competence breeds un-trust, as the person that does not have means or skills needed to support our case will not be able to do it. However, to create distrust one should demonstrate 'negative competences' – competences that will allow him or her to interfere and disrupt our preferred scenarios. If, for example, our goal is to have a secure computer network, we may trust someone who demonstrates skills needed to make such network secure. We may also distrust someone who has broken into the network, as he or she has demonstrated skills that negate skills that are required to support our goal.

This example shows clearly a potential problem with negative competences. It is too often that negative and positive competences do not differ too much. Knowledge of medicines and knowledge of poisons go hand in hand. Power and monetary resources can both support and destroy our case. Breaking the system is a part of exploring its security and so forth. White hat and black hat are sitting next to each other.

Note, however, that there are not latent competences that are at stake here, but demonstrated ones. Even though we may know that poison and medicine are close, we can easily discern between demonstrated evidence of healing and killing, attributing potentially trust to the former and distrust to the latter.

## 6.7.3 Motivation

While, the two previous classes of evidences may require a certain effort to create their negation, the 'negative motivation' is one of the easier to grasp. Trust requires evidence of motivation, i.e. evidence that Bob has certain reason to support Alice, mostly by encapsulating her interest in his interest [Hardin2004a]. Symmetrically, distrust requires evidence to the contrary: that Bob has a reason to harm and damage Alice's case, e.g. that the destruction of her case is encapsulated in his interest.

For example, if Alice is competing with Bob to win a large contract, it is evident that Bob's best interest as a competitor is not to allow her to have this contract, so that the destruction of her case is encapsulated in his case. Even if Bob is bound by rules to play fair, Alice should be aware that he may be motivated to hurt her. Certainly, this should lead to distrust, rather than to the simple lack of trust.

What we have demonstrated here is that three classes of evidence that support trust have their mirror-negative counterpartis that support distrust – each class having its own negative image. This line of reasoning follows Figure 6.2 established above: evidence regarding distrust is placed along negative parts of axes associated with different classes.

# 6.8 The Area Between

The world would be a simple place if evidence from the same dimension of trust comes only from the positive or only from the negative side (i.e. supporting only trust or only distrust). In many cases, we are presented with a mixed set of evidence where some evidence may support trust and others distrust – while yet others may support the lack of trust or the lack of distrust.

Let's consider four bits of evidence that have been developed in the course of a business relationship between two companies. First, both companies have a history of successful cooperation in standardisation groups. Second, both companies are competing now in a limited marketplace. Third, they are not interested in the joint sponsorship of a cultural event. Fourth, they share the same office in one foreign country. Should they trust each other or not?

The answer may be mixed, as evidence points in both directions. Successful cooperation in the past could have been attributed to trust, but the current competition suggests distrust, reinforced by the lack of desire to co-sponsor the event. Sharing the office requires a certain level of trust, but it might have been driven more by necessities of the local rental market than by choice.

The actual outcome seems to be determined by the dynamics of trust. The way events appear in time, combined with their direction and intensity, builds the history of relationship. Such history leads to certain current levels of trust or distrust through the process of building the relationship. Not going into details, from the perspective of this analysis it is enough to say that evidence of the same class (whether positive or negative) can be consolidated in time into the single perception of trust or distrust.

The more interesting case is when one dimension indicates trust while another leads towards distrust, i.e. when we have conflicting consolidated evidences in different dimensions (Figure 6.3). As they are generally incompatible (being from different dimensions), they cannot be consolidated through their historical analysis. Figure 6.3 shows that points from the remaining grey six cubes (the majority of the space discussed here) refer to such cases. For example, Bob may demonstrate desired competence but evil motivation; or he may be positively motivated but his perception of continuity significantly differ from Alice's.

Those cases form the space of mix-trust (this new term has been created in the absence of any better description of this phenomenon), i.e. the space where trust and distrust compete and where no simple answer can be provided, as there is no consolidation mechanism that can work across different dimensions.

The state of mix-trust is undesired from the complexity point of view (see the discussion later) so that there are certain ad-hoc methods to handle mix-trust. First, we can set arbitrary rules that will reduce the space of mix-trust into trust or distrust. The principal approach may be to use distrust on any axis to override any signs of trust on other axis, so that if Bob is distrusted in one dimension, he is distrusted in all of them. A more tolerant approach may, for example, allow certain distrust along one axis if offset by trust along other ones. As such rules are set in an arbitrary manner (potentially in response to social standards), it is hard to generalise them, except for the fact that they demonstrate the desire to resolve the case of mix-trust, even at the expense of a personal perception of justice.

Another social method to resolve mix-trust is the identity split. The identity of a single person can be socially split into several ones (e.g. private identity at home and public one at work). Each identity, as it deals with a fragment of a personal activity, attracts only a fraction

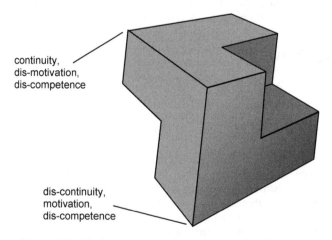

**Figure 6.3**  Mix-trust areas and some characteristic nodes

of evidence, so that it is possible to produce the coherent perspective whether to trust or distrust the person within the context of a single identity. 'The trusted father but unreliable worker' (or vice versa) is a good example of such approach. Again, the identity split is no more than an escape route to deal with mix-trust and it demonstrates that mix-trust may not be socially bearable.

## 6.9 Trust, Distrust, Mix-Trust and Complexity

The primary role of trust is to contain complexity of the future by assuming on the basis of available evidence that the other party will act to our benefit. Let us now consider whether distrust can act as a similar complexity reductor, becoming the functional equivalent of trust [Luhmann1979].

The way trust decreases complexity is by removing a number of possible future scenarios Alice should be dealing with. Future development scenarios that depend on Bob are reduced only to those that are beneficial for her – removing those where Bob is not helping her. If Bob is an essential part of Alice's preferred futures, this approach significantly increases subjective chances that a positive scenario will be realised.

Interestingly, distrust is working to a great extent in the same way. By distrusting, Alice assumes that Bob, the 'dis-trustee', will act against her. Therefore, all the scenarios where he is supposed to support her can be removed, as she subjectively believes that those scenarios will not be realised. Certainly, if Bob is the critical element of Alice's preferred scenarios, those scenarios are unlikely to happen. However, the goal has been achieved: she deals with a simpler future where certain scenarios have been eliminated. Having the complexity reduced, she can make the decision how to proceed: whether to abandon her goal, whether to neutralise Bob or whether, for example, to find a different goal. Whatever she chooses, the understanding of the future behaviour of the dis-trustee eliminates complexity and enables her to concentrate on working towards her goal.

Mix-trust, on the other hand, becomes the problem from the perspective of complexity. As Alice has no firm opinion about Bob, she cannot remove either set of scenarios from her future: Bob may act to her benefit or against her and she should be prepared for both. As the reduction does not happen here, Alice may be actually willing to engage in one or more strategies to artificially introduce the reduction of mix-trust to trust or distrust, as discussed above.

Note that here we can see a difference between risk assessment and trust-based approach. From the perspective of risk, mix-trust is better than distrust (and worse than trust). Mix-trust implies that there is an increased chance of the beneficial scenario happening, compared to distrust. For risk assessment, complexity is not an issue, as available resources and instrumentation to study the future can be expanded to satisfy needs. Therefore resourceful organisations can embrace risk assessment and mix-trust while individuals (or small organisations) must resort to a choice of trust or distrust as their main strategy, as it is the reduction of complexity that drives them.

## 6.10 Dis-control

While considering distrust it is worth checking whether there is a meaning in dis-control. Such a word ('dis-control') cannot be found in any dictionary and this may raise suspicion that we are talking here about a void construct that cannot be properly identified or maybe does not even exist at all.

We are not going into a complete analysis of the subject, but as the logic of the model may evoke the concept of dis-control, this construct should be at least briefly explored. First, let's differentiate between dis-control and un-control (where the latter can actually be found in dictionaries). In a manner similar to differentiating between dis-trust and un-trust, we can provisionally define un-control as the lack of evidences of control and dis-control as the presence of evidence to the contrary.

## 6.10.1 Dis-control and Un-control

Recalling the discussion about control (and controllability), Alice can gain confidence along the 'control path' if there is evidence that demonstrates that certain instruments of control work on Bob, i.e. that he is controllable through such instruments (assuming that Alice is confident in those instruments). Contrary to trustworthiness, controllability cannot be defined as Bob's internal property, but only as Alice's perception of how Bob responds to control. We can reasonably expect that everybody can be controllable, assuming that Alice can afford the right instruments – even though Alice will not always benefit from such excessive control. However, Alice is interested only in controlling Bob within the limits of her affordable complexity, and it is here where Bob can be more or less controllable.

What we discuss here is therefore Alice's perception of her ability to control Bob through instruments she can afford. Similar to her perception about trusting Bob, Alice should distinguish between three different cases: for a given instrument: (1) she may have evidence that Bob can be controlled with it; (2) she may not have sufficient evidence regarding Bob and the instrument; and (3) she may have evidence that Bob cannot be controlled with this instrument. Intentionally, we have excluded the case of ignorance, where Alice has simply no evidence regarding Bob and the considered instrument.

The first case has already been covered by the model: Alice can use an instrument to gain additional confidence regarding Bob. The second case is more interesting: Alice has evidence but they are inconclusive. This can be attributed to only two things: either the instrument is not relevant to Bob or Bob is indifferent to it. For example, if we consider reputation-based systems, Bob may not be aware of such a system or he may simply ignore its rating. In either case, what Alice receives is a kind of random readings, showing that Bob sometimes seems to respond to control, and sometimes does not.

This leaves the last case: Alice has evidence that Bob cannot be controlled by an instrument. What is a difference between 'random readings' described in the second case and inability to control discussed here? The difference is in Bob's attitude. In the second case Bob simply ignores the instrument, in this case he is proactively working against the instrument. Returning to the reputation-based system, Alice may have evidence that Bob is quite aware of the system but is actually able to use the system to his advantage, e.g. by manipulating his rating. This is the class of evidence of dis-control: Bob actively fighting the control system.

From Alice's perspective, there is a huge difference between those three cases. In the first case she knows that by using an instrument she can gain higher confidence regarding Bob. In the second case (un-control), she knows that an instrument is useless with regard to Bob – she is not gaining anything by using it, while potentially spending some of her complexity on it. Finally, in the case of dis-control, she knows that the use of a particular instrument will actually lower her confidence, as Bob is willing to act against her.

## 6.10.2 Example

Let's assume that Alice would like to buy something from Bob and what she needs is confidence that Bob will deliver as promised. She may look at Bob's past performance (using the 'knowledge' component) to see what is Bob's usual behaviour. In case of un-control, she may find Bob's behaviour erratic and unpredictable: sometimes he performs brilliantly, sometimes he fails. In the case of dis-control she may find evidence that Bob has actually manipulated his performance ratings. For example, Alice may find the rating that is too perfect or too regular so that it indicates Bob's sinister manipulation (rather than Bob's consistent performance).

Alice may also verify whether she has an instrument of influence. For example, she may want to be able to call the customer service (in the hope that this will speed up the delivery). In case of un-control, her experience with customer service will be of a mixed nature – sometimes she may receive help and sometimes not, sometimes delivery will be accelerated and sometimes not. In case of dis-control she will see that Bob is intentionally 'punishing' customers who call the customer service – possibly in the expectation that customers will realise this, stop calling and save him the trouble of running this service. For example, whenever she calls the customer service, the delivery is delayed.

Finally, Alice may want to gain some assurance, for example, by verifying that Bob is bound by appropriate laws and regulations that can be disputed in the court. In case of un-control, Alice is unlikely to receive any evidence of this or maybe the jurisdiction selected by Bob is too exotic for Alice – so that she is unlikely to pursue the dispute anyway. In the case of dis-control, Bob may offer a legal recourse on condition that he (or e.g. his dependent company) will represent Alice in the court – so that Alice is not only unlikely to win, but she will actually pay Bob's legal expenses.

Concluding this short analysis of dis-control (and un-control), we can see that the construct of dis-control (similar to the construct of dis-trust) follows the logic of the model and can find its justification in a real-life experience. The complete analysis of dis-control probably requires a separate discussion, but we can see at least that it is justifiable to have such discussion.

## 6.11 Conclusion

We deal here with constructs of distrust and dis-control, which are understood as a negation of trust and control, respectively, not just the mere lack of them. By alleviating some of the ethical issues that are usually associated with distrust, we have established distrust as a valid partner to trust. From the perspective of the complexity-based model of confidence, distrust is a logical complement of trust and can be explored along the same three dimensions; while it is the mix-trust (mixed trust and distrust) that brings additional problems as it increases complexity. Similarly, dis-control can be interpreted within the framework proposed by the model.

By discussing distrust and dis-control within the framework defined by the model we can bridge the gap between the social perception of distrust and dis-control and their theoretical underpinnings and technical implications. The ability to address both trust and distrust (as well as control and dis-control) within the single model is potentially very attractive, specifically it does not lead to the significant increase in the complexity of the model while it can provide the uniform reference framework, where the actual computation of trust and distrust may be possible.

## Bibliography

[Abdul-Rahman2005] Alfarez Abdul-Rahman: A Framework for Decentralised trust Reasoning. PhD thesis. Available at: http://www.cs.ucl.ac.uk/staff/F.AbdulRahman/ docs/thesis-final.pdf. 2005.

[Barber1983] Bernard Barber: *The Logic and Limits of Trust*. New Jersey: Rutgers University Press. 1983.

[Bolton2004] Gary E. Bolton, Elena Katok and Axel Ockenfels: How Effective are Electronic Reputation Mechanisms? An Experimental Investigation. *Management Science*, 50(11): 1587–1602. 2004.

[Cofta2006] Piotr Cofta: Distrust. In *Proc. of Eight Int. Conf. on Electronic Commerce ICEC'06*, Fredericton, Canada. pp. 250–258. 2006.

[Fukuyama1996] Francis Fukuyama: *Trust: The Social Virtues and the Creation of Prosperity*. New York: Touchstone Books. 1996.

[Gans2001] G. Gans, M. Jarke. S. Kethers and G. Lakemeyer: Modelling the Impact of Trust and Distrust in Agent Networks. Available at: http://www-i5.informatik.rwth-aachen.de/~gans/tropos/dokumente/AOIS01.pdf. 2001.

[Giddens1988] Anthony Giddens: *The Consequences of Modernity*. Cambridge: Polity Press. 1988.

[Grandison2003] Tyrone Grandison: Trust Management for Internet Applications. PhD thesis, University of London. 2003.

[Hardin2004a] Russel Hardin: Distrust: Manifestation and Management. In Russell Hardin (ed.): *Distrust*. New York: Russell Sage Foundation, pp. 3–33. 2004.

[Josang2004] Audun Josang and David McAnally: Multiplication and Comultiplication of Beliefs. *Int. J. of Approximate Reasoning* 38/1: 19–55. 2004.

[Kamvar2003] Sepander D. Kamvar, Mario T. Schlosser and Hector Garcia-Molina: The Eigentrust Algorithm for Reputation Management in P2P Networks. In: *WWW2003*, May 20–24, Budapest, Hungary. 2003.

[Kern1998] H. Kern: Lack of Trust, Surfeit of Trust: Some Causes of the Innovation Crises in German Industry. In C. Lane (ed.): *Trust Within and Between Organizations*, Oxford. 1998.

[Larson2004] Deborah Welch Larson: Prudent, If Not Always Wise. In Russel Hardin (ed.): *Distrust*. New York: Russell Sage Foundation, pp. 34–59. 2004.

[Lewicki2003] Roy J. Lewicki and Edward C. Tomlinson: *Distrust*. 2003. Available at: http://www.beyondintracability.org/m/distrust.jsp. 2003.

[Luhmann1979] Niklas Luhmann: *Trust and Power*. Chichester: John Wiley & Sons. 1979.

[Lumsden2006] Jo Lumsden, Lisa MacKay: How Does Personality Affect Trust in B2C e-Commerce? In *Proc. of Eight Int. Conf. on Electronic Commerce ICEC'06*, Fredericton, Canada. pp. 471–481. 2006.

[Markoczy2003] Livia Markoczy: Trust but Verify: Distinguishing Trust from Vigilance. Available at: http://www.goldmark.org/livia/papers/socint/socint.pdf. 2003.

[Marsh1994] Stephen Marsh: Optimism and Pessimism in Trust. Technical Report CSM-117. In *Proc. IBERAMIA94/CNAISE'94*. Ed. J. Ramirez. New York: McGraw-Hill. 1994.

[Marsh2005] Stephen Marsh and Mark R. Dibben: Trust, Untrust, Distrust and Mistrust – An Exploration of the Dark(er) Side. In P. Herrmann (ed.): *iTrust2005*, LNCS 3477, pp. 17–33. 2005.

[McKnight1996] D. Harrison McKnight and Norman L. Chervany: The Meanings of Trust. In University of Minnesota, http://www.misrc.umn.edu/wpaper/wp96-04.htm. 1996.

[McKnight2003] D. Harrison McKnight, Chuck Kacmar and Vivek Choudhury: Whoops … Did I Use the Wrong Concept to Predict E-Commerce Trust? Modelling the Risk-Related Effects of Trust versus Distrust Concepts. In: *Proc. of the 36th Hawaii Int. Conf. on System Sciences (HICSS'03)*. Available at: http://csdl2.computer.org/comp/Proc./hicss/2003/1874/07/187470182b.pdf. 2003.

[McKnight2006] D. Harrison McKnight and Vivek Choudhury: Distrust and Trust in B2C E-Commerce: Do They Differ? In *Proc. of Eight Int. Conf. on Electronic Commerce ICEC'06*, Fredericton, Canada. pp. 482–491. 2006.

[Shafer1976] G. Schafer: *A Mathematical Theory of Evidence*. Princeton, NJ: Princeton University Press. 1976.

[Smith2005] Carole Smith: Understanding Trust and Confidence: Two Paradigms and their Significance for Health and Social Care. *J. of Applied Philosophy*, 22(3). 2005.

[Ullmann-Margalit2004] Edna Ullmann-Margalit: Trust, Distrust and In Between. In: Russel Hardin (ed.): *Distrust*. New York: Russell Sage Foundation, pp. 60–82. 2004.

[Valsiner2005] Jaan Valsiner: Civility of Basic Distrust: A Cultural-psychological View on Persons-in-society. Paper presented at Symposium Risk, Trust, and Civility. Toronto, Victoria College. May. 6–8. 2005.

[Webb1996] Eugene J. Webb: Trust and Crisis. In Roderick M. Kramer and Tom R. Tyler: *Trust in Organizations*. London: Sage Publications. 1996.

[Yan2003] Zheng Yan and Piotr Cofta: Methodology to Bridge Different Domains of Trust in Mobile Communications. In *Proc. The First International Conference on Trust Management (iTrust'03)*, LNCS 2692/2003, pp. 211–224. 2003.

[Yao2003] Walt Teh-Ming Yao: Fidelis: A Policy-Driven Trust Management Framework. In *Proc. of First Int. Conf. on Trust Management iTrust 2003*. LNCS 2692, Berlin: Springer. pp. 301–317. 2003.

# Part II
## Confidence and Technology

# 7

# Impact of Digital Technology

That question is not what will the computer be like in the future but what will we
be like?

(Sherry Turkle, *The Second Self*)

## 7.1 Introduction

It may come as a surprise but one can have a comfortable life and not be in touch with anything
natural or unprocessed – including air, water or soil. Even though we may still love our
gardens, we gradually are spending more and more time in controlled environments, eating
pre-processed food and using tools developed by someone else. Technology has become the
new nature, pervasive to the extent of invisibility. Digital technologies, despite their relative
juvenile status, are responsible for an enormous part of our technical universe. Their impact
is specifically visible in the area of information processing, including communication and
entertainment, even though they are present almost everywhere – from watches to washers.
They are intermediaries in almost everything we do. Being so intensively pervasive, they
have profoundly altered our perception of trust and control – our ability to be confident. They
have also brought new, intriguing questions of our relationship with technology: confidence
in technology and confidence through technology.

Giddens [Giddens1991] rightfully observes that we have an instrumental relationship with
nature when our direct experience is replaced by a mediated one, sequestrated by science. It is
not only that we often do not have direct access to experiences related to nature, but it is also
that such experiences are re-defined and re-positioned by the development of technology.
Specifically, the extent, nature or selection of experiences are increasingly being defined by
our technical capabilities, most notably by our communication capabilities. One effect of
television in 'mashing' our world experience has been widely documented but more currently
the Internet is a defining instrument of new relationships between us and the world around
us. It is not unknown for two persons in the same room to exchange e-mail or that a person
at the railway station wirelessly consults the remote website for local train information.

This chapter explores different facets of the impact that digital technology (specifically
digital communication) has on our understanding of confidence. From the perspective of
the book, it is a watershed chapter: in earlier chapters Alice and Bob were presumably

*Trust, Complexity and Control: Confidence in a Convergent World*   Piotr Cofta
© 2007 John Wiley & Sons, Ltd

human, and technology has been playing only second fiddle. In all subsequent chapters Alice and Bob will become agents, either human or technical (digital) and technology will be our primary interest. This chapter is organised as follows. We start with an analysis of the main driving forces behind the digital technology to see what their potential impact on confidence is. Next, we discuss two types of confidence, contrasting what is considered human confidence with what is considered to be the technical one. Interestingly, we will see that, despite differences, both can be unified. The differentiation between what is 'human' and what is 'digital' allows us to conveniently structure the discussion area, highlighting different approaches to confidence. From there, we move to investigate digital equivalents of constructs from the model of confidence.

## 7.2 Drivers

Three drives seem to have a major impact on our relationships with digital technology: (1) disembodiment and the erosion of reality: (2) removal and increased transparency of our actions; and (3) intentionality combined with high expectations. The following is a short discussion of those drivers.

### 7.2.1 Disembedding

Giddens [Giddens1991] uses the concept of abstract systems to explain how ready-made solutions (such as institutions, social concepts, means of transport, etc.) can rapidly populate the sphere of our social interactions, introducing new intermediaries on a technical as well as on a conceptual level. Once we accept such intermediaries, we disembed our relationship with others so that it becomes managed by such abstract systems. Communication technologies (even such old ones as paper, writing and letters) can be blamed here for such growing disembedding. However, it is only recently that digital technologies increasingly have made communication also de-personalised.

From the perspective of confidence creation, such dis-embedding means that personal trust (i.e. 'trust' according to our model) is gradually being replaced with trust in abstract systems and institutions (confidence in control instruments). As such, we can see that the dis-embedding actually increases the need for control while decreasing the ability to trust. As we cannot have a rich relationship with an individual, we – following the model – choose to trust external institutions in the expectation that they will control those invisible and frequently anonymous entities at the other end of digital communication link.

We trust our monetary systems, our doctors, the Internet – not exactly because we decided to trust, but because often we have little choice. We are not in a position to control them, because our knowledge (and associated complexity of the process) do not allow it. It is not, however, a confident behaviour expressed out of despair [Deutsch1973], but the rational (or at least rationalised) decision necessitated by the structure of our society. Note that even though we have no choice but to trust, we still have a choice whom to trust.

### 7.2.2 Removal and Transparency

Our digital identity is facing two parallel processes, both of profound importance. First, we increasingly remove ourselves from interactions, hiding behind digital proxies: websites, browsers, messaging, avatars, etc. In this process we can shape our presentation [Turkle1997], slicing it into convenient pieces, introducing intermediaries, creating disposable ones, etc. From the perspective of confidence, removal erodes one of the key enablers – physical identity. Apparently, the development of trust is slower in teams that cannot rely on face-to-face communication [Alexander2002].

The phenomenon of the 'small world' is the best example of another form of removal – we are close to everybody, but almost never close enough to make personal contacts. Opinions, recommendations, reputation, gossip are the ways to make (or break) our social standing. Increasingly, the time is coming when our digital rating is more important than our real one.

At the same time, we are being watched all the time. The amount of digitally-processed information about us (the 'digital footprint') is growing at an increasing rate. We have digital witnesses that continuously collect evidence. Such evidence can be used to make our identities transparent, reconciling presentations with our true self. Even though this process threatens our privacy, if controlled, it may be beneficial for the development of confidence.

### 7.2.3 Intentionality

The notion of intentionality has been discussed several times throughout this book, so that it is recalled here only briefly. At certain level of complexity, technical devices are too complex to be perceived as designed, so that we adopt the intentional stance [Dennett1989] in our dealings with them – we attribute human-like intentions to them. Digital technology, combining wizard-like robustness with a flexibly designed outlook, is a primary candidate for such human-like (or even super-human-like) perception.

Expectations are set high, though. While people in general are reasonably happy relying on digital devices and are eager to attribute human-like properties to them, they are even more eager to withdraw from the relationship if devices are not entirely human and fall short in fulfilling even an unreasonable expectation [Numan1998].

This finding is related to the similarity enabler – the perception of similarity is essential for the proper interpretation of signals and even the slightest trace of an 'alien' behaviour invalidates all assumptions, thus effectively terminating the relationship. It seems that making devices more intelligent, predictable and accommodating comes at a price: once above a certain threshold, they are expected to be human-like – or risk being rejected. There may be no smooth path leading from 'dumb' device to the one that satisfies the Turing test, but rather an abrupt transition from knowledge-based to trust-based relationships.

## 7.3 Two Types of Confidence

Of the two components of confidence: control and trust, the latter generates significantly more interest and confusion when the impact of digital technology is considered. The notion of control is usually well understood and is transferable between human and technical domains. Apart from the essential and existential question of 'who is controlling whom', the problem of control seems to be settled. However, the discussion of digital trust (as a contrast and complement to human trust) encounters problems because two different meanings of trust are investigated simultaneously by different research areas, in a manner that is far from interdisciplinary [Cofta2006c]. In fact it may seem that apart from sharing the word 'trust' both areas pursue totally different agendas.

### 7.3.1 Two 'Trusts'

We can talk about human trust, the relationship that exists between people and that is explored by psychology, sociology and other social sciences. Then there is digital trust, the management concept applicable to complex, mostly digital, systems. Despite having identical names and despite leading quite often to very similar observable behaviour, those two meanings of confidence bear only limited similarity when it comes to fundamentals that drive them.

Both human trust and digital trust share several items in common. Vulnerability, the need for knowledge (evidence), dependence of context and limited predictability of the trustee are equally visible in both definitions. Both agree well with the notion of trust being a part of confidence, whether human or technological.

There are also differences. Human trust is a subjective belief and the exact way of developing trust may be different for each person. One cannot experience trust of another by direct exploration of their minds, but only through observation of external behaviour, introspection or assumed similarity of beliefs. Contrasting, digital trust is objective in a sense that it is programmable and can be fully observed, explored (and modified) through the code. Human trust is driven by the necessity, mostly by the overwhelming complexity of the future. Such trust can be seen as a psychological protective mechanism. In contrast, digital trust is one of several possible management strategies that has been selected on the basis of its fitness to purpose and expected merits. Digital trust benefits certain categories of systems [Cofta2004b] but is not the universal panacea for the management of complex systems.

Finally, human trust assumes the best effort of the trustee without setting exact expectations while digital trust relates to the fulfilment of expectations. For example, the taxi driver is trusted to drive us to the desired place but if he did not do so, he may still be trusted due to possible excuses – diversion, congestion, etc. On the contrary, the trust-based routing (e.g. [Jensen2006]) is trusted to deliver packets but the non-delivery of the packet is considered to be a breach of trust. Breach of human trust is caused by the disbelief that the best effort was tried, breach of digital trust is the result of not meeting expectations.

Note that the root cause of the misunderstanding between different concepts of trusts is in fact that they lead to very similar observable behaviour – no surprise if one considers that digital trust has been designed to mimic human trust-based behaviour. The observer may notice a similar pattern: important local decisions being made on the basis of limited evidence, despite vulnerabilities, but he will not notice whether they are the result of the human operator's decision or the outcome of trust-based decision-making process.

## 7.3.2 Unification

The proposition is therefore as follows. The technical and human confidences ('trusts') are two manifestations of the same phenomenon: the desire to be reassured about the future behaviour of the other party, whether human or technical, in the light of uncertainty and limited by complexity. They therefore serve the same purpose and are expected to work fundamentally in the same way. However, they are different at the surface, as they need different evidence, develop in a different context and at different speed.

Starting with the dissenting note, Marcella [Marcella1999] and Solomon [Solomon2001] noticed that trust should be reserved for our relationship with other people, so that we cannot trust machines. This is further re-stated in [Kracher2005] where a distinction is proposed between technology trust (understood as trust in technology, in technical safeguards, including technical security); technology-mediated trust (where the trustee is an object that is communicated with through some technical means) and website trust where trust is related to the technical representation of the trustee, not to the trustee himself (in the majority of cases trust is related to the website of the trustee, hence the name).

Supporting the notion is the proposition of intentionality [Dennett1989] where any device or system, once it crosses a certain threshold of perceived complexity, is better explained in terms of its intentions than in terms of its structure and operation. That is, the complex system (whether technical or not) is being attributed human properties of intentionality, regardless of intentions of its creator.

Numan [Numan1998] noted that there is sufficient evidence to confirm that trust between people and digital devices exists as an observable phenomenon. However, he did not distinguish between confidence (that may have elements of control) and pure trust. It is worth noting that trust is greater among technically-savvy people than among general public, suggesting that either the feeling of 'being in control' or familiarity (knowledge of expectations) is the main driver here and also suggesting that this relationship may have more to do with control than with trust.

The observation that people refer to technology as if it were other people comes also from Reeves and Nass [Reeves2002]. Experimental psychological studies of how people relate to technology and technology-mediated communications demonstrate the striking similarity between our interaction with technology and our interaction within the society. Devices and technically-mediated information are related to as if they were people, including the perception of certain free will, intentions, etc.

Even though there is a significant amount of evidence that people treat computers as other people and are willing the extend the belief of confidence to them, this unification come with a caveat. Despite all the technological advances, the ability to contact live people is the most important fact that enables confidence in websites to be built [Lumsden2006]. This can be interpreted as a sign that people are still concerned that the face value presented by the technology may not match its real trustworthiness. Even though they are otherwise willing to over-interpret clues conveniently placed by web designers (in the absence of other evidence), they are not ready to surrender their ability to assess trust in other person – or possibly they are trying to look beyond what they perceive as a mask hiding the 'real' machinery of the Internet. The unification is not yet over and the technology is not yet accepted as an equal partner.

From now on, we will refer to Alice and Bob (and the rest of the group) as agents, stressing that they may be either human or technical. Unless necessary, we will not distinguish between those two types of agents. We should, however, distinguish between intentional agents [Dennett1989] and cognitive agents [Castelfranchi2000]. The intentional agent is the one that is believed to have intentions, i.e. the one that can alter its behaviour in a way that is unpredictable by other parties. Cognitive agent is the one that can reason about other agents. As we can see, Bob is an archetypal intentional agent while Alice is a cognitive agent.

## 7.4 Structuring the Area

Even though we can assume the similarity of concepts of confidence (in both aspects of trust and control) between humans and technology, it is worth keeping the differentiation between human and technology a bit longer and use it to structure the research area of trust and confidence. As the research into those two forms of confidence has been separated for quite a long time, we can see that by separating human entities and technical entities and by exploring different relationships between them, it is possible to identify interesting relationships between the subject discussed here and several related areas (see e.g. [Rea2001]). For an in-depth discussion, please refer to [Cofta2004a].

We will use a simple notation, referring to our initial configuration: Alice trying to be confident in Bob (see Figure 7.1). We assume that Alice and Bob populate the 'human domain', i.e. the area where agents are human and adhere to the human notion of confidence. However, Alice can have her representation in the technology domain, *(Alice)*. For example, *(Alice)* may be Alice's computer, her website, mobile phone, etc. Similarly, Bob may have a technical representation, *(Bob)*. Both *(Alice)* and *(Bob)* populate the technology domain, but they are somehow connected to human Alice and Bob.

Alice can establish her confidence in Bob entirely within a human domain (by direct contact or possibly though other people) – the relationship that is marked as (1). In the technology domain, *(Alice)* can establish her confidence in *(Bob)* – similarly, either directly or

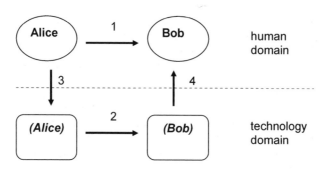

**Figure 7.1**  Domains and primary relationships

through some intermediating technical components. This relationship is marked as (2). The relationship (3) is Alice's assessment of her confidence in *(Alice)* while (4) is the assessment of *(Bob)* of his confidence in Bob.

Figure 7.1 allows us to integrate several research areas on 'trust' into the common framework. The following discussion provides only a glimpse into the otherwise complex relationship between different areas of trust-oriented research. Most of them have been or will be discussed throughout the book.

### 7.4.1  Human to Human (1)

The creation of confidence between humans has been widely studied across disciplines such as psychology, sociology, management, economy, etc. The way confidence is assessed and developed between people is used as an example, illustration and reference. Not entering into a discussion, let's just note that on-line relationship weakens traditional trust-building elements. Missing identities and personal characteristics are important barriers to trust [Nissenbaum1999]. However, predispositions to deceive, behave trustworthily, trust and distrust are preserved in quasi-virtual environment where traditional signals of trust are not available and the available communication channel is very narrow [Bacharach2002]. Initial chapters of this book cover this area in greater details.

### 7.4.2  Technology to Technology (2)

The remaining part of the book will make extensive use of the concept of the 'technology to technology' confidence, specifically of the concept of agents and networks. Autonomous agents (whether hardware-based, software-based, mobile or otherwise) have their behaviour programmed but what is of major concern and interest is not their individual behaviour (that is usually predictable), but the emergent behaviour of large societies of such agents. Specifically, the ability of such a society to establish cooperation on the basis of emergent confidence (rather than embedded control or pre-defined trust) has been the subject of extensive research.

Developments in autonomous agents face a formidable philosophical challenge. We can replicate our social relationships inside the technology domain – with all the problems that it brings, with the ability to choose non-cooperative strategies, to cheat, to monopolise markets (e.g. [Sen2006]), to free-ride the welfare society, to build technical equivalents of organised crime, etc. Such a technical system will be no better than our own society but at least we will be able to recognise it and we will know how to deal with it. Another option is to build the system that follows our desires regarding perfect societies – the technological Utopia. This may, however, lead to the system that is not understood by us (as it may be potentially driven by principles that we do not fully comprehend) so that it may lead us into an area where we do not want to be. Unfortunately, it seems that there is no middle way here: either replication of our problems or alien perfection.

### 7.4.3 Human to Technology (3)

Making humans confident in technology has mostly been the areas of human computer interaction (HCI), and – to certain extent – security (e.g. trusted path). Even though the model of confidence can potentially contribute to the HCI, it is not discussed separately in the book. Several comments regarding the use of human-like agents, usability, etc. are addressed while discussing other subjects.

In general, confidence in technology comes at two different levels: the first is the consideration regarding the immediate interface between human and technology: website design, etc. Assuming that the underlying operation is trustworthy (which unfortunately is not guaranteed), the research is concerned with convincing the user of such trust-worthiness (see e.g. [Egger2000] for the toolkit approach). Even very simple elements of social presence in web pages, easily achievable within existing technology, facilitate an increase in trust [Hassainen2004]. Further, there is some evidence that people respond positively to animated social actors, e.g. in decision support tools [Al-Quaed2006]. Naturally looking and naturally behaving interaction agents greatly improve relationships and build an equivalent of interpersonal trust [Bickmore2001]. Similarly, embedding human social rituals in devices allows them to be perceived as more trustworthy [Cassell2000]. Finally, confidence in technical systems can be seen as an intersection of privacy, security and reliability [Camp2003].

Alternatively, there should be methods to present the true extent of trustworthiness of underlying technology, in a way that is apparent to the user. The task here is much more challenging as it spans the design and operation of actual systems, well beyond their user-facing components. Some comments regarding this approach are included later in this book, while describing Trust-Enhancing Technologies in Chapter 11.

### 7.4.4 Technology to Human (4)

This is the rich area that has several direct links to security, in the form of authentication, authorisation, identification, access control, etc. The number of different methods by which the device can gain certain confidence about humans is enormous (and actually exceeds the number of methods humans can gain confidence in technology): passwords, fingerprints, smart cards, breath analysers, etc. There are even methods to recognise human emotional states.

Authentication as such is not discussed here, leaving that to security-oriented discussion. Within this book, there are some considerations regarding trust-based access control (TBAC). In Chapter 14 we address this particular aspect of confidence with a discussion of Privacy Preserving Authentication, the method to deliver high confidence regarding identity while not endangering privacy.

### 7.4.5 Other Configurations

Figure 7.1 allows several other configurations to be considered. Two more are discussed in the book. The first is the digitally-mediated confidence between humans, the configuration where technology delivers a communication channel between people. Even though it is widely understood that the introduction of a digital intermediary decreases opportunities to trust, we can demonstrate later in this book (while discussing Trust-Enhancing Technologies), that the proper use of technology can be beneficial to inter-personal communication.

Trust management is yet another configuration, discussed in a separate chapter, where both Alice and Bob delegate their confidence to the digital domain so that their respective agents replicate the existing relationship between them. Trust management is an important part of computer security and an inherent element of digital communication.

## 7.5 Digital Equivalence

The original model of confidence was built on the basis of social sciences and assumed that Alice and Bob were humans. If we assume that we are discussing the creation of confidence between *(Alice)* and *(Bob)* instead, without any help from humans, the question is whether there are already some digital equivalents of constructs known from our model that can be used by *(Alice)* to determine her confidence in *(Bob)* in a manner that follows what Alice could have done. In order to demonstrate that such confidence can be built entirely in a digital domain, we should be able to show the design components of agents or of protocols that can reflect various social confidence-building constructs in a technical way, i.e. technology signals that can be treated as evidence of trust and control between agents.

Information exchange that happens between agents is an exchange of signals rather than exchange of goods. Certainly, inter-connected agents are in a position that is comparatively worse than humans engaged in an interaction. The amount of available information is significantly limited (as it is restricted by the protocol), so that deducing from such information is harder. From this perspective, some classes of evidences may look irrelevant, as they can be artificially produced whenever needed. Even worse, resources available to particular agents cannot be reliably determined by another agent, so that the relative value of different evidence is hard to determine. Still, the proposition is that certain properties of the protocol that is used by both agents, as well as certain properties of agents themselves, can facilitate confidence.

In order to identify digital equivalences, we will briefly re-visit different constructs of the model, using Table 7.1. Whenever needed, further explanation is provided. The purpose of this overview is to demonstrate that (almost) all the components of the model have their digital equivalents. We will look for examples of technology-related properties that are applicable to the interaction between technical agents (e.g. devices) and that are visible throughout the technology alone, without the potential support in social structures. Unless stated otherwise, we will think here of agents as computers connected by the network.

Some entries seem to be self-explanatory, e.g. the provision of identity by unique identifiers embedded in devices. Others may require additional explanation that is provided in notes following the table.

### 7.5.1 Similarity (Note 1)

Trusted computing [Pearson2002] can be seen as a solution that allows one device to establish the extent of similarity between another device and the desired one. If Bob (the device) is fitted with such a trusted computing platform, through cryptographic processes he can reliably attest to the other device that its hardware and software configuration is as is stated, i.e. that no uncontrolled changes has been made to the device – it can prove that Bob is exactly as Bob is supposed to be.

What does it do for Alice? From her perspective Bob delivers the evidence of similarity – actually the best evidence she could have dreamed of. If she has sufficient knowledge, she can learn exactly how Bob is constructed and how he should eventually behave. But first and foremost, she is convinced that Bob is a computing device of certain capabilities – exactly the way Alice-human can be convinced that Bob-human is a human indeed.

**Table 7.1**  Digital equivalences

| Construct | Equivalent | Note |
|---|---|---|
| Enablers | | |
| identity | unique identifier for devices | |
| | signatures for software | |
| | message digest and signature for information | |
| similarity | trusted computing as attestation of non-changeability | note 1 |
| accessibility | protocol features: heartbeat, ping, status report | |
| honesty | not guaranteed | |
| Evidence for trust | | |
| continuity | mutually accepted standards | note 2 |
| | shared norms | |
| competence | response time, quality of service | |
| | proper maintenance, security updates, etc. | |
| | global performance indicators | |
| motivation | callbacks | note 3 |
| | shared resources (e.g. memory) | |
| | inter-dependent processing | |
| Evidence for control | | |
| knowledge | direct experience, opinions, reputations | |
| | past performance | |
| influence | scripting | note 4 |
| | code hosting | |
| assurance | guarded contracts | |
| | external arbitrage, trusted third party | |
| Processing | | note 5 |
| reflexive | policies, centralised security | |
| reasoning | trust-based security | |

## 7.5.2  Continuity (Note 2)

The perception of shared standards is the strong driver for continuity, the shared future of agents. It has its pragmatic justification: if both agents do not adhere to the same standard (e.g. to the same protocol), then they are unlikely to develop any interaction in the first place. However, shared standards can influence the perception of continuity beyond this. For example, the usage of the same operating system can be interpreted as an anticipation of higher compatibility beyond current protocols, expectation of similar functionality and similarity of agents' update lifecycle, all important for the continuous cooperation in the future.

Shared norms (discussed later in this book) provide another perspective on developing the perception of shared future [Elgesem2006]. By demonstrating that both agents adhere to the same set of norms (meta-protocol), agents may develop confidence despite occasional incompatibilities or errors.

## 7.5.3  Motivation (Note 3)

The strongest evidence of motivation is an encapsulation of interest, where one agent's welfare depends on another agent's, thus capitalising on an interdependence of agents' fate. In the technology domain, callbacks and various forms of inter-dependence bring a similar notion of motivation. Callback is a control flow paradigm: if Alice asks Bob to perform some action, she may request him to call her back (hence the name) while performing such action. He cannot complete without her and she cannot complete without him. His performance depends on hers, her resource utilisation depends on his. They are both motivated to cooperate to complete an action.

Shared resources or shared processing demonstrate a similar concept: Alice and Bob must cooperate to proceed. Neither agent can do it alone, so that they must both wait for each other (thus decreasing their individual performance) to complete the task. Similar to callbacks, they are both motivated to complete at least the shared part of the task. Note that some mechanisms of parallel programming such as monitors, semaphores, etc. have a similar effect.

### 7.5.4 Influence (Note 4)

Digital agents are not like subjects of enforcement: they do not yield to pressure, they are not afraid of physical abuse and generally are not responsive to persuasion. However, certain technologies deliver the equivalent of an enforcement. The availability of such technologies can be used as evidence of enforceability of a behaviour. In general, all personalisation and remote execution technologies fall into this category.

Let's assume that Bob allows Alice to send a code to be executed within his context. For example, Alice is a web server and Bob is a web browser accepting scripts. Of course, he may take some precautionary measures (e.g. by sandboxing such code), but in fact he allows Alice to directly control his behaviour – whoever interacts with Bob will see Alice as a part of his functionality. It is not even necessary for Alice to actually control Bob to increase her confidence in him, as the fact that she is able to control him (e.g. Bob is willing to accept the script from her) may be sufficient.

### 7.5.5 Confidence Processing (Note 5)

Giddens [Giddens1991] has noticed that the rules that we use to assess confidence fall into two categories. Reflexive rules form an automatic behaviour that delivers a satisfactory assessment rapidly and with low cognitive complexity. In contrast, reasoning rules allow us to handle more complex cases at the expense of complexity and time.

Technology has developed two very similar solutions to handle the creation of confidence. Reflexive rules are implemented as policies – 'if-then' statements that can be used to rapidly interpret the situation. Policies are used, for example, to determine what kind of access rights should be granted to Bob – the action that apparently has a direct relationship to the confidence vested in him. Policies became one of the cornerstones of modern computer security [Bishop2005] and the creation, management and verification of policies are the fundamental tasks of any security or configuration manager.

However, Alice-computer may resort to more complex algorithms (that represent the machine equivalent of reasoning rules) if time and situation permit. The area of autonomous agents is an excellent example of technical solutions that are applicable in such situations. We can even see the development of solutions that to certain extent exhibit both behaviours (e.g. [Quercia2006]) where the device works reflexively according to the relatively simple policy until it encounters a situation that requires more detailed reasoning, at which point it switches itself to solutions that consume more time and resources but that may eventually lead to the isolation of problematic elements and the restoration of reflexive behaviour.

## 7.6 Conclusion

The main drivers behind digital technology – disembodiment, removal and intentionality – are introducing new challenges to the way confidence has been established between people, but they are also introducing new opportunities as the intermediate technology can facilitate the development of new relationships. Technology becomes a valid partner in confidence, both as an intermediary and as an object and subject of confidence. Alice and Bob are no longer only humans, but they can be agents, attributed with intentional and reasoning capabilities.

It is possible to think of human and technical confidences as two manifestations of the same phenomenon, addressing similar needs and employing similar mechanisms. It is even possible to identify technical equivalents of almost all constructs from the model of confidence, where such equivalents operate only among devices. However, it is still possible to think of a duality of human confidence and technical confidence when it comes to structuring the research area. Different relationships of those two confidences reflect different approaches and define different compartments of the large area related to trust and confidence. The structuring itself allows differences and similarities between various areas of interest to be demonstrated.

The unification of human and technical confidence is cemented by the analysis of digital equivalents of psychological and social constructs that has been postulated by the model. By demonstrating that almost all constructs have their digital equivalents in existing protocols and architectures, we can indeed accept technical and human confidences as being very similar.

# Bibliography

[Alexander2002] Patricia M. Alexander: Teamwork, Time, Trust and Information. In *ACM Int. Conf. Proceeding Series*; Vol. 30. *Proc. of the 2002 Annual Research Conf. of the South African Institute of Computer Scientists and Information Technologists on Enablement Through Technology*. 2002.

[Al-Quaed2006] Faisal Al-Quaed and Alistair Sutcliffe: Adaptive Decision Support System (ADSS) for B2C ECommerce. In *Proc. of ICEC'06*. 14–16. August Fredericton, Canada. 2006.

[Bacharach2002] Michael Bacharach: How Human Trusters Assess Trustworthiness in Quasi-virtual Contexts. In: *AAMAS2002 Workshop on Deception, Fraud and Trust in Agent Societies*. 2002.

[Bickmore2001] Timothy Bickmore, Justine Cassell: Relational Agents: A Model and Implementation of Building User Trust. *Proc. Human Factors Computing Systems (SIGCHI 01)*, New York: ACM Press. 2001.

[Bishop2005] Matt Bishop: *Introduction to Computer Security*. Reading. MA: Addison-Wesley. 2005.

[Camp2003] L. Jean Camp: Designing for Trust. In: R. Falcone et al. (eds): *AAMAS 2002 Workshop on Trust, Reputation . . .* LNAI 2631, pp. 15–29. 2003.

[Cassell2000] Justine Cassell and Timothy Brickmore: External Manifestations of Trustworthiness in the Interface. *Communications of the ACM*, 43(12): 50–56. 2000.

[Castelfranchi2000] Cristiano Castelfranchi and Rino Falcone: Trust is Much More than Subjective Probability: Mental Components and Sources of Trust. In *Proc. of the 33rd Hawaii Int. Conf. on System Sciences (HICSS2000)*. Vol. 6. Available at: http://www.istc.cnr.it/T3/download/Trust-more-than-probability.pdf. 2000.

[Clark2007] David Clark: The End-to-end Argument in Today's World: Do We Have to Destroy the Principle to Save it? Presentation, in BT UK, 28 Feb. Suffolk. 2007.

[Cofta2004a] Piotr Cofta: Phenomenon of a Borderline Trust: Position Paper. Available at: http://piotr.cofta.eu. 2004.

[Cofta2004b] Piotr Cofta: Applicability of Digital Trust. Available at: http://piotr.cofta.eu. 2004.

[Cofta2006c] Piotr Cofta and Hazel Lacohee: Trust or There and Back Again: A Cautionary Tale of an Interdisciplinary Cooperation. Paper presented at First Research Meeting of Centre for Systems and Services Sciences (CS3), Grenoble, France. 2006.

[Dennett1989] D.C. Dennett: *The Intentional Stance*. Cambridge, MA: MIT Press. 1989.

[Deutsch1973] Morton Deutsch: *The Resolution of Conflict: Constructive and Destructive Processes*. New Haven, CT: Yale University Press. 1973.

[Egger2000] Florian N. Egger: From Interactions to Transactions: Designing the Trust Experience for Business-to-Consumer Electronic Commerce. PhD thesis, Eindhoven University of Technology, The Netherlands. ISBN 90-386-1778-X. 2000.

[Elgesem2006] Dag Elgesem: Normative Structures in Trust Management. In K. Stolen et al. (eds): *iTrust 2006*, LNCS 3986, pp. 48–61. 2006.

[Giddens1991] Anthony Giddens: *Modernity and Self-identity: Self and Society in the Late Modern Age*. Cambridge: Polity Press. 1991.

[Hassainen2004] Khaled S. Hassanein and Milena M. Head: Building Online Trust Through Socially Rich Web Interfaces. In *Proc. of Second Annual Conf. on Privacy, Security and Trust*. 13–15 October. Fredericton, New Brunswick, Canada. 2004.

[Jensen2006] Christian D. Jensen and Paul O Connell: Trust-Based Route Selection in Dynamic Source Routing. In K. Stolen et al. (eds): *iTrust 2006*, LNCS 3986, pp. 150–163. 2006.

[Kracher2005] Beverly Kracher et al.: A Foundation for Understanding Online Trust and Electronic Commerce. *Info., Comm. and Ethics in Society*, 3: 131–141. Available at: http://www.troubador.co.uk/image/J.s/3-3-4-Kracher6651.pdf. 2005.

[Lumsden2006] Jo Lumsden and Lisa MacKay: How Does Personality Affect Trust in B2C e-Commerce? In *Proc. of Eight Int. Conf. on Electronic Commerce ICEC'06*. Fredericton, New Brunswick, Canada. pp. 471–481. 2006.

[Marcella1999] A. J. Marcella: *Establishing Trust in Virtual Markets*. Altamondate Springs, Florida: The Institute of Internal Auditors. 1999.

[Nissenbaum1999] Helen Nissenbaum: Can Trust be Secured Online? A Theoretical Perspective. *Ethics and Politics*, 1(2). 1999. Available at: http://www.units.it/~etica/1999_2/index.html. 1999.

[Numan1998] John Han Numan: Knowledge-based Systems as Companions. Theses on Systems, Organisation and Management. University of Groningen. Available at: http://dissertations.nb.rug.nl/faculties/management/1998/j.h.numan. 1998.

[Pearson2002] Siani Pearson et al: *Trusted Computing Platforms: TCPA Technology In Context*. Englewood Cliffs. NJ: Prentice-Hall. 2002.

[Quercia2006] Daniele Quercia et al.: STRUDEL: Supporting Trust in the Dynamic Establishment of Peering Coalitions. In *Proc. of SAC'06, 2006*. Dijon, France. 2006.

[Rea2001] Tim Rea: Engendering Trust in Electronic Environments. Roles for a Trusted Third Party. In Cristiano Castelfranchi and Yao-Hua Tan (eds): *Trust and Deception in Virtual Societies*. Dordrecht: Kluwer Academic Publishers. pp. 221–236. 2001.

[Reeves2002] Byron Reeves and Clifford Nass: *The Media Equation*. Stanford, CA: CSLI Publications. 2002.

[Sen2006] Sandip Sen and Dipyaman Banerjee: Monopolizing Markets by Exploiting Trust. In *Proc. of Fifth Int. Joint Conf. on Autonomous Agents and Multiagent Systems (AAMAS06)*. 2006.

[Solomon2001] Robert C. Solomon and Fernando Flores: *Building Trust in Business, Politics, Relationships and Life*. Oxford: Oxford University Press. 2001.

[Thibadeau2006] Robert Thibadeau: Trusted Computing for Disk Drives and Other Peripherals. *IEEE Security & Privacy*, 4(5): 26–33. 2006.

[Turkle1997] Sherry Turkle: *Life on the Screen: Identity in the Age of the Internet*. New York: Simon & Schuster. 1997.

# 8

# Digital Security

Your computer is at risk.

(too often seen on the screen)

## 8.1 Introduction

The relationship between security (specifically digital security, i.e. computer security, network security, etc.) and confidence deserves some detailed discussion. In fact, it is quite surprising that we have come so far in the discussion about trust, control and technology and not yet mentioned digital security. Definitely, it is time to catch up with this subject.

Digital security is about minimising the probability of improper behaviour, i.e. events that are deemed inappropriate but technically possible. For example, if the security scheme is used to deny Alice access to a particular file, it is understood that the purpose of such security is to minimise (eliminate in the best case) chances that she will gain access to this file. It is also understood that such a file may be present and she has the capability to access it, as there is no point in denying access to something that cannot exist or denying access to someone who cannot gain access anyway.

Confidence in preventing improper behaviour can be gained in several different ways, depending on the structure of the system. For example, Alice may pledge (or sign an agreement) that she will never access the particular file. Alternatively, by using technologies such as DRM (Digital Rights Management) we can make sure that the file itself may refuse to disclose its contents to Alice. Finally, we may place both Alice and the file in a context where Alice's request to open the file will be declined – implementing certain access control. Our analysis (and associated risks, threats and vulnerabilities, etc.) may include only Alice, only the file, all people, all files, all systems, etc. – they become actors in our security scenarios, part of the transactional scope of the confidence assessment.

Security is not only about preventing certain actions, but also about requesting that certain features are available (e.g. that the web server responds to incoming service requests). In fact, confidence in a context of digital security can be seen as being closely related to reliance and dependability, i.e. expectations that the system will exhibit certain properties and deliver certain services across a wide range of circumstances such as faults, errors, accidents or attacks [Powell2003].

Consolidating, digital security is about being confident that all identified actors (users, programs, files, etc.) will behave according to given rules and expectations, within the scope and architecture of the system and across the wide range of potential scenarios (including hostile attacks). We know that the issue of confidence in an actor can be addressed either by control or by trust: we can build the elaborate control mechanism to force actors to behave properly or we can trust those actors (or at least some of them) to behave. This duality can be seen, e.g. in Flowerday [Flowerday2006], where assurance is used to build trust and control is used to support trust.

As the problem can be addressed by control or by trust, we can see that both approaches can be applicable – sometimes as equally valid alternatives. Let's consider an example of gaining confidence that the given code will behave according to prescribed rules – e.g. that it will never reach beyond the memory area that belongs to it. One method is to build the 'cage' (or the sandbox) around the code so that the behaviour of the code will be closely monitored and every violation of the rule will be reported (e.g. by virtualising the platform). This approach definitely follows the control path, as in fact we enforce certain behaviour on a code and we ignore its trustworthiness. The code has no other way but to behave properly.

However, it is also possible (and actually it is sometimes the only feasible method) to let the code run without any constraints – to let it run on trust. Specifically, if there is no support for control or if the code is time-critical, we may find out that control is not possible or not feasible, and that trusting the code is the only option. We certainly may implement some technical trust-building measures (e.g. we may want to sign the code) so that we know that original intentions of the creator of the code have not been altered, or that the code has been properly reviewed. In such cases the digital signature serves only as a technical proxy for the trust that has been placed in actors that contributed to the code: developers, reviewers, verifiers, etc. – it does not provide any control of the code itself (even though it may provide control over the distribution process).

## 8.2 Reductive Approach: Security through Control

The first of two approaches to security presented here is the reductive one, where security is achieved mainly through control. The modern computer security stresses control over trust, trying to create elaborate enforcing structure where trust is contained and minimised. This is best seen in the NSA's operational definition of trustworthy element (see [Anderson2001]) that stipulates restricting trust to a very limited number of actors because trusted actors are those whose failure will invalidate security. Interestingly, such a definition affirms what we know from the model – that instruments of control must be trusted and that there is an element of trust even in the deeply embedded chain of control. The definition is also pragmatic, operational, and first of all – prescriptive. It does not say what trust (or trustworthiness) is, but it suggests that trust must be isolated and reduced to well-defined design elements.

The approach presented here is popular not only for digital security. For example, [Bons1999] seeks the improvement in trust (in e-commerce) in developing a formalised and verifiable procedure of exchange of electronic documents prior to and during the transaction, i.e. in introducing additional control and minimising areas where trust is needed. In accord with our earlier considerations, control seems to take the upper hand at the expense of trust.

The question of trust in particular elements (as such trust is always needed) is addressed by techniques that are outside of the scope of a system itself. For example, the technical system may resort to socially embedded methods of attestation [Bishop2005] to demonstrate that there is credible evidence that a component meets its given requirements. It is important that such trust is not generated within the system itself, but it is provided from the outside, by instruments that are not available to agents that operate within the system. For example, while the banking card can be attested to work properly by the procedure that has been preformed in the lab, such a procedure is not available to the ATM, that can only trust that the card has not been compromised.

Absolute trust in platform cannot always be guaranteed. However, through the process of assurance, the extent of such trust can be assessed. The assessment highlights the difference between the pure control mechanism and confidence that can be attained through trust, where non-technical elements such as an expert's opinion, development process, design criteria, etc. are taken into consideration. Examples are TCSEC (Trusted Computer System Evaluation Criteria (Orange Book)), FIPS 140, Common Criteria and similar assessment schemes.

The outcome of such design decisions is that the way the security is built resembles a sandwich where layers of security and trust interleave, connected by an upward spiral of dependence, with the topmost layer of control and with an irreducible trust at the bottom (Figure 8.1). Each layer of control assumes that all the layers below are trusted – specifically that all the irreducible 'roots of trust' are trusted with no proof within the scope of the system. Each layer of trust (except for the lowest one) is then built on the basis of some control mechanisms that operate outside the scope of the system defined by the upper layers. The concept of layered security, where trust in lower layers is required for the upper layers to operate, is fundamental to modern security (e.g. [Bishop2005]) and has also been noted e.g. in [Flowerday2006].

## 8.2.1 Ontological Security

There is a specific kind of trust that is worth mentioning here, as it directly reflects the layered structure of trust and control. It is a software trust in its immediate environment. Assuming that Alice is a system (e.g. a software autonomous agent), she is willingly and implicitly trusting the foundation of its existence – hardware, software, network connections, etc. If Alice is a Java application, she is not in a position to question the integrity of Java Virtual Machine; if she is a distributed application, then she trusts the communication between different parts of herself.

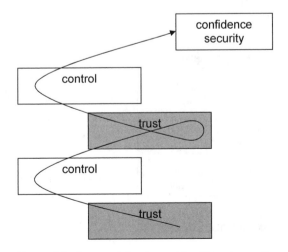

**Figure 8.1**  Layered structure of traditional security

The development of such trust has been widely studied in psychology, demonstrating that this kind of trust is essential in the creation of ontological security [Giddens1991] that enables the initial separation of self and the world. A child does not question the surrounding world, accepting it 'as is'. He also does not question the integrity of himself. Should he doubt the world around him or doubt himself, then his development would stop there. One wonders whether this trust is trust indeed, or whether this is a psychological defensive mechanism, but it enables and preconditions our existence – and the existence of software agents. Again, the social and technical concepts of confidence converge here.

## 8.2.2 Trusted Computing

Trusted computing [Pearson2002] is the area of research, standardisation and development that promises to increase confidence in software integrity by adding specific hardware and re-developing the basic software of the computer. Trusted computing is becoming quite popular as the concept has grown from its initial proposition aimed at desktops and servers to laptops, mobile devices and even peripherals [Thibadeau2006], [England2003].

This short study explores trusted computing from the developer's perspective. The user's perspective may paint quite a different picture. Note that there is some discussion whether 'trusted computing' has in fact anything to do with trust, apart from using this catchy phrase for its own benefit. Not wishing to engage in this discussion, the analysis below demonstrates that trusted computing indeed deals with trust – but probably in a manner that was not expected. Trusted computing requires trust (actually quite a significant amount of trust) to operate, but it does not create trust.

The implementation of trusted computing starts with the hardware component that is activated immediately upon the computer's start-up. The component conducts the verification of the immutability of the basic software embedded in the computer (BIOS) and then passes control to this software. The software, in turn, verifies the integrity of the operating system and passes control to it. Finally, the operating system verifies the integrity of the application.

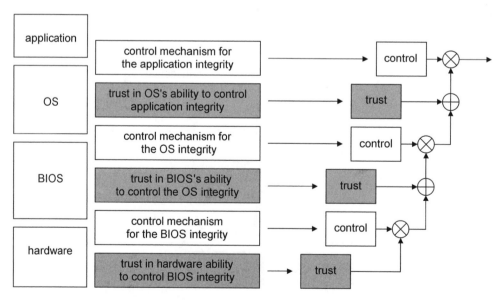

**Figure 8.2**  Layered structure of trusted computing

From the perspective of the developer it is obvious that confidence in application integrity is reached by the combination of trust and control, with stress on the latter (Figure 8.2). Such confidence is based on trust in the ability of the operating system to control application integrity in the first place, combined with the control that the operating system has not lost its own integrity. Such trust is delivered by elements that are outside the trusted computing model, such as brand, opinion of experts, experience, etc. The control is provided by the underlying software in BIOS. Note that the trust in the application itself is not the element of the model: trusted computing does not guarantee that the application works correctly (or e.g. that the system cannot be attacked by worms) [Oppliger2005], but only that it has not changed between verifications.

Similar reasoning is repeated in relation to the operating system, BIOS and finally to the hardware. At each level (except for the hardware) confidence is built through control provided by the next lower element and is influenced by the trust in the implementation of the control mechanism at the given level. At the hardware level no evidences of control are available so that confidence must be gained entirely through trust. Trust in the hardware element (Trusted Platform Module, TPM) is the irrevocable foundation of the trusted computing concept.

There is an expected similarity between the layered structure of trust and control and the way the complexity-based model of confidence is allowed to recursively invoke itself to assess confidence with regard to control instruments. The model can be used to provide an explanation of how layers of trust and control contribute to the overall confidence in application integrity.

### 8.2.3 Trusted Foundations

Looking at the wide area of modern digital security we can actually see that there are some constructs that are trusted, i.e. constructs that are the foundations of control and elements whose failure may invalidate the whole chain of control. Interestingly, quite often system developers assume trust in those constructs without any discussion. Those are as follows:

- *Root trust.* It has been already mentioned that the reduction of the need to trust (and the replacement of trust with control) cannot be done indefinitely, but it must rest on certain architectural elements that are implicitly trusted. Currently, it seems that the role of such trusted elements is assumed by hardware rather than by software, in the form of trusted computing platforms [Pearson2002] or smart cards [Rankl2003]. Either solution is built on the concept of 'unbreakable' architectural element that is trusted without any further proof – whether such element is a chip on the main board or the separate card that can be inserted into the device.
- *Cryptography.* It is surprising that cryptography appears in a context of trust (rather than in a context of control). However, modern cryptography is yet another element that is trusted rather than controlled – even though it is naturally used as the foundation of several instruments of control. The cryptography, by relinquishing the idea of 'security by obscurity', allowed public trust in cryptographic algorithms (with some notable exceptions e.g. [Schneier2004]) by disclosing them to public scrutiny by the research community. Such algorithms are trusted to be correct (and to be correctly implemented) – they are not controlled. It is only the continuous public review process of cryptographic solutions (together with other evidence of trust) that guarantees trust in cryptography – despite obvious evidence that it is fallible [Wang2005].
- *Rational user behaviour.* The assumption of rational user behaviour is probably the weakest element of computer security – but it is hard to imagine how this assumption can be modified. The significant number of successful security attacks are carried out by insiders [Keeney2005] – by people who have right to access information that is being attacked, by

people who are trusted to follow policies and regulations. Even the best control system cannot prevent this from happening. However, challenging the assumption of rational behaviour (and removing the user from the decision-making process) is also not the answer, as users do not trust technology that they cannot control [Lacohee2006].

## 8.3 Holistic Approach: Security through Trust

The fallibility of modern 'reductionist' computer security is strongly associated with its design. By building on control and allowing a very limited amount of trust, the security system builds a world that is nearly perfect but at the same time vulnerable to undetectable errors in its critical elements. If even one of the trusted elements fails, the whole control structure is rendered useless – they became single points of failure of the whole system. Even worse, the system does not know that it has been compromised.

The discussion and critique of trust through security are well expressed in [Nissenbaum1999]. Two main arguments are provided. The first is that using security both increases the price of interaction and restricts choices, hence even though the status quo may be preserved, the ability to grow is stifled due to the lack of resources and breathing space. Second (as already mentioned here), security can deliver protection only if we trust the instruments of security – so that security is not a panacea as it engages us in an endless loop of ever more sophisticated control instruments.

An alternative approach calls for the elimination of those single points of failure by distributing the foundation of trust across all (or the majority of) actors within the system and by continuously monitoring such a foundation. Security thus becomes not a designed-in feature but the potential emergent property that realises itself only if agents are proactively interested in it. Those systems are not designed to eliminate all risks, or to deliver immediate reassurance, but they are rather designed for long-term resilience. Even if one agent fails (i.e. becomes untrustworthy), the whole system will be still operational, possibly at a slightly lower level of overall security.

Such a system does not deliver the promise of near-perfection. As a whole, it will possibly deliver reasonable security, but it may be harder to measure it. While layers of trust and

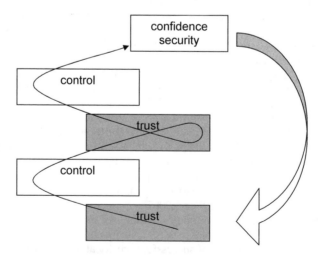

**Figure 8.3** The holistic security

control are still present (as they form the foundation of the majority of system designs), the trust at the bottom is built on the basis of the confidence itself (Figure 8.3). This makes the system self-sufficient, in that it does not require any evidence coming from outside the system to realise its security.

### 8.3.1 Applicability

Not all systems potentially benefit from holistic security. The list of features that a given digital system should fulfil in order to use and benefit from a trust-based approach to security [Cofta2004b] apparently identifies several modern computing systems, from grid to pervasive and agent-based ones. As a minimum requirement, the system should be reasonably fault-tolerant, stable, evidence-rich and resourceful. In order to significantly benefit from trust-based security (in contrast to traditional security), the system should make security-related decisions locally, probably use weak identities, have imperfect policy enforcement or already be trust-based. Trust-based security (as it is often called) seems to be the preferred solution for various categories of systems, that may include applications of a large variety such as social networks, information sharing, remote computation, etc.

Instead of policies imposed by the central management, trust-based security assumes self-governance of individual units (devices, agents, sub-networks, groups, etc.). Decisions regarding security are made and executed locally on the basis of available evidence. Agents should be flexible and tolerant of unintentionally faulty information and at the same time be able to avoid fraudulent information from malicious or unreliable agents [Barber2002b]. For example, [Gray2002] discusses the trust-based admission control where authorisation is based on the reputation of an entity and demonstrates how such a system can be applied to collaborative ad-hoc networks while the EU SECURE program (e.g. [Cahill2003]) uses trust between peers to overcome the limitations of traditional security, specifically in systems where uncertainty is high such as ad-hoc networks, pervasive computing, etc.

Note that such systems do not give away control, they are only using the mix of trust and control in a different way, linking the bottom trust back with the topmost confidence. Improvements in security and in control instruments are welcome and there are plenty of them: security frameworks for grid computing [Uszok2004], use of trusted computing [He2002], security for autonomous agents [Poslad2003], supervised interactions [Kollingbaum2003], certified communication patterns [Weigand2001], improved PKI scheme [He2001], etc.

Holistic security schemes (e.g. [Krukow2006]) are still relatively new and they radically depart from traditional notions of security to present the security scheme that is built entirely on trust (see e.g. [Barber2002a] for a discussion of research challenges in multi-agent environments). Propositions for alternative frameworks (e.g. for pervasive computing [Butler2003] or peer-to-peer networks [Caballero2006]) are available. Trust (confidence) has been identified as the most important aspect of security for the semantic web [Hu2003]. It permeates all facets of the semantic web and forms the foundation of alternatively structured security.

### 8.3.2 Trust-based Access Control

Trust-based access control (TBAC) is probably the most popular function suitable for the holistic approach. TBAC implements the traditional function of access control, with the difference that the decision regarding access is made locally, on the basis of current reputation rather than on the basis of authentication and pre-defined access control list. Figure 8.4. shows an example of TBAC where Alice decides whether to allow Bob to access certain resources on the basis of opinion about Bob that comes from Carol, Dave (through a central reputation rating system) and from herself. Note that the use of reputation means that Alice may alter

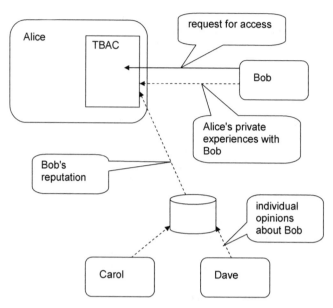

**Figure 8.4** Trust-based access control

her decisions in time, depending on the current status of Bob's reputation, even if Alice is not in contact with any central authority.

An alternative approach to the trust-based access control is presented in [Levien2002]. The Advogato trust metrics (named after the website where it was originally implemented) has been found to be significantly better in protecting trust in a group's shared identity than trust in individuals. Such an approach shifts the perception of access control from the usual authentication/authorisation towards identity/entitlement, where the group protects its shared identity by limiting access to the benefits it brings. Unfortunately, normal control instruments such as PKI have been found not to easily support such a notion of group identity [Levien1998].

### 8.3.3 Problems with the Holistic Approach

Trust-based holistic security faces certain problems of its own. The most important are the uncertainty about the identity of other parties, the unwillingness to contribute and the problem of a stranger. In order to resolve the identity-related problem, [Jordan2004] envisages the formation of an Augmented Social Network where identity and trust will be intrinsically interconnected, by relative persistence of (reputation-based) identities combined with reputation-based networking. Relationships and identities will be meaningful only if they are trusted so that they will be brokered by semantically-aware dictionaries (thus resolving the issue of similarity). The similar concept is discussed later in this book.

Regarding unwillingness to cooperate (free riding), [Huyngh2006] proposes a interesting yet simple mechanism that may overcome it. Self-interested agents may not be willing to report their experience, as they see no benefit from it, considering that providing reputation is currently not rewarded. However, if an agent whose reputation is being assessed (Bob) is able to present certified reputation ratings about himself, he may develop a compensation mechanism that will deliver incentives for agents to provide such ratings. [Jurca2002] considers the similar problem: why agents should report their experience and why such reports should be trustworthy. Understanding that existing security concepts are not easily applicable to large, distributed, multi-agent networks, he proposes the reputation-based system (further improved in [Jurca2006]) where reports on reputation are bought and sold, with pricing benefiting proactive and honest agents.

The problem of a stranger goes beyond trust-based security, as it addresses the question how one agent can be confident (trust) another agent with no prior knowledge of such an agent. Generally, the creation of confidence under those circumstances requires additional arrangements, e.g. [Ho1999] discusses the process of building trust among strangers, using the modified game of trust. Results show that while strangers are usually untrustworthy and untrusting, the emergence of a cluster of trusting persons can facilitate the process of building trust. This is similar to findings from the social sciences [Hawthorn2000].

## 8.4 Trust and System Design

The final aspect of the relationship between security and trust originates from an observation [Clark2007] that the end-to-end paradigm that has been a foundation of a modern communication (specifically the Internet) should in fact be called the trust-to-trust one. As entities cannot meaningfully communicate if they do not trust each other, trust is not only a prerequisite of the communication, but it also determines the actual architecture of a system. Specifically, agents tend to build their 'domains of trust' out of components that they trust (possibly because they have such components under their exclusive control), leaving untrusted spaces between them.

Assuming that two agents have found the means to trust each other, they are able to establish communication over the untrusted network (that may include other agents). If they are not able to trust each other, the technical embodiment of a trusted third party may facilitate such communication (see also [Yan2003]). This concept well explains the structure of Internet applications, modularity of a design and common practices such as the use of tamper-resistant tokens (e.g. smart cards) as a surrogate for the necessary trust.

From the security perspective, the trust-to-trust concept provides a strong principle that can be applied to the system design. While able to explain to the principles of good design and modularity, it does not guarantee security. We can see that trust between parties allows confidentiality and integrity to be addressed (e.g. through cryptography), but it may not be able to address availability. The untrusted network can easily mount a denial of service attack, denying any communication between parties. If the trust-to-trust principle is indeed the foundation of modern networks, then they are all susceptible to the denial of service attacks.

Certain support for the concept of a trusted path between parties across the network comes from the security principles of the next generation of networks, specifically, from the principle of separating control and data planes (the principle that has been implemented e.g. in GSM radio access networks from the very start). By moving the control flow out of bounds for 'regular' agents, the standard proposes the shared trusted path that interconnects all agents that are responsible for the network.

## 8.5 Conclusion

Modern digital security is at the crossroads. Traditionally, it has consolidated the need for trust into the smallest element and has relied on its reliable operation to construct several layers of control. However, the new approach that replaces or supplements the trusted element with system-wide confidence is emerging and is finding its applications in new architectures.

The complexity-based model of confidence can be used to distinguish between both approaches: the reductive, control-based and the holistic, trust-based. Both are modelled on social structures – but different ones. The first follows the rigid structure of large companies (and armies), with its strict reporting process, swift implementation of decisions and limited number of trusted decision-makers. The latter replicates the phenomena of peer groups, democratic decision-making, voluntary participation, etc. – phenomena well known in our social life, but not necessarily in our business or technical life.

There is a significant number of technological solutions that may benefit from the trust-based holistic approach to security. The main driver that may force the adoption of such security remains the same: complexity. It seems that the control-based security may have its limits defined by how much we can control – and what is the price of failure. As computing systems grow in size and number, controlling them becomes a task in itself. At the same time the single failure may have the increasingly damaging effect, as it may affect the security not only of a single computer, but also of a whole network.

## Bibliography

[Anderson2001] Ross Anderson: *Security Engineering: A Guide to Building Dependable Distributed Systems*. Chichester: John Wiley & Sons, Ltd. 2001.

[Barber2002a] K. Suzanne Barber, Karen Fullam and Joonoo Kim: Challenges for Trust, Fraud and Deception Research in Multi-agent Systems. In R. Falcone et al. (eds): *AAMAS 2002 Workshop on Trust, Reputation*... LNAI 2631, pp. 8–14. 2003.

[Barber2002b] K. Suzanne Barber and Joonoo Kim: Soft Security: Isolating Unreliable Agents from Society. In R. Falcone et al. (eds): *AAMAS 2002 Workshop on Trust, Reputation*... LNAI 2631, pp. 224–233. 2003.

[Bishop2005] Matt Bishop: *Introduction to Computer Security*. Reading, MA: Addison-Wesley. 2005.

[Bons1999] Roger W. H. Bons et al.: A Formal Specification of Automated Auditing of Trust-worthy Trade Procedures for Open Electronic Commerce. In *Proc. of the 32nd Hawaii Int. Conf. on System Sciences*. 1999.

[Butler2003] Michael Butler et al.: Towards a Trust Analysis Framework for Pervasive Computing Scenarios. In *Proc. of Sixth Int. Workshop on Trust, Privacy, Deception, and Fraud in Agent Societies (AAMAS'03)*. 2003.

[Caballero2006] Alberto Caballero, Juan A. Botia and Antonio F. Gomez-Skarmeta: Trust and Reputation Model Based on WSMO. In *Proc. of Fifth Int. Conf. on Autonomous Agents and Multiagent Systems AAMAS-06. Workshop WS8 – Ninth Int. Workshop on Trust in Agent Societies*. Hakodate, Japan. 2006.

[Cahill2003] Vinny Cahill et al.: Using Trust for Secure Collaboration in Uncertain Environments. *IEEE Pervasive Computing*, July–September.: 52–61. 2003.

[Clark2007] David Clark: The End-to-end Argument in Today's World: Do We Have to Destroy the Principle to Save it? Presentation, in BT UK, 28 Feb. Suffolk. 2007.

[Cofta2004a] Piotr Cofta: Phenomenon of a Borderline Trust: Position Paper. Available at: http://piotr.cofta.eu. 2004.

[Cofta2004b] Piotr Cofta: Applicability of Digital Trust. Available at: http://piotr.cofta.eu. 2004.

[England2003] Paul England et al.: A Trusted Open Platform. *IEEE Computer*, July: 55–62. 2003.

[Flowerday2006] Stephen Flowerday and Rossouw von Solms: Trust: An Element of Information Security. In IFIP Int. Federation for Information Processing, S. Fischer-Hubner, K. Rannenberg, L. Yngstrom, and S. Lindskog, (eds): *Security and Privacy in Dynamic Environments*, vol. 201. Boston: Springer. 2006.

[Giddens1991] Anthony Giddens: *Modernity and Self-identity: Self and Society in the Late Modern Age.* Cambridge: Polity Press. 1991.

[Gray2002] Elizabeth Gray et al.: Towards a Framework for Assessing Trust-Based Admission Control in Collaborative Ad Hoc Applications. Technical Report 66, Department of Computer Science, Trinity College, Dublin. 2002.

[Hawthorn2000] Geoffrey Hawthorn: Three Ironies in Trust. In Diego Gambetta (ed.): *Trust: Making and Breaking Cooperative Relations,* electronic edition, Department of Sociology, University of Oxford, Chapter 7, pp. 111–126. Available at: http://www.sociology.ox.ac.uk/papers/hawthorn111–126.pdf. 2000.

[He2001] Qi He: Security Infrastructure for Software Agent Society. In Cristiano Castelfranchi and Yao-Hua Tan (eds): *Trust and Deception in Virtual Societies.* Dordrecht: Kluwer Academic Publishers, pp. 139–156. 2001.

[He2002] Qi He, Pradeep Khosla and Zhongmin Su: A Practical Study of Security of Agent-Based Ubiquitous Computing. In R. Falcone et al. (eds): *AAMAS 2002 Workshop on Trust, Reputation . . .* LNAI 2631, pp. 194–208. 2003.

[Ho1999] Teck-Hua Ho and Keith Weigelt: Trust Building Among Strangers: An Experimental Investigation of Mutual Trust and Trustbuilding. WP 99–01. A Working Paper of the Reginald H. Jones Center, The Wharton School. Philadelphia, PA: University of Pennsylvania. 1999.

[Hu2003] Yuh-Jong Hu, Se-Ting, Chen, and Min-Huei, Yang: Trust on the Semantic Web Pyramid: Some Issues and Challenges. *Int. Semantic Web Conf. 2003 (ISWC 2003).* 2003.

[Huyngh2006] Trung Dong Huynh, Nicholas R. Jennings, Nigel R. Shadbolt: Certified Reputation: How an Agent Can Trust a Stranger. In *Proc. of AAMAS'06.* 8–12 May. Hakodate, Japan. 2006.

[Jordan2004] Ken Jordan, Jan Hauser, and Steven Foster: The Augmented Social Network: Building Identity and Trust into the Next-Generation Internet. Available at: http://www.firstmonday.dk/issues/issue8_8/jordan/. 2004.

[Jurca2002] Radu Jurca and Boi Faltings: Towards Incentive-Compatible Reputation Management. In R. Falcone et al. (eds): *AAMAS 2002 Workshop on Trust, Reputation . . .* LNAI 2631, pp. 138–147. 2003.

[Jurca2006] Radu Jurca and Boi Faltings: Using CHI Scores to Reward Honest Feedback from Repeated Interactions. In *Proc. of AAMAS'06.* 8–12 May Hakodate, Japan. 2006.

[Keeney2005] Michelle Keeney et al.: Insider Threat Study: Computer System Sabotage in Critical Infrastructure Sectors. Available at: http://www.cert.org/archive/pdf/insidercross051105.pdf. 2005.

[Kollingbaum2003] Martin J. Kollingbaum and Timothy J. Norman: Supervised Interactions: A Form of Contract Management to Create Trust between Agents. In R. Falcone et al. (eds): *AAMAS 2002 Workshop on Trust, Reputation . . .* LNAI 2631, pp. 108–122. 2003.

[Krukow2006] Karl Krukow: Towards a Theory of Trust for the Global Ubiquitous Computer. PhD dissertation, University of Aarhus. 2006.

[Lacohee2006] Hazel Lacohee, Stephen Crane, and Andy Phippen: *Trustguide.* Available at: http://www.trustguide.org. 2006.

[Levien1998] Raphael L. Levien and Alex Aiken: Attack-Resistant Trust Metrics for Public Key Certification. In *Proc. of 7th USENIX Security Symposium.* San Antonio, Texas, January. 1998.

[Levien2002] Raphael L. Levien: Attack Resistant Trust Metrics. Ph.D. thesis, University of Berkeley. Available at: http://www.levien.com/thesis/thesis.pdf. 2002.

[Nissenbaum1999] Helen Nissenbaum: Can Trust be Secured Online? A Theoretical Perspective. *Ethics and Politics,* 1(2). Available at: http://www.units.it/~etica/1999_2/index.html. 1999.

[Oppliger2005] Rolf Oppliger and Ruedi Rytz: Does Trusted Computing Remedy Computer Security Problems? *IEEE Security & Privacy,* March/April: 16–19. 2005.

Available at: http://ieeexplore.ieee.org/iel5/8013/30742/01423956.pdf?isnumber=& arnumber=1423956. 2005.

[Pearson2002] Siani Pearson, et al: *Trusted Computing Platforms: TCPA Technology in Context.* Englewood Cliffs. NJ. Prentice-Hall. 2002.

[Poslad2003] Stefan Poslad, Patricia Charlton and Monique Calisti: Specifying Standard Security Mechanisms in Multi-Agent Systems. In R. Falcone et al. (eds): *AAMAS 2002 Workshop on Trust, Reputation . . .* LNAI 2631, pp. 163–176. 2003.

[Powell2003] David Powell and Robert Stroud: Conceptual Model and Architecture of MAFTIA. Deliverable D21. Project IST-1999-11583. Available at: http://www.maftia.org/deliverables/D21.pdf. 2003.

[Rankl2003] Wolfgang Rankl and Wolfgang Effing: *Smart Card Handbook*, 3rd edn. Chichester. John Wiley & Sons, Ltd. 2003.

[Schneier2004] Bruce Schneier: *Secrets and Lies: Digital Security in a Networked World.* Chichester: Wiley Publishing Inc. 2004.

[Thibadeau2006] Robert Thibadeau: Trusted Computing for Disk Drives and Other Peripherals. *IEEE Security & Privacy.* 4(5): 26–33. 2006.

[Uszok2004] Andrzej Uszok, Jeffrey M. Bradshaw and Renia Jeffers: KAoS: A Policy and Domain Services Framework for Grid Computing and Semantic Web Services. In C.D. Jensen et al. (eds): *iTrust 2004*, LNCS 2995, pp. 63–77. 2004.

[Wang2005] Xiaoyun Wang, Yiqun Lisa Yin, and Hongbo Yu: Finding Collisions in the Full SHA-1. In *Proc. of the 25th Annual Int. Cryptology Conf. Crypto 2005.* 14–18 August. Available at: http://www.infosec.sdu.edu.cn/paper/sha1-crypto-auth-new-2-yao.pdf. 2005.

[Weigand2001] Hans Weigand and Willem-Jan van den Heuvel: Trust in Electronic Commerce: A Language/Action Perspective. In Cristiano Castelfranchi and Yao-Hua Tan (eds): *Trust and Deception in Virtual Societies.* Dordrecht: Kluwer Academic Publishers, pp. 237–257. 2001.

[Yan2003] Zheng Yan and Piotr Cofta: Methodology to Bridge Different Domains of Trust in Mobile Communications. In *Proc. The First International Conference on Trust Management (iTrust'03)*, LNCS vol. 2692/2003, pp. 211–224, Greece, May. 2003.

# 9

# Confidence and Networks

Let every eye negotiate for itself and trust no agent.

William Shakespeare,
Much Ado About Nothing

## 9.1 Introduction

Up to now we have been dealing with a relatively simple relationship between Alice and Bob. Even if they did not communicate directly, there were no other people involved, only technology (if any). However, this is seldom the case in an increasingly interconnected world. We are dealing with people through other people and computers are dealing with other computers through yet another computer. Society and the Internet is a multi-layered network of relationships.

We have already noted that the introduction of any intermediary significantly modifies the relationship between Alice and Bob, potentially depriving Alice of her ability to correctly assess her confidence in Bob. Alice may not have access to the rich set of direct evidence and must rely on comments and opinions provided by others (who may or may not tell her the truth). This can potentially be compensated for by access to a much richer set of opinions: if Alice can ask thousands of people what they think about Bob, her knowledge may be much better than in the case of a face-to-face relationship.

Society itself can be modelled as a network of relationships: Alice knows Carol, Carol knows Dave, Dave knows Bob, etc. Recommendation, reputation, introduction, gossip, rumour, hysteria: all these methods are used to transfer information inside such a network, and Alice can be a beneficiary – or a victim, depending on her actions and the topology of the network. The Internet multiplies the scale and effects of our social interactions, collecting and distributing not tens but thousands and millions of opinions, automating reputation and amplifying rumours.

If Alice (or Bob) is not a person, the question of confidence holds, but requires a certain re-formulation. As the complexity of networks and interactions grows, the intentional stance [Dennett1989] becomes a plausible model for the perception of the behaviour of technical devices. For example, in ad-hoc networks, devices wake up to the unknown environment with no pre-defined relationship of trust, yet they are supposed to self-organise themselves, to do

---

*Trust, Complexity and Control: Confidence in a Convergent World*   Piotr Cofta
© 2007 John Wiley & Sons, Ltd

reasonable work, to route messages and to resist hostilities as if they were people deployed in the unknown environment. Similarly, software agents must travel the network, find partners and competitors, develop winning strategies, build relationships – all according to human understanding of such actions. Such solutions benefit from mimicking social behaviour and using the network of relationships – with all its advantages and disadvantages.

We are – as usual – interested in Alice's welfare that we understand as her ability to correctly assess the extent of confidence she should have in Bob. We are not considering here the welfare of the whole network, nor the average level of confidence or trust. We will see that depending on the distance between Alice and Bob, she can employ different strategies that in turn will impact on her ability to determine her level of confidence. The characteristic of the network is also expected to influence the quality of such assessment, and the understanding of a possible relationship between network parameters and Alice's ability to assess Bob is a core feature of this chapter.

The chapter starts with an overview of the literature. Next, we formulate two quality indicators that can be used to determine how well Alice can assess Bob. The discussion about propagation and transitivity of confidence is followed by the discussion of removed interactions to identify four basic strategies that depend on the relative position of Alice and Bob. A brief discussion of different aspects of network-based confidence (such as network topology, details of the fusion of confidence, impact of memory, etc.) closes the chapter.

## 9.2 Overview of the Research

The concept of combining the network and confidence (trust) has been explored for a long time. Early research concentrated on technologies that can automate the management of trust. For example, PGP [Zimmermann1994] explored the ability to manage trust between e-mail participants while PKI [Ellison1999] has been aiming at the creation of trust management structure that may potentially permeate the whole network. This trend continues with solutions such as WS-Trust [Anderson2005] or KeyNote [Blaze2003]. For an in-depth analysis of this (and other related) works, please refer to [Abdul-Rahman2005].

Another research stream addresses the possibility of using the social concept of trust to improve the quality of information or the network performance. This stream builds on the observation that even though trust is essentially a social phenomenon, it can be collected and processed in an automated manner and applied to the operation of the network. Concepts such as collaborative filtering (e.g. [Resnick1994]) originate from the observation that trust can be perceived as the reliance on information [Gerck2002]. The well-known concept of PageRank [Page1998] sorts information according to human perception of its relevance and trustworthiness. Trust-based routing [Twigg2003] explores the notion that the trust relationship can be leveraged to improve the information flow in the network. Social trust can be used in certain networks to achieve better resource allocation [Bearly2004].

Yet another stream explores the use of the network to establish and maintain trust between people. Specifically, a large number of works investigate different versions of reputation-based systems such as eBay [Dellarocas2003]. The flexibility of the reputation-based systems lends itself to several interesting experiments such as the creation of the derivative honesty metrics [Fernandes2004]. Specifically the use of trust in virtual organisations has gained in importance [Yao2003], [Dimitrakos2004].

Complexity and scalability have been an important part of the research in networks and trust, responding to the growing size of available networks. XenoTrust [Dragovic2003] proposes a set of servers dedicated to trust management while [Ziegler2005] explores the scalability of trust evaluation algorithms.

Several models can be used to simulate and explore the behaviour of trust in networks. For example, systems such as [Grandison2003] or [Marsh1994] allow the trust-based reasoning to be modelled within the network while ˣTrust [Branchaud2004] demonstrates the simple trust-based routing framework suitable for established virtual networks of trust. All of them assume the propagation and transitivity of trust.

When it comes to processing trust (confidence, opinions, facts), the main difference seems to lie in the way that uncertainty is handled. In the case of probability distribution [Klos2006], it is assumed that uncertainty is not present, i.e. that agents have access to known probability distribution of the behaviour of other agents and the task may be to assess the parameters of such distribution. For subjective probabilities [Gambetta2000], opinions represent expected values of the probability of behaviour while uncertainty can be either implied by assuming a certain stance or included as the orthogonal value in the opinion. Finally, uncertain probabilities [Josang2001] embed uncertainty as an inherent part of all opinions.

Opinion processing may exhibit anomalies if the opinions of agents are not independent, specifically if a given agent is able to deliver its opinion more than once (potentially through different paths). The lack of independence not only may lead to inadequacy of theories but it may manifest itself in a form of undesired emergent behaviour such as hysteria [Josang2006] or echo [Burt2001] where Alice gradually listens to her own opinion. Unfortunately, it is relatively hard to assure independence without having a repository as it require methods that are expensive in terms of local memory, communication and processing needs.

## 9.3 Quality Indicators

The main question discussed here is not how Bob decides to behave (as an intentional agent he can freely change his mind), but how well others are able to detect and assess his behaviour: for example, how long does it take to correctly assess Bob and how long does it take to adjust such estimate if Bob decides to change his behaviour? If other agents (including Alice) can reliably assess (estimate) Bob, then they can decide whether to rely on Bob. We are specifically concerned with Alice and her ability to reliably track Bob's behaviour even if Alice may not be always in a position to directly interact with Bob (he may not be her direct neighbour).

We do not discuss here the development of relationship, Alice's strategy, reciprocity or Alice's reaction to Bob's misbehaviour. The only thing we are interested in is how well Alice follows changes in Bob's behaviour. We expect Alice to react swiftly and accurately to changes, so that we define two indicators that are related to the quality of her reactions: delay and volatility. We expect that Alice will benefit by minimising both.

Bob as a source of information about his behaviour can provide his neighbour with a simple response: within the scope of a given transaction either Bob delivers evidence of trust or not, either he delivers evidence of control or not. If he can be neither trusted nor controlled, then confidence is low. If he is trusted and controllable, then confidence is high. Otherwise it is somewhere in the middle. Observed by the outsider over the longer period of time, averaged out, Bob's behaviour can be estimated to be at the certain level that can be understood as a subjective probability [Gambetta2000] of Bob's performance (a level of confidence in Bob). Bob's real intentions are not available to observers, but we can assume here Bob's willingness to perform can be at any moment estimated by a single number.

Let's consider (Figure 9.1) hypothetical Bob's intentions and observable behaviour (assuming one interaction per one time period). For illustration only, we can assume that within 150 observable time periods, Bob's intentions are initially at the level of 0.8, then drop to 0.2 to return to 0.8 later. For example, Bob is initially willing to perform, but makes some errors. Then, after 50 cycles, Bob becomes overwhelmed with other tasks and his intention to perform here drops (note that we do not assume any malicious intentions: Bob may be

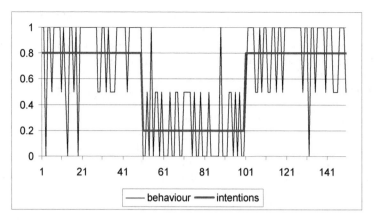

**Figure 9.1**   Bob's intentions and his behaviour

willing, but what he offers is of lower quality). Finally, Bob recovers from such temporary overload and his performance returns to normal.

For reasons of simplicity, we have assumed here that a single interaction ends up with only one of three outcomes: (1) Bob did not produce evidences of trust or control (e.g. he was unwilling and did not deliver); (2) Bob produced some (e.g. he was willing but did not deliver); and (3) Bob produced both (e.g. he was willing and he delivered). Other analysis may use less levels (e.g. only two: Bob performing or not performing) or more levels (e.g. judging the strength of different evidences). As a result Bob's observable behaviour is very uneven: it oscillates between 1, 0 and 0.5, depending whether there is evidence for trust and control, no evidence or evidence for only one of them.

When Bob rapidly changes his intentions (and consequently his behaviour), we would like Alice to change her assessment of her confidence in Bob. Figure 9.2 shows an example of the

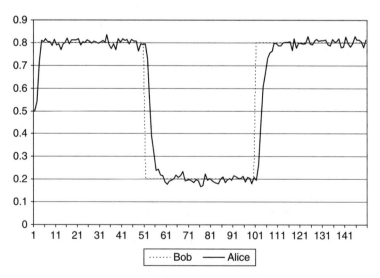

**Figure 9.2**   Bob's intentions and Alice's desired assessment

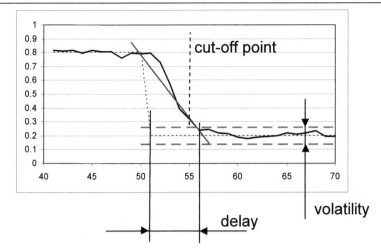

**Figure 9.3** Delay and volatility

desired transition. Initially, Alice swiftly moves from her original uncertainly (0.5) to follow Bob's intentions at the level of 0.8, with only slight deviations. When his intentions change, she reacts swiftly and then stays focused on the new level. Each change in Alice's assessment can be modelled as a two-phase process: Alice first notices the change in Bob and alters her assessment accordingly. After that Alice retains her assessment at the certain set level despite the variability in Bob's behaviour.

Accordingly, we will introduce two quality indicators: delay that will tell how fast Alice is able to react to a change and volatility that will capture how focused she stays once she has reacted to the change. Note that the way delay and volatility are calculated are somehow arbitrary and are valid within the context of this analysis only. Figure 9.3 shows the close-up of the first transition, at the 50th cycle: within one time period Bob's intentions drop from 0.8 to 0.2.

The calculation of quality indicators starts with determining the coarse cut-off point, i.e. the moment where, after the change, Alice's assessment of Bob's intentions are roughly within the new range. The value of such a range is calculated as a standard deviation of all of Alice's estimates from the moment Bob has changed his mind. The purpose of such a cut-off point is to separate the rapid slope of the transition (that contributes to delay) from its horizontal tail (that contributes to volatility).

*Volatility* captures imperfections in Alice's tracking of Bob's intentions over the 'tail' and is determined as a standard deviation $\sigma$ for Alice's estimates past the cut-off point. Note that the volatility is not artificially inflated with Alice's transition, but captures only the stability of Alice's long-term assessment. Such volatility may originate from several different sources of it, e.g. the 'noise' of the network where old estimates may linger for a long time, the gradual improvement in the estimate as time progresses, etc.

*Delay* captures the extent of Alice's delay in noticing Bob's change in intentions. Delay is measured as the extent of time that passes between Bob's transition and Alice's estimated reaction, the linear regression of the slope. The value of delay is calculated at the level of $\sigma$ (standard deviation) above (or below) Bob's intentions

## 9.4 Propagation and Transitivity of Confidence

The key problem in analysing confidence and networks is the question of the propagation and transitivity of confidence. Indeed, if confidence cannot propagate, then there is no reason to discuss its behaviour in networks. In systems that use out of band trust management (e.g. PGP [Zimmermann1994], PKI [Ellison1999], WS-Trust [Anderson2005] or KeyNote [Blaze2003]) this problem is not present, but if confidence can be established by opinions of other agents, propagation and transitivity become an important element. For example, research into autonomous agents (e.g. [Fullam2005]) as well as research into the use of social networks (e.g. e-commerce [Dellarocas2003], information filtering [Resnick1994], e-mail filtering [Golbeck2004] or trust-based routing [Twigg2003]) concentrate on systems where one agent can assess its confidence in another agent without the complete pre-existing trust structure, but by inferring trust from the observable behaviour and opinion of other agents.

Problems of propagation and transitivity are different. Propagation is concerned with the way confidence is distributed and used within the network in general while transitivity is a specific mathematical property of confidence itself. Let us consider a simple example (Figure 9.4) where Carol is an intermediary between Alice and Bob. The question of propagation asks how Alice can learn about $conf^{subject}_{Carol \to Bob}$ and what she can do with such information. The question of transitivity can be formulated differently: given that there is $conf^{subject}_{Alice \to Carol}$ and $conf^{subject}_{Carol \to Bob}$, can Alice reasonably assume that there is $conf^{subject}_{Alice \to Bob}$?

It is widely understood that confidence can propagate in networks by various mechanisms such as reputation, trusted parties, recommendation, etc. Regarding specifically the transitivity of trust, some researchers are content with the observation that trust is not transitive ([Castelfranchi2000], see also [Jones2001] for an analysis from the logical standpoint). Others (see [Josang2006] for an interesting discussion) discuss different forms of the transitivity of confidence (trust) and develop tools to calculate the level of confidence. The problem of transitivity can be also discussed in the context of using historical data to predict future behaviour [Mui2002].

The complexity-based model of confidence can be used to consolidate different views of the problem of transitivity of confidence. The proposition presented here draws on [Huang2006] (the similar proposition can be found in [Dimitrakos2003]) and can be summarised as follows. Neither trust nor control is transitive in the same frame of discourse. However, if we consider enablers, we can see that confidence can be formed remotely which gives an impression of pseudo-transitivity. The proposition is discussed below, separately for trust and for control.

It is proposed that trust is not transitive, i.e. from the fact that there is $trust^{subject}_{Alice \to Carol}$ and $trust^{subject}_{Carol \to Bob}$, Alice cannot reason that there should be $trust^{subject}_{Alice \to Bob}$. The reason for it is as follows. Trust is a belief and Alice has instant access to the first belief ($trust^{subject}_{Alice \to Carol}$), as it is her own internal state of mind. However, she can learn about the second belief ($trust^{subject}_{Carol \to Bob}$) only if either Carol reports it or if Alice (or someone else) deduces it from observations. Regardless of whether Alice has access to actual evidence or only opinions about Bob, Alice

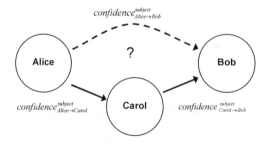

**Figure 9.4** The problem of transitivity of confidence

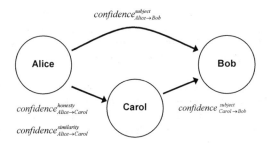

**Figure 9.5** The pseudo-transitivity of confidence

does not have access to a key bit of information: whether she can be confident that opinions or evidence are reported honestly and are meaningful. Unfortunately, Alice's trust in Carol ($trust^{subject}_{Alice \rightarrow Carol}$) does not help, as such trust remains within the frame defined by the *subject*, not by Carol's ability to report evidence or opinions. So, even though Alice can potentially use information about $trust^{subject}_{Carol \rightarrow Bob}$ to establish $trust^{subject}_{Alice \rightarrow Bob}$, the way she can do it has nothing to do with $trust^{subject}_{Alice \rightarrow Carol}$. Alice simply needs a different confidence: confidence that Carol is honest and has a similar mindset, $conf^{honesty}_{Alice \rightarrow Carol}$ and $conf^{similarity}_{Alice \rightarrow Carol}$.

Alice may be tempted to use the causal relationship to bypass the reasoning presented above: if there is $trust^{subject}_{Alice \rightarrow Carol}$ and if Alice is willing to consider Carol as a 'whole person', she may infer that it is quite likely that Carol will be also honest (as positive traits tend to go together), implying that there is $conf^{honesty}_{Alice \rightarrow Carol}$ so that Carol's opinion about her trust in Bob ($trust^{subject}_{Carol \rightarrow Bob}$) may be useful. However, the link from $trust^{subject}_{Alice \rightarrow Carol}$ to $conf^{honesty}_{Alice \rightarrow Carol}$ may be misleading: Carol's competence in the subject may not translate into Carol's competence in assessing people or in expressing herself.

Following a similar reasoning, we propose that control is not transitive as well, so that from $control^{subject}_{Alice \rightarrow Carol}$ and $control^{subject}_{Carol \rightarrow Bob}$, Alice cannot conclude that there is $control^{subject}_{Alice \rightarrow Bob}$. However, control can be made to look transitive if we consider that Alice can use Carol as an instrument to control Bob (e.g. to gather knowledge about Bob). This, however, implies a different configuration of relationships and a different type of control: Alice should be able to control Carol regarding Carol's ability to control Bob, not regarding the subject. If e.g. Alice wants to use Carol to collect knowledge about Bob, Carol's ability to process such knowledge will be the key factor.

To summarise, confidence does not seem to be transitive if we restrict ourselves to the single frame of discourse defined by the subject, i.e. from $conf^{subject}_{Alice \rightarrow Carol}$ and $conf^{subject}_{Carol \rightarrow Bob}$ Alice cannot reason that there is $conf^{subject}_{Alice \rightarrow Bob}$. However, if Alice can somehow resolve the problem of $conf^{honesty}_{Alice \rightarrow Carol}$ and $conf^{similarity}_{Alice \rightarrow Carol}$, then she can use $conf^{subject}_{Carol \rightarrow Bob}$ to build her own $conf^{subject}_{Alice \rightarrow Bob}$, as shown on the Figure 9.5. Even though it is not exactly transitivity (as Alice's reasoning is conducted in two or even three frames of discourse), at least confidence can propagate. To summarise, confidence can therefore propagate on the foundation of honesty and similarity.

## 9.5 Taxonomy of Interactions

We propose and extend here the taxonomy from [Patrick2002] and [Battison2006] where it has been demonstrated that there are important differences in the way trust (confidence) develops depending on the level of removal, i.e. the number of intermediate agents placed between Alice and Bob along the shortest connecting path between them.

Non-removed interaction happens when Alice can directly observe Bob. In such a situation she may not be concerned with the opinions of other agents and network properties may have no effect on such an interaction that is entirely on an inter-personal level. Confidence can be built rapidly, but Alice must develop a mechanism to cope with the extreme volatility of Bob's behaviour.

Once-removed interaction assumes that both Alice and Bob have a set of 'common friends' that can relay their first-hand observations about Bob to Alice. If Alice disregards any other sources of information, there is an increased (but fixed) delay, as friends must gather and relay information. At the same time Alice has potentially access to the larger amount of information, which should result in lower volatility. However, this gain can only be realised if Alice and Bob indeed have a large group of common friends.

Twice-removed interaction happens where there are two or more intermediate agents. In this kind of interaction there is a significant influence of opinions coming from other agents, as there is no direct relationship between Alice and Bob. Information about Bob arrives to Alice through different, sometimes unpredictable paths. It can be integrated, discounted, amplified, looped, etc. Alice must develop some coping mechanism to determine the quality of information delivered by different neighbours.

Level-removed interactions use the separate repository to store all available information about Bob. All Bob's neighbours can store their experience and Alice has access to the outcome of the evaluation of such experiences. The interaction is called 'level-removed' as it requires an additional level of confidence where all agents implicitly trust the repository that does not belong to the network. Level-removed interactions are characteristic of modern technology-augmented confidence-based systems such as electronic marketplaces (e.g. [Dellarocas2003]).

We will discuss below those four different types of interactions to see how the distance between Alice and Bob affects Alice's ability to assess her confidence in Bob. For simplicity, we will restrict our discussion only to two frames of discourse: the main subject and honesty, assuming that all agents are similar, that their identity is guaranteed and that there is sufficient flow of evidence throughout the network.

Further, we will use a simple simulation to estimate and compare some values of quality indicators. Even though such values are valid only within a context of this analysis, they provide additional reassurance regarding quantitative changes that we expect. It has been simulated that Bob is changing his intentions in a way presented in Figure 9.1. The simulation has been performed for the ER network of 1000 agents, where each agent has 10 neighbours on average (unless stated otherwise). The simulation has been run at least 100 times and results have been averaged, for all runs and for both changes in Bob's intentions. No agent has used any memory so that any memory-like effect results from the properties of the network, not the properties of agents.

## 9.5.1 Non-removed Interactions

The first and the simplest interaction is the non-removed one where there is a direct connection between Alice and Bob. Ignoring the presence of other members of the network, this simple relationship between Alice and Bob can be as shown in Figure 9.6. If Alice ignores all the other agents, this interaction takes us back to our considerations from earlier chapters and can serve as a reference point to see whether the existence of a network actually improves Alice's estimations.

Alice can use all the evidence that she is able to gather about Bob to estimate her confidence in him. We can see that Alice is in a position to explore both trust and control with regard to Bob. Specifically, we will see later that evidence of trust is generally restricted to this kind of direct (non-removed) relationships [Wang2003].

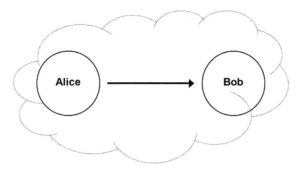

**Figure 9.6** The non-removed interaction

What happens when Alice assesses her confidence in Bob simply follows the model (Figure 9.7). Alice is not concerned with anyone's honesty, as evidence is gathered directly by her. Similarly, she is not concerned with the question of similarity, as she is the only one to interpret the evidence. If she has been using any instruments of control, they are also under her complete control. Consequently, whatever evidence she has, it very strongly influences what she thinks about Bob.

The price that she pays is the direct exposure to the volatility of Bob's behaviour. In order to dampen such volatility and smooth her assessment, Alice can only rely on her own memory. The longer and better is Alice's memory, the lower the volatility. However, at the same time the delay of Alice's reaction grows, as new evidence can only gradually overcome the remembered ones. The impact of memory will be discussed later in this chapter.

If we run the simulation, such a simulation sets the base reference for future comparisons. Assuming that Bob is changing his intentions in a way presented earlier in this chapter, and assuming that there is a non-removed interaction between Alice and Bob, the delay is 0.25 and the volatility is 0.28. We assume further that Alice has only been listening to Bob, as discussed above, i.e. that she has not been collecting indirect opinions from others about Bob.

Simulation results confirm our expectations: the delay is very short: 0.25 means that Alice has been able to track changes in Bob's intentions immediately in the majority of cases: quite an achievement considering the limitations of Bob's responses. However, the volatility of 0.28

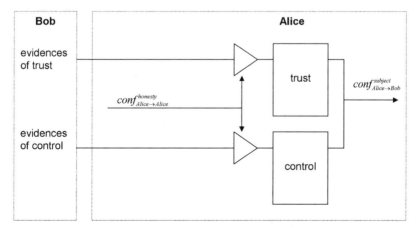

**Figure 9.7** Confidence evaluation for non-removed interaction

is very high and Alice cannot be certain about her estimates. In fact, as she is lacking any memory, her volatility is identical with Bob's. We can expect that even though she can detect rapid and significant changes in Bob's behaviour, she may have problems with smaller and gradual ones.

### 9.5.2 Once-removed Interactions

If Alice is not Bob's immediate neighbour, she may be able to gather information about Bob from their common neighbours, i.e. those agents that are immediate neighbours of Bob and Alice. In such case we can speak about once-removed interaction. Figure 9.8 shows an example of a network with once-removed interaction between Alice and Bob: Alice is a neighbour of Carol and Dave and both Carol and Dave are neighbours of Bob.

Compared to non-removed interactions, the once-removed brings some significant changes. First, Alice is no longer able to collect evidence directly, so she must develop an appropriate strategy to deal with Carol and Dave. For the first time the question of the honesty of other agents become important, as Alice is vulnerable to malicious actions of her neighbours: if either Carol or Dave are dishonest, then Alice will be unable to correctly determine her confidence in Bob.

Alice has two strategies to choose from: she can either ask her neighbours to forward evidence that they have or she can ask them to forward their opinions about Bob. Those strategies may potentially lead to very different results so that they deserve a separate discussion.

If Alice is using her common neighbours to forward evidence about Bob, she in fact overcomes her inability to interact with Bob: her neighbours act as her remote 'eyes and ears' and the interaction strategy folds back into the one that is very similar to the non-removed one. They hopefully only report what they have observed, not their beliefs. In such a situation, Alice is able to use both trust and control to develop her confidence in Bob. The main difference is that evidence is coming from several sources and that each source may be of different quality (specifically honesty). Figure 9.9 demonstrates how information may flow if Alice chooses such a strategy. Note that Carol and Dave here are no more than simple

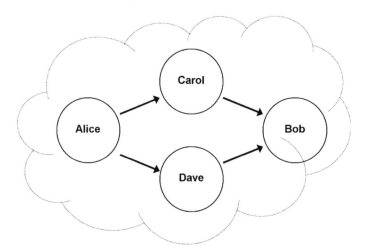

**Figure 9.8**  The once-removed interaction

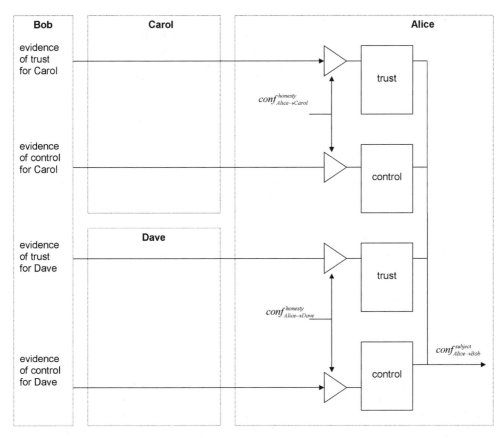

**Figure 9.9** Evidence-forwarding strategy

conduits of evidence. The price that Alice pays is complexity: she must process much more evidence than in the case of non-removed interaction or in the case of opinion-gathering.

Another price that Alice pays (that is inherent to all interactions that are somehow removed) is the additional complexity of assessing the honesty of her information sources. For simplification, we are not discussing here the methods that Alice uses to determine Carol's and Bob's honesty as it is performed within a different frame of discourse.

Note that the evidence-forwarding strategy is feasible in once-removed interactions mostly because the number of common neighbours is small so that neither Alice nor any of her neighbours is overwhelmed by the task (Figure 9.9). The added benefit of this strategy is the fact that neighbours gather and forward information about Bob along independent paths, i.e. what Carol knows about Bob does not interfere with Dave's knowledge. The strategy does not scale, though. If there is a large number of neighbours or if they interact with each other in forming their opinion about Bob, they are unlikely to forward raw evidence. In such a case Alice should employ the other strategy: to collect opinions.

From the perspective of the quality of assessment, we can expect that evidence-forwarding will decrease the volatility of Alice's assessment, as she will have access to a larger number of bits of evidence (assuming that there is more than one common neighbour). As with all forms of once-removed interactions, the delay will be longer than in the case of non-removed interactions, but it will be fixed and predictable.

The simulation demonstrates that the delay has indeed increased to 0.62 while the volatility has remained the same at 0.28. This can be explained by the parameters of the network that has been used in simulations: with the average number of neighbours equal to 10, there is little chance for more than one common neighbour between Alice and Bob, so that in fact she is not gaining anything by being once removed. Only when the average number of neighbours grows, does volatility go down – but at the slightly higher expense of delay. With 50 neighbours, the delay is 0.72 and the volatility is 0.19.

The second possible strategy that Alice can apply is to ask her neighbours for their opinion about Bob. An opinion is the processed statement from an agent about its confidence in another agent, but without a detailed description of the process that has led to such a conclusion. Contrary to the first strategy of evidence-forwarding, opinions are opaque: Alice knows an opinion but does not know how an agent came to such opinion. While using opinions, Alice is not only concerned with the honesty, but also with the similarity of agents: if Carol or Dave have a different perception of confidence, their opinion, even though it is honest, may be useless.

Opinions are generated by collecting evidence of trust and control, but they are later processed entirely as evidence of control. Opinions are qualified by the confidence in the honesty of a source of opinions, i.e. whether Carol or Dave are honestly reporting their opinion. If there is doubt about similarity, then an additional qualification is needed. Figure 9.10 demonstrates how opinions can be collected and processed. Similarly to the previous strategy, we are not concerned here with the process of determining honesty (or similarity) of Alice's neighbours.

From the perspective of the quality of Alice's assessment, the situation should be similar to the previous strategy: the delay should grow and the volatility should decrease. Simulation confirms this, delivering 1.10 for the delay and 0.25 for the volatility – not a great improvement. Similarly, the growth in network density improves volatility and introduces additional delays. The difference between strategies does not seem to be significant with regard to the quality, but it is quite significant with regard to the distribution of the workload. The first strategy gives significant work to Alice: she has to evaluate all evidence, both for trust and for control. The second strategy distributes the work: her neighbours are generating opinions and she is only consolidating them. Because of this, the opinion-based strategy scales better.

It may come as a slight surprise that delays reported here can be below 1. Indeed, as it takes as additional one time slot for common neighbours to forward evidences or opinions, so we should expect delays to be above 1. The answer is in the way quality indicators are measured. In the case of high volatility, delay is shorter than for lower ones, i.e. a regression line that approximates as Alice's reaction crosses the line defined by a standard deviation faster if the deviation is high.

## 9.5.3 Twice-removed Interactions

Up to now we have assumed that Alice has already had some facts about Bob or that she has been able to find, in her immediate vicinity, agents that has been in contact with Bob. However, as the network grows, this assumption may be less realistic. The phenomenon of the 'small world' [Milgram1967] calls for up to six degrees of separation, much more than the once-removed strategy can handle. Alice must be ready to deal with the network where agents may know other agents only through a chain of intermediaries.

We now move to networks where Alice has no personal experience with Bob and has no personal experience with anyone who may have such an experience. This is the situation where the interaction is classified as twice-removed: Alice knows someone who knows someone who knows Bob. Note that in fact the interaction may be removed more than twice,

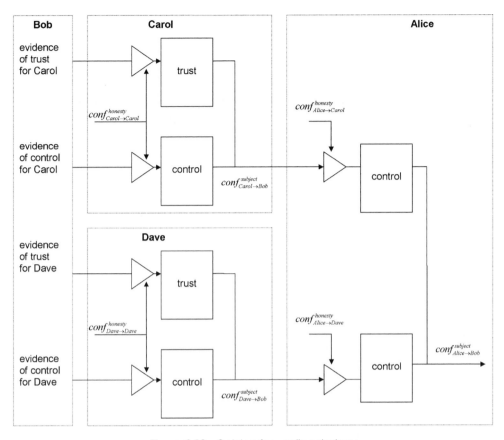

**Figure 9.10** Opinion-forwarding strategy

with a long chain of intermediaries. However, it is at least twice removal that is characteristic here so that we will use the name 'twice-removed interaction' for all these interactions.

Let's consider the simple chain of relationship, as demonstrated in Figure 9.11. Alice can contact Carol who can contact Dave who knows Bob. In fact, in more complex networks there can be several chains of this kind, potentially partly overlapping or forming circles. Implications of this fact (as compared to the simple structure of once-removed interactions) are significant. For the simplicity of discussion, we will restrict ourselves for the moment to the single chain.

In order to be confident about Bob, the only viable strategy is opinion-gathering: Alice asks her immediate neighbours about their opinion about Bob, they ask their neighbours, etc. At a certain moment this process reaches someone who knows Bob and, through possibly a convoluted link, returns this assessment back to Alice.

Considering only the displayed path (Alice–Carol–Dave–Bob), this leads to the characteristic chain that spans from Bob to Alice. Dave uses both trust and control to develop his confidence in Bob; Carol uses Dave's confidence to build her confidence in Bob and finally Alice uses input from Carol. Figure 9.12 shows details of such a chain. Note that Carol is concerned only with Dave's honesty and Alice is concerned only with Carol's honesty (assuming similarity of agents).

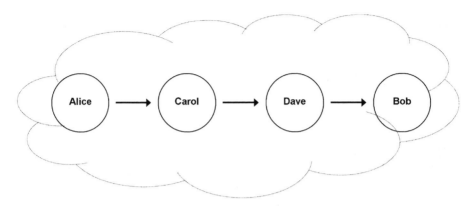

**Figure 9.11**   Twice-removed interaction chain

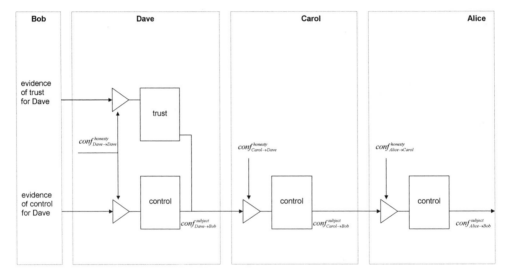

**Figure 9.12**   Confidence evaluation in twice-removed interaction

The most important difference between once-removed and twice-removed is the growth in opaqueness. Even though the confidence is known, the method that has been used to evaluate it is not known. Alice does not know all the possible interaction chains between herself and Bob. She is not aware how long is the shortest link between her and Bob, how many agents interacted with each other, whether the same opinion has been used more than once, how the honesty of such agents have been established, etc. Even though the chain presented on Figure 9.11 looks deceptively simple, the network of relationships is much more complicated.

Simulation results produce 3.46 for delay and 0.12 for volatility. Even though there is a significant improvement in volatility, it has been paid for with an enormous growth in delay. The expected delay depends on the average path length in the network and for ER networks it can be expressed as $\log_m n$ (where m is the average number of neighbours and n is the number of agents). For Alice, the problem may be that this is an average – given Bob, she does not know what to expect, even if she knows properties of the network.

The positive impact on volatility comes from the fact that the network itself acts as a convoluted memory. Even though each agent can hold its opinion only for one period of time, they collectively form a complex system that continuously processes and re-processes opinions, so that the same opinion can arrive at Alice several times, combined with time-shifted other opinions – as if it has been stored in Alice's memory. Unfortunately, this is potentially also the source of problems, as twice-removed interactions are subject to emergent behaviours where the network artificially and unexpectedly attenuates certain opinions.

## 9.5.4 Level-removed Interactions

Repositories are the most popular forms of opinion-processing architectures, implemented e.g. in eBay or in Google [Page1998] and in several other computer-mediated social groups. Their architecture usually features two characteristic elements: a central hub – repository – where all agents report their opinions, and the common reputation evaluating algorithm executed in such a hub. Alternative architectural solutions may call for the ring of peers continuously agreeing upon their opinions and delivering their estimate.

The existence of a central repository for the first time permits us to introduce the notion of reputation – the commonly agreed expectation regarding Bob's intentions and behaviour. Alice can here use the common shared experience of the whole network regarding Bob. Reputation can be (and usually is) used by Alice as the only evidence of her confidence in Bob. She is of course free to combine global reputation with any other strategy (e.g. with her direct experience).

Confidence evaluation schemes available in such systems are based in general on the assumption of the similarity of agents, that allows global reputation to be used. The reputation-based interaction may be incorrectly perceived as once-removed as there is only the central repository between Alice and Bob. However, the repository is not a part of the network of agents and Alice has no means within the network to build confidence in the repository. From this perspective, she is no match for the central repository – the repository can assess her while she cannot assess it. She is therefore level-removed from the repository – her confidence in the repository must arrive at a different level from the rest of the interaction (Figure 9.13).

Simulation demonstrates that level-removed interactions produce the best results: at the delay of 0.75, the volatility is 0.09. Increasing network density (which is equivalent to decreasing the number of reports on Bob) predictably decreases volatility even further while slightly increasing delay. Note, however, that those simulation results are not directly compa-rable: in all the previous types of removal we have been considering ER networks, but the existence of a central repository implies that the network is not ER. It changes the topology of ER network into something that resembles highly centralised scale-free networks. This has significant implications on other parameters: available connections (every agent must have a direct connection to the repository), structure of the traffic (repository is heavily loaded with a traffic), fault resistance (repository is a single point of failure), etc.

The popularity and relative quality of repository-based systems should not obscure the fact that they suffer from some significant shortcomings. First, they employ algorithms that may be different from what Alice would like to consider. As long as Alice shares her perception of confidence with other members and with the repository, she can put up with the opinion provided by the repository. However, if for whatever reason she would like to assess Bob by herself, she is not in a position to do so. In order to circumvent this shortcoming, some repositories provide direct access to facts, so that Alice can freely explore them and make up her mind – at the expense of her complexity.

The second shortcoming is that there must be a mutual relationship of honesty between the repository and all members. The repository will use its understanding of the honesty of

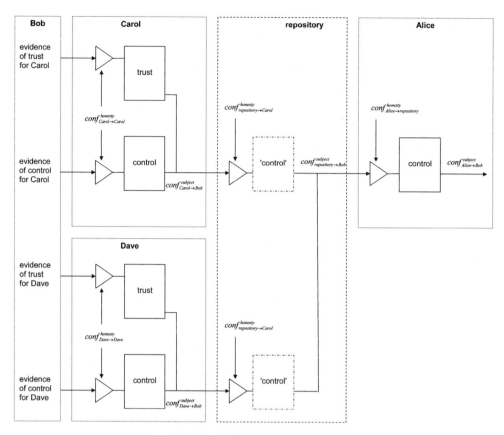

**Figure 9.13** Confidence in level-removed interaction

members to assign a proper weight to opinions reported by them while members will use their assessment of the honesty of the repository to assess the importance of the opinion it produces. Note, however, that the dishonest repository endangers the whole system as there is no means of bypassing it.

## 9.6 Additional Considerations

The propagation of trust, control and confidence in networks is a very complex area with several aspects that are worth separate discussion. Up to now, this chapter has presented how the model applies to situations where Alice and Bob are parts of the network, rather than individuals, in the context of four different removal models. Below there is a brief discussion of some aspects of confidence evaluation within the network that are common to all situations, regardless of the level of removal between Alice and Bob.

### 9.6.1 Network Architectures

There is potentially an endless number of different network architectures, that may differ in their properties such as the number and distribution of agents, connections between them, presence or absence of cycles, etc. However, when discussing confidence, we are usually concerned only with two things: the number of agents and the way they are connected. Specifically, in large networks we are concerned with the probability distribution of connections between agents: how likely it is that two randomly chosen agents are connected. Looked at from this perspective, networks come usually in one of two flavours: random or power.

## Random Networks (ER Networks)

Random networks assume that the probability of connection between two agents is constant across the whole network, regardless of any other properties of either agent. This leads to Poisson distribution of the probability that the agent has n connections to other agents. Random networks were first studied by Erdos and Renyi [Erdos1960], hence they are frequently referred to as 'ER' networks. The random network can easily be generated by creating the empty network (with no connections) and then randomly choosing missing connections and adding them to the network.

Random networks are usually characterised by only two parameters: the number of nodes-agents, m, and the average number of connections per agent, k (k is called the scale or the degree of the network). Alternatively, the network can be characterised by n and p, the probability of the connection between two agents. Random networks are useful in situations where little is known about the actual distribution, but they also well describe the structure of networks within closed communities such as universities or interest groups.

## Power Networks (Scale-free Networks)

Power networks are also random networks in the sense that we assume a certain probability distribution that determines how connections are placed, but they significantly differ from random ER networks, as some agents are significantly better connected than others (even though the average number of connections, k, can remain the same). The name 'power networks' comes from the fact that the probability that the given node is connected to n nodes has a form of $n^{-y}$, where y is the parameter of the network [Barabasi1999]. If visualised, it is possible to observe that the power network consists of a set of well-connected agents (hubs) and a majority of agents with the relatively little connections.

Power networks are normally used to describe scale-free networks (sometimes both names are used as synonyms), i.e. networks that can potentially grow indefinitely from a relatively small set of well-connected nodes. Power networks reflect the nature of social networks (e.g. collaboration networks in the Internet) where people tend to favour services that are already popular, so that some nodes attract high traffic while a significant number of nodes attract very little traffic. Interestingly, power networks can be also used to model other relationships within the society as well as the overall structure of the Internet itself. There has been also growing interest in power networks and power distribution in relation to Internet-based marketing and biological phenomena. An interesting review of relationships between nature-built and artificial networks, with special focus on different power networks is included in [Drogovtsev2003].

Random ER networks are usually used to model closed networks of reasonably stable size, such as the power grid network, highway system, closed communities, etc. Power scale-free networks are used to model open, growing communities. It has been demonstrated [Pennock2002] that the structure of the Internet is actually a combination of both: the structure of hubs and service providers can be described with the help of power scale-free networks, but established communities tend to follow random ER network structure. It has even been claimed [Clauset2005] that the perception of the power-network structure of the Internet can be partly explained by the way the actual analysis has been conducted, while the underlying network may in fact be the ER one.

## 9.6.2 Memory

Throughout this chapter we are generally considering memory-less networks, i.e. we are assuming that agents can store and process only information related to the most recent interaction. Specifically in the case of non-removed interactions, we can see that the lack of memory leads to a very high volatility, as Alice cannot do anything more than simply follow Bob's highly erratic behaviour. Even though the assumption of memory-less operation is useful in demonstrating the relationship between the network and quality of assessment, the impact of memory deserves some discussion.

The model of confidence caters for the existence of memory, allowing every agent to refer not only to current but also to past evidence. There are several possible strategies that dictate how an agent can use its memory: how many pieces of evidence it stores, how they are rated, how they are processed etc. The most popular one is to allow an agent to remember the fixed maximum number of pieces of evidence (i.e. the 'size' of the memory is fixed) and then to take into account all evidence, but to apply a certain discount factor to older ones. This strategy is supposed to simulate the rational behaviour where older evidence has less effect on the outcome and is eventually forgotten. In general, this strategy leads to significant improvements in an assessment: volatility dramatically decreases, even though delay generally grows. Figure 9.14 shows the impact of a memory size on delay and volatility, assuming the fixed discount factor. Figure 9.4 has been drafted for non-removed interactions and averaged across several simulation runs.

Note, however, that such a strategy is not the only one. For example, people attribute significant value to the first evidence, suggesting that our memory collects and re-processes opinions rather than facts. An equivalent strategy that is based on continuous re-evaluation of opinions does not require us to store all the facts, but only to store the last (newest) opinion. Once the new fact arrives, such an opinion is re-evaluated in light of the past one and the new fact, and only the new opinion are stored. Note that even though this strategy does not require significant memory, the strong impact of first few pieces of evidence cannot be easily overcome.

Alternatively, it may be enough to store only the last evidence and entirely discard all the previous ones. Even though such a strategy may look simplistic, it is worth noting that

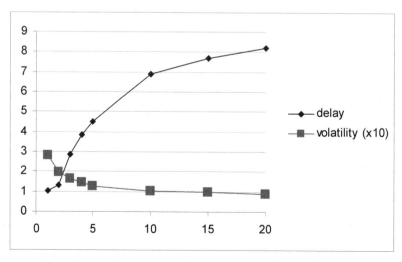

**Figure 9.14**  Performance indicators as a function of memory size

it is exactly the foundation of the winning strategy in the game of trust [Axelrod1984]. The tit-for-tat strategy is successful despite its simplicity (or possibly because of it), and it requires the memory that remembers only the last evidence. If the goal of the assessment of confidence is to develop the best interaction strategy, then the large memory may not be the best one.

### 9.6.3 Honesty and Similarity

The most important step Alice has made while moving from the non-removed to a once-removed interaction is to rely on others' experiences and opinions. The model already contains provision to include the honesty and similarity of agents in the equation.

Note that we are not discussing here the idea of malicious agents, i.e. agents that intentionally circumvent the truth (or simply lie) about their experience and opinions. The discussion about such agents rightfully belongs to attacks on confidence-based networks. We expect here that all agents are willing to provide their real experience or opinion – but not all of them are in an equally good position to deliver it. For example, some agents have significantly less interaction with Bob (or cannot reach Bob at all), so that their experience is of limited value.

The main question for Alice is therefore to determine the strategy that can let her assess the extent of honesty and similarity that she can attribute to different agents. There are at least two different strategies that can be used here.

First, Alice can take the 'whole person' approach and say that her opinion about the agent's ability to deliver an opinion is the same as the agent's ability to do the task – that there is only one level of confidence which represents the whole agent. Within society, this approach is supported by common perception that there is a single person with a single mindset represented through different activities. Some of the well-known reputation algorithms such as EigenTrust [Kamvar2003] use this approach both to simplify computation and to express their belief that the reliable peer is the one that provides the good service and the reliable opinion at the same time.

It is specifically hard to assess honesty if there is no additional independent evidence. For example, honesty can be assessed on the basis of opinions about honesty that one agent has about other agents – with no further evidence. In the equivalent of gossiping agents may try to agree which agent is the honest one and which one is not. The evaluation of honesty, if built on insufficient evidences, too often leads to the liar's paradox: if Carol is not honest, then her opinion about Dave's honesty should be ignored. Symmetrically, if Dave is not honest, then his opinion on Carol's honesty should be ignored as well. The problem is that opinion about Carol is provided by Dave while the opinion about Dave is based on Carol's reports.

In the case of level-removed interactions, the complete knowledge of all facts allows the repository to attempt to correct this paradox. It can be done in at least two different ways. The first is to convert the graph into the acyclic one by arbitrarily removing some opinions. For example, the PageRank [Page1998] introduces the notion of 'boredom' to break circular relationships as they are encountered.

The other method is to introduce an equivalent of lazy evaluation, effectively making the graph acyclical in time rather than in space. At every step the evaluation is done on the basis of information from the previous step, that is, all opinions are provided before any new fact is admitted into the repository while the trustworthiness of a new fact is assessed on the basis of historical facts only. This, again, does not remove the problem – it only makes the calculation of an opinion possible. For example, the distributed EigenTrust algorithm [Kamvar2003] uses this method to resolve circular references.

Another approach calls for the separation of different skills and separation of evidence: the ability to formulate opinion on other entities does not necessarily go together with the ability to deliver the service. This approach can be also justified by everyday observations: e.g. there are enough researchers that are poor speakers and good engineers that misjudge other people's intentions to see that it may be valid to talk about separate confidence in the ability to support us and another confidence in the ability to formulate the opinion.

## 9.7 Conclusion

If Alice moves from a direct interaction with Bob into the network-based environment, she is facing both opportunities and risks. On the positive side, she may access and utilise the wealth of facts and opinions that have not been previously accessible to her. This may definitely improve the quality of her assessment. On the negative side, she must rely on others, and this bring questions of honesty and similarity, additionally complicating the way confidence is evaluated.

The network reveals its important characteristics when we consider not the stable situation, but the transitional one, when Bob is changing his intentions. Two quality indicators can help determine the impact of different network configurations on Alice's ability to assess her confidence in Bob in such a case. Delay tells how long it takes for her to react to the change while volatility describes how well she is able to track the new level of confidence.

The concept of removals defines four types of interactions: non-removed, once-, twice- and level-removed. Each type requires a different strategy from Alice as it allows her to access different types of information and benefit from them in a different way. Those four types of interaction also affect quality indicators, their expected values, their behaviour, dependence on network parameters, etc.

Of all the network configurations, the level-removed network with a central repository and a centrally processed reputation delivers the best quality of assessment. However, it is also the most vulnerable one, requires modifications to the network and calls for additional requirements regarding honesty and similarity. Of the others, non-removed interaction benefits from very low delay while twice-removed one brings low volatility.

The key problem of networked environment is the ability to determine the honesty and similarity of other agents. Even though the process is essentially similar to the one of determining confidence in a given subject, it is either built on an alternative set of evidence or it may eventually lead to unsolvable paradoxes.

## Bibliography

[Abdul–Rahman2005] Alfarez Abdul–Rahman: A Framework for Decentralised Trust Reasoning. PhD thesis. Available at: http://www.cs.ucl.ac.uk/staff/F.AbdulRahman /docs/thesis-final.pdf. 2005.

[Anderson2001] Ross Anderson: *Security Engineering: A Guide to Building Dependable Distributed Systems*. Chichester: John Wiley & Sons, Ltd. 2001.

[Anderson2005] Steve Anderson et al.: Web Services Trust Language (WS–Trust). Available at: http://specs.xmlsoap.org/ws/2005/02/trust/WS–Trust.pdf. 2005.

[Axrelrod1984] Axelrod R. M. *The Evolution of Cooperation*. New York: Basic Books, 1984.

[Barabasi1999] Albert–László Barabási, and Albert Reka: Emergence of Scaling in Random Networks. *Science*, 286: 509–512. 1999.

[Battison2006] Stefano Battison, Frank E. Walter and Frank Schweitzer: Impact of Trust on the Performance of a Recommendation System in a Social Network. In *Proc. of Fifth Int. Conf. on Autonomous Agents and Multiagent Systems AAMAS-06. Workshop WS8 – Ninth Int. Workshop on Trust in Agent Societies*. Hakodate, Japan. 2006.

[Bearly2004] Terry Bearly and Vijay Kumar: Expanding Trust Beyond Reputation in Peer–To–Peer Systems. In *Proc. of the Database and Expert Systems Applications, 15th Int. Workshop on (DEXA'04)*. pp. 966–970. 2004.

[Blaze2003] Matt Blaze et al.: Experience with the KeyNote Trust Management System: Applications and Future Directions. In *Proc. of First Int. Conf. on Trust Management iTrust 2003*. Berlin: Springer. LNCS 2692, pp. 284–300. 2003.

[Branchaud2004] Marc Branchaud and Scott Flinn: xTrust: A Scalable Trust Management Infrastructure. In *Second Annual Conf. on Privacy, Security and Trust (PST 2004)*. Fredericton, New Brunswick, Canada. 14–15 October, pp. 207–218. NRC 47411. 2004.

[Burt2001] Ronald S. Burt: Bandwidth and Echo: Trust, Information, and Gossip in Social Networks. In James E. Rauch and Alessandra Casella (eds): *Networks and Markets*. New York: Russell Sage Foundation. 2001.

[Castelfranchi2000] Cristiano Castelfranchi and Rino Falcone: Trust is Much More than Subjective Probability: Mental Components and Sources of Trust. In *Proc. of the 33rd Hawaii Int. Conf. on System Sciences (HICSS2000)*. Vol. 6. Available at: http://www.istc.cnr.it/T3/download/Trust–more–than–probability.pdf. 2000.

[Clauset2005] Aaron Clauset and Cristopher Moore: Accuracy and Scaling Phenomena in Internet Mapping. *Phys. Rev. Lett.* 94, 018701. 2005.

[Dellarocas2003] Chrysanthos Dellarocas and Paul Resnick: Online Reputation Mechanisms: A Roadmap for Future Research. Available at: http://ccs.mit.edu/dell/papers/symposiumreport03.pdf. 2003.

[Dennett1989] D.C. Dennett: *The Intentional Stance*. Cambridge, MA: MIT Press. 1989.

[Dimitrakos2003] Theo Dimitrakos: A Service–Oriented Trust Management Framework. In R. Falcone et al. (eds): *AAMAS 2002 Workshop on Trust, Reputation . . .* LNAI 2631, pp. 53–72. 2003.

[Dimitrakos2004] Theo Dimitrakos et al.: Towards a Trust and Contract Management Framework for Dynamic Virtual Organisations. Available at: http://epubs.cclrc.ac.uk/bitstream/701/E2004-305-TRUSTCOM-OVERVIEW–FINAL%5B1%5D.pdf. 2004.

[Dragovic2003] Boris Dragovic, Steven Hand, Tim Harris, Evengelos Kotsovinos and Andrew Twigg: Managing Trust and Reputation in the XenoServer Open Platform. In *Proc. of iTrust 2003*. P. Nixon, S. Terzis (eds): *Trust Management 2003*, LNCS 2692, pp. 59–74. 2003.

[Drogovtsev2003] S. N. Dorogovtsev and J. F. F. Mendes: *Evolution of Networks: From Biological Networks to the Internet and WWW*. Oxford: Oxford University Press. 2003.

[Ellison1999] Carl Ellison et al. SPKI certificate theory – IETF RFC 2693. Technical report. *The Internet Society*, September. 1999.

[Erdos1960] P. Erdos and A. Renyi: *Random Graphs*. Publication of the Mathematical Institute of the Hungarian Academy of Science, 5, pp. 17–61. 1960.

[Fernandes2004] Alberto Fernandes et al.: Pinocchio: Incentives for Honest Participation in Distributed Trust Management. In C.D. Jensen et al. (eds): *iTrust 2004*, Berlin: Springer. LNCS 2995, pp. 63–77. 2004.

[Fullam2005] K. Fullam, T. B. Klos, G. Muller, J. Sabater, A. Schlosser, Z. Topol, K. S. Barber, J. S. Rosenschein, L. Vercouter, and M. Voss. A Specification of the Agent Reputation and Trust (ART) Testbed: Experimentation and Competition for Trust in Agent Societies. In *AAMAS*, pp. 512–518. 2005.

[Gambetta2000] Diego Gambetta: Can We Trust Trust? In Diego Gambetta (ed.): *Trust: Making and Breaking Cooperative Relations*, electronic edition, Department of Sociology, University of Oxford, Chapter 13, pp. 213–237. 2000. http://www.sociology.ox.ac.uk/papers/gambetta213–237. 2000.

[Gerck2002] Ed Gerck: Trust as Qualified Reliance on Information. In *The COOK Report on Internet*, x(10). ISSN 1071–6327. Available at: http://nma.com/papers/it–trustpart1.pdf. 2002.

[Golbeck2004] Jennifer Golbeck and James Hendler: Reputation Network Analysis for Email Filtering. In *Proc. of the First Conf. on Email and Anti-Spam*, 30–31 July. Mountain View, California. 2004.

[Grandison2003] Tyrone Grandison: Trust Management for Internet Applications. PhD thesis, University of London. 2003.

[Huang2006] Jingweni Huang and Mark S. Fox: An Ontology of Trust – Formal Semantics and Transitivity. In *Proc. of Eight Int. Conf. on Electronic Commerce ICEC2006*. Canada, pp. 259–270. 2006.

[Jones2001] Andrew J. I. Jones and Babak Sadighi Firozabadi: On the Characterisation of a Trusting Agent – Apects of a Formal Approach. In Cristiano Castelfranchi and Yao-Hua Tan (eds): *Trust and Deception in Virtual Societies*. Dordrecht: Kluwer Academic Publishers. pp. 157–168. 2001.

[Josang2001] Audun Josang: A Logic for Uncertain Probabilities. *Int. J. of Uncertainty, Fuzziness and Knowledge–Based Systems*, 9: 279–311. 2001.

[Josang2006] Audun Josang, Stephen Marsh and Simon Pope: Exploring Different Types of Trust Propagation. In K. Stolen et al. (eds): *iTrust 2006*, Berlin: Springer. LNCS 3986, pp. 179–192. 2006.

[Kamvar2003] Sepander D. Kamvar, Mario T. Schlosser and Hector Garcia-Molina: The Eigentrust Algorithm for Reputation Management in P2P Networks. In: *WWW2003*, 20–24 May. Budapest, Hungary. 2003.

[Klos2006] Tomas Klos and Han La Poutre: A Versatile Approach to Combining Trust Values for Making Binary Decisions. In: K. Stolen et al. (eds): *iTrust 2006*, LNCS 3986, Berlin: Springer. pp. 206–220. 2006.

[Marsh1994] Stephen P. Marsh: Formalising Trust as a Computational Concept. PhD thesis, University of Stirling 1994. Available at: http://www.nr.no/~abie/Papers/TR133.pdf. 1994.

[Milgram1967] S. Milgram: The Small World Problem. *Psychology Today* 1, (61). 1967.

[Mui2002] Lik Mui et al.: A Computational Model of Trust and Reputation. In *Proc. of 35th Hawaii Int. Conf. on System Science (HICSS)*. IEEE Computer Society Press, USA. 2002.

[Page1998] Larry Page, Sergey Brin, R. Motwani and T. Winograd: *The PageRank Citation Ranking: Bringing Order to the Web*. Stanford Digital Library Technologies Project. 1998.

[Patrick2002] Andrew S. Patrick: Building Trustworthy Software Agents. *IEEE Internet Computing*, Nov/Dec: 46–53. 2002.

[Pennock2002] David M. Pennock, Gary W. Flake, Steve Lawrence, Eric J. Glover, and C. Lee Giles: Winners Don't Take All: Characterizing the Competition for Links on the Web. In *Proc. of the National Academy of Sciences*, 99(8): 5207–5211. Available at: http://www.pnas.org/cgi/reprint/99/8/5207.pdf. 2002.

[Resnick1994] Paul Resnick et al.: GroupLens: An Open Architecture for Collaborative Filtering of Netnews. In *Proc. of the 1994 ACM Conf. on Computer Supported Cooperative Work*. Chapel Hill, North Carolina, pp. 175–186. 1994. Available at: http:// delivery.acm.org/ 10.1145/200000/192905/p175–resnick.pdf? key1=192905&key2=8948 234411&coll=portal &dl=ACM&CFID=68878040&CFTOKEN=74261221. 1994.

[Twigg2003] Andrew Twigg: A Subjective Approach to Routing in P2P and Ad Hoc Networks. In *Proc. of First Int. Conf. on Trust Management iTrust 2003*. Berlin: Springer. LNCS 2692, pp. 225–238. 2003.

[Wang2003] Yao Wang and Julita Vassilieva: Trust and Reputation Model in Agent–to–Agent Networks. Available at: http://www.cs.usask.ca/grads/yaw181/publications/120_wang_y.pdf. 2003.

[Yao2003] Walt Teh–Ming Yao: Fidelis: A Policy–Driven Trust Management Framework. In *Proc. of First Int. Conf. on Trust Management. iTrust 2003*. Berlin: Springer. LNCS 2692, pp. 301–317. 2003.

[Ziegler2005] Cai–Nicolas Ziegler: Scalable Trust Propagation Models. *Information Security Bulletin*, July (10): 223–232. 2005.

[Zimmermann1994] Phil Zimmermann (ed.) *PGP User's Guide*. Cambridge, MA: MIT Press. 1994.

# 10

# Attacks on Confidence

Nusquam tuta fides
(Nowhere is trust assured)

Virgil, *Aeneid*

## 10.1 Introduction

Not all of Alice's neighbours are wearing white hats [Bono2000]. Some may actually try to alter her assessment of her confidence in Bob in a specific way – increase, decrease, modify the type of reasoning, etc. If we want to understand the risks that Alice is facing, we should enter the area of security of confidence-based systems ('security of confidence') where we may consider potential attacks and their implications.

The problem of attacks (and more generally of deception) can be addressed and structured in many ways. For example, [Falcone2001a] identifies the rich set of deceiving practices in competitive games while some types of attacks are listed, e.g. in [Quercia2006a]. Here we will combine the structure given by the complexity-based model of confidence with an approach known from computer security. We will identify assets that can be compromised. Threats (potential attacks of different feasibility) will be categorised depending on protection categories: we would like to protect integrity, confidentiality and availability of assets. Finally, some common or interesting forms of attacks will be presented, to illustrate possible vulnerabilities.

Security of confidence should not be confused with the basic security of Alice or of the network of her neighbours. If the system is confidence-based, then there is a new exposure area with vulnerabilities characteristic of the process of the confidence assessment. Compromising basic security is not discussed here, even though attacks against such security may lead to compromising confidence. For example, a compromised neighbour can be used to submit a false reputation rating while a compromised communication link may allow evidences to be monitored and altered. However, if the particular threat works by legally injecting dishonest members into the community to skew Alice's expectations, then regular security measures will not help. This chapter concentrates on the latter.

*Trust, Complexity and Control: Confidence in a Convergent World*   Piotr Cofta
© 2007 John Wiley & Sons, Ltd

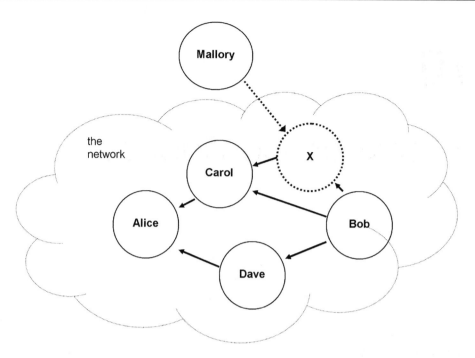

**Figure 10.1**   Hypothetical attack configuration

The taxonomy of threats and the list of attacks presented here apply equally well to human societies, to humans communicating through digital media and to societies of autonomous agents. There may certainly be differences in implementation, in feasibility of particular forms of attacks or in their range and impact. Examples freely mix different forms of networks, from the entirely social to the entirely digital. Common threads and re-usage patterns can easily be identified.

In this chapter we are introducing a new dramatis persona: Mallory. Mallory is an attacker, a manipulator who from the outside alters the way Alice, other agents and the network works. Mallory lives outside of Alice's network (such a network includes Bob, Carol, etc.), but he (or his allies) may join the network if allowed to do so. Mallory's intention is to modify Alice's assessment of her confidence in Bob – not necessarily improve or decrease it, but potentially also e.g. prevent Alice from having any opinion about Bob at all. Figure 10.1 demonstrates the hypothetical configuration where Mallory, himself outside the network, has managed to insert his subordinate (agent X) as an alternative connection between Carol and Bob.

This chapter is divided into two parts. First, the taxonomy of threats is introduced to deliver a structure that may guide the analysis of threats. The second part contains descriptions of various attacks that are facilitated by the introduction of the assessment of confidence.

## 10.2 Taxonomy of Threats

The structure of the system that will be considered here is dictated by the complexity-based model of confidence and is presented in Figure 10.2. Key assets are of course Alice and Bob: an agent that is assessing confidence and the one that is being assessed. The configuration of a network, including all the other agents is an asset as well, as modifications may alter Alice's

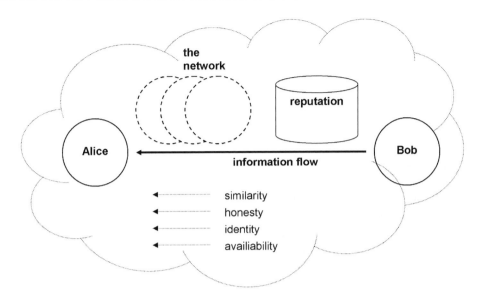

**Figure 10.2** The architecture of assets

assessment. Similarly, the repository, that is present in some architectures, is yet another asset. Finally, the information flow represents yet another asset that should be protected on five different levels that are related to different frames of discourse, as specified by enablers of the model.

Figure 10.3 combines different security-related constructs. The system is presented through the structure of its assets (as discussed above), while its security protects integrity, confidentiality and availability of those assets. This approach allows us to structure threats in relation to assets and protection categories.

The following is a detailed analysis of threats for different assets and different protection categories. The analysis concentrates on demonstrating how threats to particular assets can impair Alice's ability to correctly assess her confidence in Bob. The wide scope of this taxonomy allows us to discuss threats that might not usually be included in the analysis.

### 10.2.1 Alice

Alice's ability to reason (about confidence) is the critical asset – if Alice is unable to come to a decision about confidence in an unconstrained way, then the whole process of gathering evidence, opinions, building reputation, etc. is meaningless. From this perspective Alice is the single most important point of vulnerability, and potentially the most interesting point to attack.

Attacking Alice's rational thinking may at first seem futile, as the reasoning is an algorithm that works inside her (in her brain, her processor, etc.). Apart from impairing her mental activity (e.g. through selective intoxication) there may seem to be little that can be done from the outside. Even though such impairment is a feasible option (see e.g. bot attacks or some considerations presented at the end of this book), we are more concerned here with remote attacks that can be done through existing information channels.

*Integrity* of Alice's operation is a major concern. If Mallory is able to undermine her reasoning, or to alter the way she conducts such reasoning (alter her internal algorithm), he can set her thinking on the course he may benefit from. Because the reasoning is

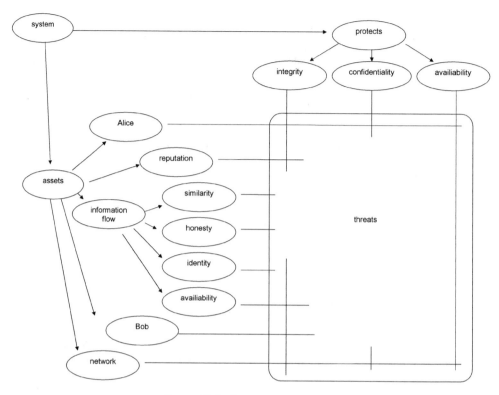

**Figure 10.3** Taxonomy of threats

conducted within Alice, Mallory must either re-program Alice (e.g. by taking over the control of Alice-agent or by instilling some teaching on Alice-person) or he must put Alice into the situation where rational thinking is impossible. For example, if Mallory can instil in Alice a doubt about the nature of confidence (e.g. demonstrating that the concept itself is useless), Alice may entirely abandon the reasoning and possibly decide to blindly trust someone.

Alice can reasonably assume that her decision regarding confidence (and the way she came to such decision) remains confidential. Threats against Alice's *confidentiality* reveal either the outcome or the whole reasoning process to Mallory. Revealing the outcome is a feasible threat while revealing the process may not be feasible, unless Mallory has close access to Alice (so that he can e.g. question her about her decision or inspect the trace of her code). If Alice is unaware of the disclosure, Mallory seems to be a major benefactor, as he can learn about Alice and incrementally discover more feasible vulnerabilities. If Alice is aware of the disclosure (or the possibility of disclosure), she may alter her decision (being afraid of social pressure or reciprocity from Bob) or she may entirely withdraw from acting on the outcome of her reasoning.

Alice's *availability* is probably the easiest thing to be compromised, at least for Alice-human. Complexity is already a major driver of and limitation on her reasoning about confidence, so that increasing the complexity of a situation beyond Alice's capabilities (beyond her disposable complexity) is a very feasible threat. Note that Alice's reaction may differ: under heavy attack she may have no choice but to withdraw (which may not be what Mallory wants her to do), but under moderate attack she may choose to trust rather than to control.

## 10.2.2 Reputation

Alice's reasoning about Bob is conducted within the main frame of discourse. For example, if Alice would like to buy goods from Bob, the main frame of discourse revolves around the problem of being confident in Bob as a seller: whether he will deliver the ordered goods as agreed, how he will handle exceptions and problems (faulty goods, returns), what quality of after-sales service he can provide, whether he will respond to correspondence, etc.

Reputation can be thought of as a perception that an agent creates through his past actions about his intentions and norms [Ruohomaa2005]. It is a collective knowledge about an agent. Reputation is extremely important and quite often is central to the concept of the confidence-based system. By altering Bob's reputation, Mallory can influence Alice's assessment. Reputation is Bob's important asset that facilitates transactions and delivers added value.

Bob's reputation is not within the domain of his direct control but it stays with the community of agents. Reputation has been externalised and detached from Bob. Mallory can compromise Bob's reputation without the need to deal with Bob at all.

Alice decides on her level of confidence on the basis of evidence that arrive from Bob, either directly or indirectly (in the form of opinions from other agents). Specifically, Alice may use Bob's reputation as one (possibly the most important) evidence. Mallory may attack the flow of opinions with the intention to alter the outcome of the process – to make Alice believe in the reputation that is not true.

The use of reputation does not imply that there is a central database in place. There are several possible methods to implement the reputation processing. The most popular one indeed includes a central repository of recommendations from where an algorithm (such as EigenTrust [Kamvar2003]) derives a standardised reputation rating, available to all agents. The repository does not have to be centralised but may be implemented in a distributed version (e.g. [Ingram2005]) and an algorithm does not have to be standardised, but each agent may employ its own, that reflects its own perception of reputation.

Threats to reputation have been studied in depth, specifically in light of the fact that reputation-based systems are widely used in several commercial systems and that reputations have a significant influence on profit margin. Those threats are built on vulnerabilities that are inherent to the reputation and those that are specific to particular algorithms, storage methods and processing. In [Dellarocas2004] there are four variants of attacks on reputation: ballot stuffing (unfairly high rating for Bob), bad-mouthing (unfairly low rating for others – that indirectly increases Bob's rating), negative discrimination (Alice is victimised by a sub-standard service without significant loss to Bob's rating) and positive discrimination (Alice receives preferential treatment in the expectation she will improve Bob's rating). Attacks on reputation may lead even to the monopolisation of the market [Sen2006].

If the reputation-based systems is used as a basis for transactions that have a monetary value, several new forms of threats are viable, specifically if combined with weak identities [Gavish2006]. For example, Ingram [Ingram2005] lists several possible attacks on reputation such as a newcomer attack (where Mallory creates a one-time identity and cheats on transaction), a waiting attack (Mallory builds a reputation and then discounts it in a final transaction), etc. Note that such attacks are not aimed at the reputation as such, but they explore vulnerabilities of the link between the reputation and a financial value of a transaction. In a waiting attack, for example, the reputation as such is not attacked (i.e. it is correctly calculated and processed), but it is used to convince Alice to enter into the fraudulent transaction.

Information **integrity** is compromised if Mallory is able to alter the reputation according to his needs. Mallory may e.g. collude with other agents (dependent on him) to achieve his goal. As the reputation usually assumes independence of agents, collusion alters the dynamics of the reputation system by creating a relatively large group of agents that – in a way concealed from others – decide on their opinions. For example, Mallory may perform the self-recommendation Sybil attack. Mallory may also collude with honest agents (unbeknown to them), by selectively amplifying naturally fluctuated opinions if they support his goal. For example, Mallory and his cronies may repeat (thus amplify) every negative opinion about Bob and sit quietly while hearing positive opinions. In the long run Bob's reputation will suffer even though Mallory did not state anything that is not true.

In several systems there is no threat to information **confidentiality** because information related to confidence is public (at least to certain extent). There are, however, some good reasons why it should not be always public and why Mallory may be interested in accessing this information. First, he can create much better targeted attacks if he knows the information flow – e.g. without this knowledge he cannot amplify opinions that suit him and cannot identify highly connected nodes that are his primary targets. Second, agents may alter their opinions if they know that those opinions might become public – due to a fear of reciprocity, social conformance, etc. Finally, by discovering and then promoting the good reputation of some agents, Mallory may direct to them unexpectedly high traffic, potentially shutting them down.

In general, threats to **availability** result in Alice having less than an optimal amount of information, that leads to her increased uncertainty. If insufficient information about the subject is present, Alice may interpret it as e.g. a lack of competence, inexperience, lack of knowledge, etc., leading to lower perceived confidence in Bob. If Mallory cannot remove information about Bob from the system, he may be able to obscure it by flooding the system with information about others. As Alice's assessment is quite often made on the basis of relative performance, experience, knowledge, etc., Alice may lower her confidence in Bob only because she has relatively less information about him.

### 10.2.3 Information Flow

Threats against the information flow are the classical ones (from the perspective of computer security) – Mallory attempts to alter evidence that reaches Alice, in the expectation that such a differently shaped set of evidence will result in the desired change in her confidence in Bob.

Here, classification of threats naturally follows the classification of protection categories: integrity, confidentiality or availability of information can be attacked. Information flow can, however, be attacked within different frames of discourse (as defined by the model), depending on the outcome that Mallory wants to achieve – forming different classes of threats. If e.g. he is interested in attacking Bob's identity, he may selectively alter information related to his identity only. Different objectives influence the interpretation of threats. Therefore, even though all threats against information flow are quite similar on a technical level (data are altered, added or removed), they are discussed below separately for each frame of discourse.

### Enabler – Similarity

Similarity means that Alice may apply various standards of expectations depending on the perceived category of subjects: e.g. Alice is differently confident about Bob-person and about Bob-software agent, Bob-individual and Bob-representative of a large company, etc. Specifically, she is more likely to be confident about someone that is similar to her than about someone entirely different.

Threats to similarity alter Alice's perception of Bob's similarity by altering the flow of information that allows Alice to make a judgement about such similarity. Those threats are not aimed at altering information within the main frame of discourse, i.e. they do not change what Bob says with regard to the subject, but it may change how Bob said it or they may alter the context of it. By altering interpersonal clues that arrive from Bob (as well as other's opinions of such clues), Bob is perceived as a different entity, with Alice being more or less confident in him.

Threats to *integrity* of similarity aim at altering the perceived category of an agent, e.g. Bob-person is presented as Bob-agent (by removing clues about his human origin) or possibly he is presented as a private person while he is an employee of a large company. The digitally-mediated communication specifically facilitates those threats, by obscuring and restricting available communication channels. As 'on the Internet nobody knows you're a dog', it is relatively easy for an agent to pretend that he is someone else.

*Confidentiality* threats work by revealing Alice's expectations regarding similarity. As the determination of similarity is the element of Alice's reasoning, disclosing particularities of this reasoning makes Alice more vulnerable to attacks. For example, by deploying agents of different profiles, Mallory can determine which types of agents are mostly trusted by Alice and then use such agents to instil false trust in Alice.

By de-personalising communication Mallory can create threat to *availability*. Information flow that is devoid of personal clues prevents Alice from determining similarity, thus making it potentially easier to conduct other attacks. If – for example – Alice's communication with Bob-person misses individual personal clues, Bob can be easily replaced by someone else (e.g. by Mallory).

## Enabler – Honesty

When it is not possible for Alice to directly interact with Bob, she must rely on information that she receives from others. As long as the other agents are honest, Alice has only to cope with natural problems created by the network architecture: increased delay, undesired noise, possibility of hysteria, information decay, etc. However, if Mallory (or a group of colluding agents directed by Mallory) decide to behave in a dishonest manner, Alice faces a much more serious problem: she cannot fully rely on the information she receives. An attack on honesty becomes an attack on the quality of information.

The main purpose of attacks on honesty is the desire to influence Alice's perception of honesty to the extent where she will be more confident about the honesty of an attacker than about other agents. Assuming that Alice makes her decision on the basis of the perceived level of honesty, this may lead to Alice choosing to listen to Mallory – at her peril. Attacks on honesty do not aim to change evidence or opinions about Bob or Mallory with regard to the main subject but they try to change Alice's perception of the value of agents as sources of information.

Attacks on *integrity* allow Mallory to alter Alice's perception of the honesty of a particular agent. If, for example, Mallory can create a perception that such an agent is particularly honest (e.g. by manipulating other's opinion about him), Alice may be willing to attentively listen to opinions provided by such an agent. If this agent is colluding with Mallory, then Mallory has already succeeded in creating an efficient channel to influence Alice.

*Confidentiality* of honesty does not seem to require protection at all – usually one's honesty is a piece of public information that can be used by agents to select opinions from honest agents. However, there may be reasons why agents would prefer to keep their honesty confidential and why the breach of confidentiality may be damaging. The most important reason is the potential lack of resources to handle requests for opinions. As honest agents (e.g. Carol) are expected to attract more traffic than dishonest or unrated ones, Carol faces the burden of handling such traffic – or refusing to give an opinion. In the case of some systems (e.g. [Seigneur2005]), providing an opinion may actually cost Carol some of her reputation and it may be considered too expensive. However, blank refusal is not an answer as it decreases the perceived level of trust in Carol. On balance, Carol may be willing to keep her honesty rating semi-private and disclose it to the limited number of agents. Making them public may ruin Carol.

By selectively removing or deferring information related to honesty, Mallory attacks *availability* of honesty. Alice receives less information about honesty so that her assessment of the honesty of others is created on a less than optimal basis. In an extreme case, Mallory can withhold all information related to honesty so that Alice will perceive all agents as similarly honest (or dishonest, depending on her stance).

## Enabler – Identity

Attacks against identity are not only associated with confidence-based systems. In fact, such attacks are an unpleasant fact of everyday life, mostly in the form of identity theft. Those attacks can be directed against different agents – Alice, Bob, other members of the network, but their common characteristic is the fact that they aim at identity, not at other properties of an agent.

From the perspective of confidence-based systems, the value of identity is strongly associated with reputation and attacks on identity can be used as substitutes of attacks on reputation. For example, if the evaluation of reputation is held centrally, then it may be more feasible to attack the identity of a highly reputable agent than to attack the central repository. The goal of the attacker may be to take control of the identity of a good reputation and possibly to capitalise on potential financial gain [Resnick2006].

*Integrity* of identity is preserved when it is only the agent that has created the identity (his presentation) who is controlling it. Attacks may come in several forms that decrease or eliminate this control: identity can be copied (so that two agents use the same identity), it can be altered (so that the original agent is not longer the only one that controls it), hijacked (so that Mallory disowns the original agent from his identity), multiplied (if Mallory creates several identities that are linked to him), etc. The most feasible attack is to copy a presentation identity ('identity theft'), as presentations are routinely exposed to outsiders. If there is a reputation attached to such identity, Mallory can use such a reputation for his own gain.

Attacks on *confidentiality* of identity are similar to attacks on privacy: the connection between different presentations is revealed or the nature of a 'true self' is no longer protected by presentations. As a result of such an attack, reputation may be damaged (if one of the cross-linked identities has a significantly lower reputation), or an agent may withdraw from one or more of his identities (to protect privacy), thus abandoning the already developed reputation.

Hiding an agent against his will represents an attack on his *availability*. Availability of identities can be attacked e.g. by removing Bob from the directory service (if there is one) or by altering results of search queries. If Bob is hidden from Alice's view, then Alice, unaware of his existence, may turn to other actors, possibly less reputable but more visible (e.g. to Mallory).

The link between weak identities and threats to confidence-based systems is well established. In fact [Ingram2005], if identity is not adequately protected, then the system is always susceptible to Sybil attack. Certifications, almost-permanent identities, entry fees, etc. can be used to add protection to identities or decrease incentives to create multiple identities.

## Enabler – Availability

The purpose of attacks against availability is to increase Alice's uncertainty by decreasing the number of available evidence. Attacks are not aimed at altering Alice's assessment of confidence, similarity, honesty or identity, but at making Alice less certain about the outcome of such an assessment. If the level of certainty is used subsequently by Alice to determine confidence (e.g. by influencing the weighting of different opinions), then of course the final outcome will suit Mallory's needs – to change Alice's confidence in Bob.

In a simple case, compromising the *confidentiality* of evidence provides Mallory with information that he can use to plan attacks while compromising the *integrity* allows him to arbitrarily alter data or inject false data. For example, Mallory may conduct a random attack, distributing random information in the system. The existence of such randomness will decrease Alice's certainty regarding any information available. However, the more interesting attack on confidentiality comes from Alice's reaction on knowing that data have been compromised – regardless whether they have been indeed compromised. Alice may reject such data or treat them with suspicion in the expectation that they are available to potential attackers or that other agents may not have been entirely sincere in their opinions knowing that data may become public or altered. Such behaviour is in fact a version of a Byzantine problem (e.g. [Bishop2005]) known in computer security.

Attacking information *availability* is the simplest method to decrease certainty (not surprising, considering that the name of the enabler and the name of the risk are identical). Less available data should make Alice more suspicious about the level of certainty that she can achieve. If, for example, Mallory is able to remove the majority of opinions about Bob, then Alice will end up being significantly less certain about her confidence. The similar impact on uncertainty can be also achieved if data are intentionally discounted before reaching Alice, e.g. by delaying delivery.

## 10.2.4 Network

Alice is a part of a network. Even if Alice and Bob are non-removed (they can communicate directly with each other), Alice's decision quite often depends on opinions, recommendations or evidences provided by other agents. Attacks against the network (discussed e.g. in [Abdul-Rahman2005]) are aimed at the topology and operation of the network of agents, not at the operation of individual agents. Those attacks do not directly alter information that flows within the network, but they attempt to shape this information (for the benefit of an attacker) by altering the configuration of a network.

A network can be attacked regardless of its topology, even though certain topologies facilitate different forms of attacks and offer different levels of resistance. For example, [Crucitti2003] scale-free networks are extremely vulnerable to attacks on highly connected nodes, compared to random networks. Attacks against the central repository can also be treated as a category of attacks on the network and are conceptually no different from attacks on a distributed repository or attacks where agents individually collect reputation ratings.

Attacks on network *integrity* modify the network topology without necessarily decreasing the type or amount of information that travels between agents. Mallory may e.g. introduce the large number of agents (as an initial step in Sybil attack), alter the structure of the network (e.g. make it more similar to a scale-free network in expectation of taking over control of one of the well-connected nodes), alienate particular agents (so that e.g. Alice will be surrounded by Mallory's agents), etc. Alterations to network topology may suit his further needs (so that they may be the prelude to other attacks) or may be the goal itself, if they make Alice reconsider her opinions about Bob.

Disclosing the topology of the network is a typical attack on network *confidentiality*. There is a general assumption that even though global network parameters (number of nodes, average density, probability distribution of connection) are public, the detailed topology is not. Discovering details about the topology aids other attacks, either directly or by allowing Mallory to hide himself for an extended period of time. For example, in power networks there are usually only a few well-connected (and highly influential) agents while the majority of agents contribute very little to information flow. Focusing attacks on those agents may lead to significant damage at relatively low cost.

Weakening and destruction [Abdul-Rahman2005] of the network attack its *availability*. By decreasing the confidence of the agents in the network itself, Mallory may reach the stage where agents are removed (or remove themselves) from the network, as they see no value in participating in the network. With sufficient number of agents leaving, the network will eventually collapse. In a simple attack, Mallory may join the network and then, through abusive behaviour, subvert the perceived value of the network. Even though he may be eventually identified and removed from the network, it may be too late to save it from collapse.

### 10.2.5 Bob

Directly attacking Bob is an alternative to attacks on Alice or on the network: if Bob, under attack, becomes untrustworthy (or becomes perceived as untrustworthy), then Alice is likely to spot this and change her assessment. If Bob is attacked, a properly working network and a rational Alice are allies of Mallory: he needs them to convey information about changes that have happened to Bob.

Attacking Bob's *integrity* seems to be slightly confusing, because 'integrity' means both: (1) being an undisturbed whole; and (2) being firm on moral principles. Certainly, if Bob's second integrity is compromised (e.g. Mallory corrupts him) then Alice is quite right to stop trusting Bob. With some social engineering this may be quite a feasible attack, specifically that Mallory does not have to compromise Bob's integrity – it is sufficient to make him behave as if it has been compromised. However, Mallory may also decide to attack the first meaning of integrity (which is the one usually associated with the analysis provided here): he may interfere with the way Bob behaves and thinks. For example, Mallory may become a devious teacher that alters Bob's perception of the world (thus his intentions) or – if Bob is a software agent – Mallory can re-program Bob.

Disclosing information about Bob should benefit Alice so that attacking Bob's *confidentiality* does not seem to be a viable threat. However, there is one type of information that Bob should keep to himself: information that may lead Alice not to assume the intentional stance, to stop believing that Bob has intentions and acts upon them. If Mallory provides Alice with evidence that ruins this perception of Bob, then Alice should no longer employ confidence as a method to deal with Bob. For example, if Mallory demonstrates that Bob behaves randomly (e.g. that Bob is insane or that Bob's algorithm is flawed), then Alice is no longer in a position to trust or control Bob. Similarly, if Mallory is able to deliver the plausible and simple explanation of Bob's behaviour, then Alice can assume a simpler stance, such as a physical or design one [Dennett1989].

Alice can see only part of Bob's overall activity – the part that is somehow related to her future. By attacking Bob's *availability*, Mallory can decrease the size of this part. For example, Mallory can ask Bob to do some unrelated work or he may impair his working conditions or – if Bob is a software agent – may run several simultaneous processes on the same platform, thus decreasing Bob's share of the processor. Whatever the case, Alice will observe a delayed response time, a potentially lower quality of responses or maybe even a reduction of services – all these can be interpreted at least as signs of incompetence (thus affecting trust) or signs of unsatisfactory experience (affecting control).

## 10.3 Attacks

The list of attacks is non-exhaustive and is used mostly to illustrate differences between attacks on confidence and attacks on core security features. Attacks are not provided in any specific order and the selection of attacks intentionally goes for variety, not for an in-depth analysis. A certain overlap of categories may be visible, as an attack seldom follows only one threat or exploits only one vulnerability. Attacks may require social engineering or technical expertise, depending on the nature of the system that is being attacked, but the nature of a particular attack remains unchanged.

Whenever possible, references are provided to a more exhaustive analysis of particular attacks, and (even more important) to methods to protect against such attacks.

## 10.3.1 Information Overload

There are several forms of denial of service attacks that are possible within the system. The following is an information overload attack on Alice that can easily be performed using social engineering. The attack uses information overload to compromise Alice's availability and potentially her integrity (depending on the severity of attack).

Complexity is a driver of and a limitation on Alice's decision about confidence and increasing the complexity of a situation is a viable threat. Alice can 'trade' between control and trust depending on her disposable complexity. This can be used by Mallory if his intention is not to let Alice use control instruments – indirectly impairing the integrity of her reasoning and preventing her from using some of the available evidence.

Let's assume that Alice talks directly with Mallory (there are no intermediaries). Mallory can pretend to be trustworthy (e.g. by faking evidence of trustworthiness) but is afraid that control may reveal his real nature. For example, Mallory-salesman dresses smartly and talks nicely but his verifiable public reputation is poor (e.g. he cheated his previous customers). His intention therefore is to drive Alice to trust him rather than control. For that purpose Mallory can artificially inflate the complexity that she is dealing with, e.g. by increasing the information load (e.g. speaking fast and providing a significant amount of information). Even though all the information that Mallory provides is honest and valid, Alice will not be in a position to process them. Alternatively, he may place her in a position to respond rapidly (again, increasing her information load).

If Alice faces increased complexity, then she may withdraw (which in effect is a variant of a denial of service attack) or she may decide to go for trust (which is associated with lower complexity). If the latter is the case, Mallory wins by modifying her behaviour in a way that suits him.

## 10.3.2 Starvation

Starvation and hiding are simple attacks against availability, aimed at increasing Alice's uncertainty. In the former case Alice is deprived of information, in the latter case Bob is unable to send evidence. Depending on the severity and details of attacks, Alice may be entirely cut off from all communication with the world or with Bob (so that she is able only to use historic data, leading to a rapid increase in uncertainty). Alternatively, she may receive a very limited set of data, or significantly delayed data, insufficient to deliver assessment at the necessary level of certainty.

Let's consider a situation where Alice and Bob exchange encrypted and signed e-mails and Mallory has been able to intercept control over the mail server. Mallory cannot alter the e-mails but he can selectively remove them or he may delay their delivery. In order to avoid suspicion, Mallory decides to delay communication between Alice and Bob within the limits of the e-mail system, so that mail is delivered next day rather than the next minute. This strategy decreases the intensity of communication within a single mail thread, thus increasing Alice's uncertainty about Bob. In addition, every piece of information provided by Bob is already delayed, hence less relevant to the current context (specifically if the context changes rapidly). This further deteriorates Alice's certainty about her confidence in Bob. Even if Mallory does not change anything in the e-mails, Alice may not be willing to cooperate with Bob.

## 10.3.3 Masquerading

The very feasible attack on similarity comes in a form of masquerading. Mallory knows that Alice is more likely to be confident about someone who is similar to her than about someone who is very different. For example, in a digital world, Alice may be more likely to be confident about another person than about a computer. This can be exploited by Mallory who can masquerade Bob as an entity that is similar to Alice (in several different ways) in order to increase her confidence (mostly trust) in him.

Masquerading is actually a quite popular practice and quite often is not considered an attack as all. Computers pretend to be human-like (e.g. by providing synthesised voice and the image of automated assistants), operators in outsourced call centres pretend to be located in the country of a caller, salesman try to form an emotional (yet false) bonding with customers, etc. It is actually quite often a recommended practice and studies show that the relationship improves indeed – customers tend to trust masqueraded agents more.

What makes masquerading an attack is the fact that it distorts Alice's reasoning process. By presenting himself as someone (or something) else, Mallory receives trust that he may have not deserved. What demonstrates that this is an actual attack, is an angry reaction upon discovering the discrepancy between pretended and real identity. The fact that it is a popular practice does not preclude the notion of Alice being attacked. There is a fine line between richer interaction and false interaction, and false interaction is a form of attack.

## 10.3.4 Misrepresentation

Alice generally assumes that a single person is not represented within a given context by more than one identity. If this is not the case (e.g. Mallory is known as Mallory and – under a different name – as Bob), then Alice may have problems correctly assessing the similarity, as she incorrectly assumes that she is dealing with two separate individuals, not with a single one. Similarly, the large company may e.g. set up a separate identity where it is pretending to be a private person (popular e.g. with advertisement-driven blogs). Again, the main identity is disassociated from an alternative one and again Alice may be driven into the belief that she is dealing with a different type of entity.

Misrepresentation can be used to obfuscate the size or the nature of the operation. For example, if eBay's small seller was promoted to PowerSeller, he may believe that this recognition may put off his clientele who expected a small collector, not a professional seller. Therefore, he may create a set of alternative accounts and – even though he is selling from the same stock – keep volumes low to preserve the casual trading attitude. However, his action deceives Alice who may have different behavioural expectations regarding casual sellers and professionals.

Note that this threat is classified here as a threat against similarity, not against identity. If the system permits several identities, then just having such identities does not count as an attack – this is only a consequence of the system's architecture.

## 10.3.5 Reputation Takeover

Reputation takeover is a popular identity-related threat, known also as identity theft. Mallory may be able to copy Bob's identity without Bob's knowledge so that he can pretend to be Bob in a relationship with other agents. For as long as Bob does not realise what has happened, Mallory can benefit from the assumed identity by using Bob's reputation for his own purposes. If Bob's reputation is high (and this is usually the reason why Bob has been attacked), then Mallory can benefit from the potential financial advantage. Alternatively, Mallory may use Bob's good reputation to perform an attack (e.g. defamation) on another agent, at the expense of Bob.

Even the simplest identity thefts (such as the credit card theft) are in fact reputation takeovers, as Mallory's primary interest is in using the credit rating (e.g. spending limit) for his own benefit, at the same time damaging Bob's reputation. Complex takeovers (e.g. fake websites in connection with phishing, brand hijacking, etc.) utilise the more sophisticated version of the same concept.

Such attacks can be identified (usually too late) either by Bob or by the party that manages his identity (e.g. bank) on the basis of a discrepancy between Bob's recollections of his own activities (his identity) and what others remember about Bob. Alternatively, others may observe and react on discrepancies between Bob's usual behaviour and Mallory's behaviour while he is pretending to be Bob (profiling), which may allow early identification of the attack.

## 10.3.6 Unfair ratings

Systems that build agent reputation on the basis of ratings (reputation, feedback, opinions) are subject to unfair ratings attacks, where Mallory can artificially alter Bob's reputation.

Successful attacks on the rating system usually require certain forms of collusion where Mallory is able to command a large number of agents. If the cost of creating new agents is low, Mallory may perform attacks on rating using e.g. Sybil attack [Douceur2002] by creating a large number of dependent hostile agents and introducing them into the network so that they disseminate his opinions in an orchestrated manner. As the weight of an opinion is quite often associated with the number of agents that deliver such as opinion, Mallory's opinion can be perceived as a strong one, overwhelming opinions from honest agents. In addition, dependent agents may issue positive opinions regarding their own honesty, further increasing the impact of their opinions on Bob's rating.

The architecture of such as attack is shown in Figure 10.4. Mallory has been able to create four dependent agents and they deliver positive opinions about Bob to the central repository

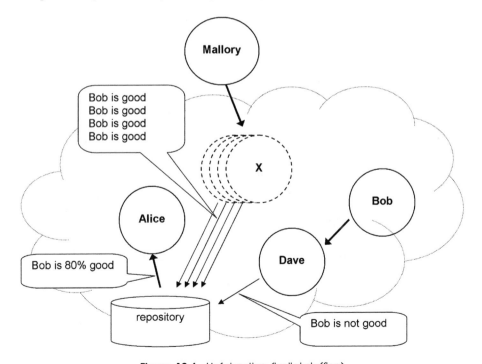

**Figure 10.4** Unfair rating (ballot stuffing)

(without actually dealing with Bob at all). Dave, the only one who really dealt with Bob, has a negative opinion, but when Alice checks Bob's rating, the result is 80 per cent of positive feedback.

There are several counter-measures that can be applied to make the system more resistant to unfair ratings. Pinocchio [Fernandes2004] is an example of a system that attempts to reduce incentives for cheating by identifying agents that consistently hold opinions that are significantly different from the average. Controlled anonymity, median filtering or frequency filtering [Dellarocas2004] can be used as well. Certain improvements to the rating mechanism (e.g. recursive verification of the credibility of the opinion provider [Xiong2004]) can increase its resistance to some variants of attacks. An alternative scheme exists where trust takes on a form of pseudo-currency, so that providing a rating is not free [Seigneur2005]. The problem of unfair ratings can also be alleviated by comparing private (non-removed) and public evidence [Zhang2006].

## 10.3.7 Dis-reputation

Dis-reputation is a directed attack where Mallory subverts Alice's perception of Bob's honesty. Even though the most efficient form of this attack requires the ability to modify information, Bob can be attacked also by a malicious agent that does not alter information provided by other agents. Further, directed dis-reputation does not require a significant number of colluding agents, but benefits from an understanding of the network topology.

Let's assume (Figure 10.5) that Mallory joins the network (through his agent X) for the purpose to attack Bob and that the reputation regarding honesty of agents is maintained by each agent separately (there is no central repository). Alice listens to Carol and Bob listens to Dave (e.g. because of the existing network topology). Mallory may attack Bob by spreading

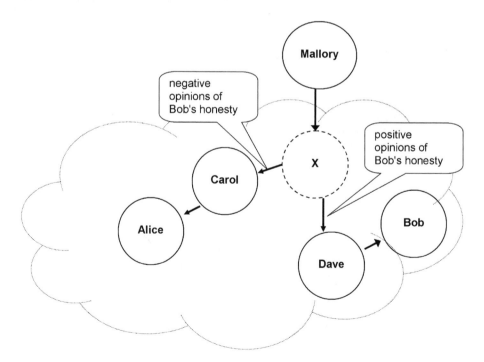

**Figure 10.5** Dis-reputation attack

false information about Bob's honesty to other agents, in the expectation that this information will reach Alice. For example, he may misinform Carol about Bob's trustworthiness. In order to delay detection, Mallory may deliver genuine information to those agents that are likely to forward it to Bob (e.g. to Dave) and false information only to those that will deliver it to Alice.

## 10.3.8 Prejudice

It is possible to create a prejudicial attack that does not interfere with any element of the network, does not require modifications to information nor direct access to agents. The only thing that is needed is Alice's willingness to listen to Mallory (initially at the low level of confidence in his honesty) and his ability to monitor her experiences with Bob. The attack is aimed primarily at compromising Alice's integrity by attacking Alice's reasoning (Figure 10.6).

Here, Mallory offers Alice an explanation of Bob's actions that is plausible (so that Alice can accept it), not falsifiable (so that it cannot be verified) and subverts his true intentions (so that it is an attack). For example, for every positive action of Bob (e.g. as reported by Carol), Mallory offers an explanation that in fact Bob builds his reputation to deceive Alice and take advantage of her confidence in future. The explanation has a sound foundation in experience, cannot be verified (if next Bob's action is positive, then the explanation holds – Bob still builds his reputation to take advantage of Alice somewhere in the future) and hopefully has nothing to do with Bob's intentions. Naturally Bob's failures (that may happen from time to time, e.g. reported here by Dave) further support the explanation. If Alice accepts such an explanation, then she is in no position to be confident in Bob, as she does not believe in continuity or motivation.

Note that even if Mallory may not be regarded initially as very honest by Alice, the belief in his honesty will grow with every failure by Bob, where her reasoning will receive perceived

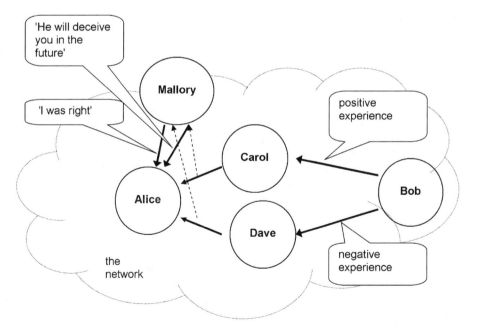

**Figure 10.6** Prejudicial attack

factual reinforcement. In a pattern of self-reinforcing behaviour, Alice may end up not only being not confident about Bob, but also being very confident about Mallory's honesty – something that he can exploit for subsequent attacks.

### 10.3.9 Free Riding

Free riding is an interesting form of attack on the network, as it subverts the perceived value of the network (thus potentially leading to its destruction), usually without breaking any rules, thus being undetected for a long time. Free riding is a behaviour where agents benefit from opinions provided by others yet do not contribute to the overall opinion pool, despite the fact that they have information that may be useful to other agents. In systems with a central repository agents may simply not submit their ratings while in distributed systems agents may not even be willing to forward information that is not useful for them, relying on others to perform this task.

If Mallory is a free rider, then he affects Alice's assessment by limiting the set of evidences and opinions that is available to her while not limiting his own options. This gives him a clear competitive advantage over Alice, both in terms of information availability and resource utilisation. The problem of free riding is specifically acute in mobile ad-hoc networks where the goodwill contribution to the confidence-based data may exhaust the resources of mobile devices. In such cases the free rider benefits not only from free service but also from significantly longer active life (survival).

As long as free riders are in an insignificant minority, the network may compensate for them (e.g. by providing alternative routing paths, distribute opinions, etc.) and Alice's assessments may be good enough. However, the network may start underperforming if there is a high level of free riders, where Alice is deprived of the significant amount of information. Even more damaging to the network is the fact that Alice may realise that free riders decrease her chances for survival, so that – for her own long-term benefit – she must withdraw from the network.

## 10.4 Conclusion

This chapter presents an overviews of threats to confidence-based systems. It provides the systematic framework that uses the model of confidence to determine assets and combines it with classes of attacks that are known from security analysis. The analysis – guided by such a framework – allows a systematic discussion of a wide collection of threats, potentially driven by social engineering, technology or a combination of both.

The list of threats is unfortunately incomplete – the creativity of attackers is quite high. However, by using the framework, we are able to have a more systematic discussion about several threats that may have been overlooked.

Preventing attacks on confidence-based systems can be quite difficult. Some attacks described here have received significant attention and there are known improvements that prevent the most drastic forms of them. However, some vulnerabilities are inherent in the construct of confidence so that only significant modification to the foundation of such systems can minimise the threat of those attacks.

It seems that copying social constructs into the technical domain comes at a price. It may be true that confidence-based systems are less prone to total failure, but it is also true that they are not easily manageable in the traditional sense of this word. Certain level of attacks, abuse and manipulations may be inherently associated with confidence, in the same way as a certain level of deception is visible in human societies. Hopefully, this will not prevent confidence-based systems from being accepted.

# Bibliography

[Abdul-Rahman2005] Alfarez Abdul-Rahman: A Framework for Decentralised Trust Reasoning. PhD thesis. Available at: http://www.cs.ucl.ac.uk/staff/F.AbdulRahman/docs/thesis-final.pdf. 2005.

[Bishop2005] Matt Bishop: *Introduction to Computer Security*. Reading, MA: Addison-Wesley. 2005.

[Bono2000] Edward de Bono: *Six Thinking Hats: Your Success in Business Depends on How Well You Think*. Harmondsworth: Penguin Books Ltd. 2000.

[Crucitti2003] Paolo Crucitti, Vito Latora, Massimo Marchiori and Andrea Rapisarda: Efficiency of Scale-Free Networks: Error and Attack Tolerance. In *Physica A 320*, pp. 622–642. Elsevier Science B.V. 2002. Available at: http://www.w3.org/People/Massimo/papers/2003/tolerance physicaa 03.pdf.

[Dellarocas2004] Chrysanthos Dellarocas: Building Trust Online: The Design of Robust Reputation Reporting Mechanisms for Online Trading Communities. In Wen-Chen Hu et al. (eds): *Advances in Security and Payment Methods for Mobile Commerce*, Chapter VII, pp. 95–113. ISBN 1591403456. London: Idea Group, November. 2004.

[Dennett1989] D.C. Dennett: *The Intentional Stance*. Cambridge, MA: MIT Press. 1989.

[Douceur2002] John R. Douceur: The Sybil attack. *In Proc. of the First Int. Workshop on Peer-to-Peer Systems, IPTPS02 Workshop*, Cambridge, MA. March. 2002.

[Falcone2001a] Rino Falcone, Cristiano Castelfranchi and Fiorella de Rosis: Deceiving in Golem: How to Strategically Pilfer Help. In Cristiano Castelfranchi and Yao-Hua Tan (eds): *Trust and Deception in Virtual Societies*. Dordrecht: Kluwer Academic Publishers. pp. 91–109. 2001.

[Fernandes2004] Alberto Fernandes et al.: Pinocchio: Incentives for Honest Participation in Distributed Trust Management. In C.D. Jensen et al. (eds): *iTrust 2004*, Berlin: Springer. LNCS 2995, pp. 63–77. 2004.

[Gavish2006] Bezalel Gavish: How Can We Reduce Fraudulent Activities on Auction Sites? Keynote speech on Networking and Electronic Commerce Research Conf (NAEC). 2006.

[Ingram2005] David Ingram: An Evidence-Based Architecture for Efficient, Attack-Resistant Computational Trust Dissemination in Peer-to-Peer Networks. In P. Herrmann et al. (eds) *iTrust 2005*, Berlin: Springer. LNCS 3477, pp. 273–288. 2005.

[Kamvar2003] Sepander D. Kamvar, Mario T. Schlosser and Hector Garcia-Molina: The Eigentrust Algorithm for Reputation Management in P2P Networks. In: *WWW2003*, 20–24 May. Budapest, Hungary. 2003.

[Quercia2006a] Daniele Quercia, Stephen Hailes and Licia Capra: TATA: Towards Anonymous Trusted Authentication. In K. Stolen et al. (eds): *iTrust 2006*, Berlin: Springer. LNCS 3986, pp. 313–323. 2006.

[Resnick2006] Paul Resnick: The Value of Reputation on eBay: A Controlled Experiment. *Experimental Economics*, 9(2): 79–101. Available at: http://www.si.umich.edu/~presnick/papers/postcards/PostcardsFinalPrePub.pdf. 2006.

[Ruohomaa2005] Sini Ruohomaa and Lea Kutvonen: Trust Management Survey. In P. Herrmann et al. (eds): *iTrust 2005*, LNCS 3477, pp. 77–92. 2005.

[Seigneur2005] Jean-Marc Seigneur, Alan Gray and Christian Damsgaard Jensen: Trust Transfer: Encouraging Self-recommendations without Sybil Attack. In P. Herrmann et al. (eds.) *iTrust 2005*, LNCS 3477, pp. 321–337. 2005.

[Sen2006] Sandip Sen and Dipyaman Banerjee: Monopolizing Markets by Exploiting Trust. In *Proc. of Fifth Int. Joint Conf. on Autonomous Agents and Multiagent Systems (AAMAS06)*. 2006.

[Xiong2004] Li Xiong and Ling Liu: Reputation and Trust. In Wen-Chen Hu et al. (eds): *Advances in Security and Payment Methods for Mobile Commerce*, Chapter II, pp. 19–35, ISBN 1591403456, London: Idea Group, November. 2004.

[Zhang2006] Jie Zhang and Robin Cohen: A Personalized Approach to Address Unfair Ratings in Multiagent Reputation Systems. In *Proc. of Fifth Int. Conf. on Autonomous Agents and Multiagent Systems AAMAS-06;* Rino Falcone et al. (eds): *WS8: Ninth Int. Workshop on Trust in Agent Societies (TRUST)*, Hakodate, Japan. 2006.

# 11

# Trust-Enhancing Technologies

Would you like my mask?
would you like my mirror?

(Loreena McKennitt)

## 11.1 Introduction

Digital communication is not conducive to those elements that convey trust and may also resist those that convey traditionally understood control. The growing dis-embedding of modern communication [Giddens1988] leads to the situation where evidence required to build trust cannot be delivered over modern communication means. We are therefore twice deserted: not only disconnected from those that we would like to relate to, but also deprived of clues that have been used for a long time to build trust.

Such a situation is the unwanted one. Current approaches tend to explore two different areas. First, we may attempt to widen the communication channel between people in the expectation that such a channel will be more conducive to traditional signals (evidence) of confidence (specifically evidence of trust). We may therefore suggest that e.g. adding direct voice communication to the trading site should result in improving trust between trading parties. Similarly, videoconferencing is supposed to improve the trust relationship in virtual teams [Panteli2001], but its usefulness does not go beyond well-structured meetings that do not require much trust anyway. We can also explore some solutions where emotional states are captured, transferred and communicated over digital media.

The problem with such an approach is that we actually do not capitalise on the strength of the digital communication – its global nature, wide participation, asynchronicity, relative anonymity or automation. Those solutions therefore do not scale. It is of course possible to video-call one trading partner but it is hard to imagine even the mid-size chat room being converted to the video conference without losing several of its obvious advantages. Some forms of interaction (e.g. avatars) are intentionally distant from any direct relationship to the real world persons so that widening the communication channel will not lead anywhere, as participants are likely to be unwilling to disclose their physical location or personality.

*Trust, Complexity and Control: Confidence in a Convergent World*   Piotr Cofta
© 2007 John Wiley & Sons, Ltd

Second, we can replace confidence between people with confidence between a person and an automated agent, e.g. the website, so that the person will interact with the (potentially) sleek and smooth automaton. The solution is a popular one, and indeed the ability to build confidence (specifically to build trust) in complex yet inanimate objects has been explored for some time, specifically in the area of e-commerce websites. This has resulted in several design recommendations and even some vigorously protected patents.

The approach is a justifiable one, in terms of its technical viability, psychological justification [Dennett1989] and acute necessity, with millions of new sites being created every day. However, this approach brings the danger of misplacing trust, with all its negative consequences. There is nothing wrong with a website being smart, sleek and smooth – as long as the person knows that this is an automaton, not a person. If the website pretends to be human ('too human'), it risks being rejected at the first glimpse of non-human behaviour. Similarly, if the site does not evoke any human attributes yet it suddenly presents some (e.g. personalises the content), its risks the alienation of its users.

It is also possible to actually abuse such technologically-driven confidence, in a same way that the unscrupulous salesman creates a false feeling of trust only to eventually rip us off. As we can learn what elements of the digital presentation (such as a website) evoke the feeling of confidence (e.g. [Egger2000]), the clever web designer can actually create the confidence-building e-commerce website for the untrustworthy and uncontrollable merchant. In a vicious cycle, users of such a site will learn not to trust certain clues and will not participate in the e-commerce, thus forcing other merchants into redesigning their websites.

However, the challenge is neither in forcing one party (Alice) to trust nor in requesting trustworthiness from another party (Bob). It is also not in controlling or being controllable. The challenge is to allow Alice to correctly judge Bob – and to see whether and how he responds to it. If Alice find ways to correctly gauge the extent of confidence, trust and control regarding Bob that is justifiable and that matches his behaviour and his trustworthiness, she will be able to determine whether she would like to proceed.

Let us consider the situation where Alice is a remote buyer of an expensive item and Bob is a seller. As Bob requires a certain up-front payment, Alice wants to understand whether she has sufficient confidence in Bob to send money without seeing the actual product (this situation is not as uncommon as one may think). As there is technology between them, they must learn how to interpret the limited set of signals that get through such technology.

However, it is not enough to overcome the limitations provided by the technology. In fact, if technology is limiting us, the simplest solution to overcome its limitation will be to abandon technology in the first place. The challenge that Alice and Bob are facing should be positioned differently: not 'how to overcome the disadvantages of technology-based communication', but 'how to take advantage of the technology to assess confidence'. This approach is the foundation of Trust-Enhancing Technologies.

This chapter contains the description of the concept of Trust-Enhancing Technologies (TET). TET defines a set of technology properties that support and improve the assessment of confidence between people (in general between agents) over digital media. TET is aimed mostly at developing the correct assessment of trust (that suffers more from the introduction of digital communication), but it can also be used to develop the assessment of control and may eventually result in increasing the level of trust over digital media. The description of TET is provided on three levels: principles, requirements and tools. Examples illustrate existing solutions that have several elements of TET.

## 11.2 The Concept of TET

TET attempts to improve the human ability to assess trust, control and confidence while interacting through technology, by defining a set of core technical properties that facilitate

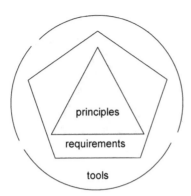

**Figure 11.1**   The structure of TET proposition

such an assessment. Instead of complaining about the destruction caused by technology, TET embraces the technology and makes it essential in the assessment of confidence. Without digital technology, TET will not be possible, as TET is not a projection of known social properties into the digital space, but it is the set of technical properties that are valid primarily in such a space only.

Therefore the proposition of TET is not about creating a false sense of trust where trust is not justified but about the properties of digital systems that facilitate the communication of new types of justified evidence of trust – or the lack of it. TET is also not a tool of enforcement or coercion that forces people into behaviour that is considered socially appropriate, but it is potentially a tool of relational education that may demonstrate the advantages of justified trust.

TET offers the solution that is significantly different from those that are currently used. TET assumes that we should benefit from the digital communication yet we should not easily be deceived by the dis-embedding that it has introduced. The concept of TET builds on the assumption that people, in a characteristically creative way, are able in the long run to estab-lish the proper interpretation of signals that are characteristic of digital communication – that society will embrace the digital communication as a proper method to deal with confidence.

However, instead of waiting for this to happen (which may take quite a long time, consid-ering the rapid development of technology), it is worth exploring this area to identify core technologies that facilitate the assessment of confidence. If such technologies are identified and subsequently embedded in various digital services and systems, their existence may greatly improve the proper assessment of confidence, eventually leading to the growth of it.

TET can be also defined as a technical environment that facilitates the assessment of confidence by selectively attenuating evidence that is relevant to such an assessment while it making costly for untrustworthy agents to mimic such evidence [Bacharach2001].

The TET proposition can be best explained through its onion-like model (Figure 11.1). Three principles of TET form its core foundation. From there, TET defines five basic requirements that should be satisfied by technology platforms. TET also offers the concept of three abstract tools that implement its requirements.

## 11.3 TET Principles

The inner most layer of TET propositions is formed by three inter-related principles of TET (Figure 11.2). All TET-based solutions must obey those three principles.

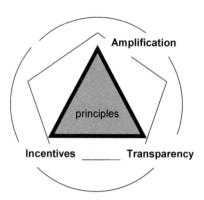

**Figure 11.2**   TET principles

These principles are defined as follows:

1. *Amplification, not alteration.* TET amplifies evidence of confidence from one or more classes of evidences, e.g. by isolating them from the noise level (e.g. filtering out from a stream of other events); aggregating them across large data collections (e.g. by building statistics out of an event log); visualising them for better communication (e.g. a special logo). TET is primarily concerned with evidence of trust but is also conducive to and supportive of evidence of control.

2. *Transparency, not blind trust.* TET does not create new evidence and does not otherwise interfere with the interaction. TET is built on the basis of existing interaction. TET is inherently accountable without the need for a trusted third party: no new evidence is introduced and there is no need to trust yet another intermediary; all evidence is within the reach of the recipient, but it may hard to discover or identify.

3. *Better assessment, not greater confidence.* TET provides reasonable incentives for the behaviour that facilitates the assessment of confidence. TET is not concerned whether the assessed behaviour leads to confidence, trust or neither: it is only the process of assessment that is of concern to TET. TET provides certain barriers to deception while adding a minimum overhead to decent users; TET is not bullet-proof and a certain level of undesired behaviour is possible within TET.

TET does not alter the content of the interaction, but it provides Alice with additional insight into it so that she can assess whether her confidence in Bob is justified. TET does not relieve her of the ultimate responsibility regarding her decision, but it provides her with a richer set of clues that she may wisely use to evaluate her beliefs.

Using the metaphor drawn from electronic engineering, TET is an amplifier and a filter that filters out what is not relevant to confidence and amplifies what is relevant, but what might have been otherwise overlooked, e.g. due to its limited visibility, inaccessibility or the relative complexity of extracting it. TET should equally well amplify signals of confidence as well as signals that relate to the lack of it. TET should, however, filter out signals that are 'fake' in a sense that someone who only pretends to be trustworthy will not be able to abuse TET.

TET does not distort the relationship of confidence. TET is neither the trusted third party nor the reputation system (even though it can be built on either technology). TET does not require the recipient to be confident in anyone (except potentially the sender) in order to benefit from TET. In fact, the recipient does not even have to trust TET, as all the signals that are amplified by TET can be found in the communication – just with significantly more effort.

Let us consider, for illustration purposes, two types of reputation-based centralised systems. Both are producing the average rating in a same way, but the first one 'consumes' feedback responses (thus leaving only the average rating) while the second one stores detailed description of all ratings in the public database. In the latter case (which is TET-conformant) Alice may browse the database and calculate the average rating by herself, even though it may be complicated and time-consuming. In the former case (non-conformant), however, Alice may have to blindly trust the system to perform the calculation correctly.

TET is considered to be a service platform, not a service itself. Users of particular domain specific services (such as e-commerce or e-health) may not see the 'trust gauge' or 'trust seal' provided by TET. Rather, services themselves will use TET in a way that unavoidably will increase the conductivity of confidence between services and users. Hence, TET is no more than the operational property of a system.

Even though TET is designed for inter-personal, digitally-mediated communication systems, it may potentially be used also in relationships between users and fully automated systems (such as Internet shops) or in general between two agents, whether human or non-human. Following earlier considerations, confidence (specifically trust) in such relationships is possible – it is just a different confidence, driven by different expectations and it requires different evidence.

TET does not take any moral stance and does not try to improve the world. It is well known that confidence (specifically trust) can have a positive impact on several areas of our lives (e.g. [Fukuyama1996]). However, technology can be potentially used for or against other people. Therefore, TET does not attempt to improve, enforce or increase the trustworthiness of either party (e.g. by enforcing certain behaviour). It does not create the law of the digital space. The only rationale for TET is to allow Alice to assess her confidence in Bob (specifically his trustworthiness). It is not within the scope of TET to 'improve' Bob or to 'convince' Alice to trust untrustworthy Bob.

Both Alice and Bob may eventually be encouraged by TET to modify their behaviour. For example, TET may encourage Bob to build his reputation through decent deeds. However, it is not the intention of TET to enforce any rules on Bob or Alice – Bob can discard his reputation if he desires so. Specifically, TET will neither force nor encourage Bob to become trustworthy – only to show his reputation so that Alice can better assess him.

Interestingly, it seems that there is a contradiction in the reasoning presented above. We have started from the demonstrated assumption that digital technology is not conducive to evidence of confidence yet we have ended with the expectation that we can build a system where such evidence can be delivered. The explanation of this phenomenon is simple: we are talking about different evidence. The evidence that has been used in a face-to-face communication is mostly lost in a digital communication and we will not attempt to recover it. What we are aiming at is evidence that is not yet widely used, that is inherent in large digital systems and that has the potential to be used in the future, as society will embrace digital communication to an even greater extent. So, a smart dress is out but the proper ISP might be in.

## 11.4 TET Requirements

Not investigating yet how TET can be implemented (or what kind of features TET can provide for services), it is worth listing certain requirements that may potentially shape TET (Figure 11.3). Those requirements are prescriptive but not definitive, as they can be implemented in several different technical manners. For example, stable identities can be

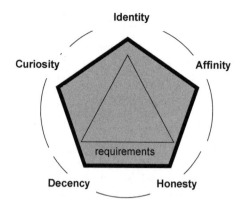

**Figure 11.3**   TET requirements

implemented through PKI schemes, validated e-mail addresses, smart cards or reputation-based identities, depending on the architecture of a particular system. Each implementation may have some advantages and disadvantages that should be compared only from the perspective of the complete set of constructs.

The requirements presented below are intermediaries between inner-circle fundamental principles that are shown above and actual technology choices. They are listed with the aim of establishing the framework where particular technologies can be discussed and their relative merits can be compared. Those requirements should always be considered jointly, i.e. the set of technological choices should satisfy all requirements, not only some of them.

## 11.4.1 Identity

The foundation of any longer-term relationship is the ability to reliably identify the other party. It seems that there is a general agreement that stable identities facilitate confidence by working both as an enabler (by allowing the collection of evidence over the period of time) and a deterrent (by increasing the exit cost and increasing chances of future encounters).

In Alice's evidence-gathering process stable identity guarantes that information about Bob collected over the longer period of time can be identified and consolidated. This applies to all types of evidence – about trust or control, direct and indirect, positive and negative. In order to be reasonably confident about Bob, Alice should be reasonably reassured that Bob remains Bob.

There is a growing research interest in digital identities and the relationship between physical and virtual ones. Anthropological observations [Turkle1997] demonstrate that the ability to create (and destroy) virtual identity is the enabler of the open communication, but it can be also used to hide misdeeds. Specifically, the expectation of short-lived identity results in an incentive to cheat.

Bob may be inclined to retain his identity for as long as he believes that Alice is confident in him (if that is what he wants), but he may be quite strongly inclined to drop (or change) his identity if he starts believing otherwise. Bob may also want to have several identities at the same time and use them for different purposes – potentially to keep some parts of his life separate or to boot his reputation by telling good things about himself under this disguise.

TET is not in the position to prevent all the possible kinds of manipulations regarding identity, but it should be able to demonstrate to Alice that Bob's identity is strong enough in the context of the system. For example, TET may demonstrate that Alice and Bob have been properly authenticated to the system (so that their identities have been verified) or that Bob has been an active member of the community for a long time (so that he has an incentive to retain his identity).

Stable and reliable identity does not imply that such identity must be permanent, that it must be the only identity of a party or that it must be attached to the real identity of a person. Understanding the desire to protect one's identity against abuse, stable identity may become no more than the guaranteed pseudonym. Currently the industry favours strong identities, that are endorsed by third parties (e.g. PKI [Adams2002]) and verified by sophisticated authentication methods, including biometrics or smart cards. The threat that such schemes represent with regard to privacy is significant, even though it can be alleviated by the use of confidence-based authentication schemes described elsewhere in this book.

It is important to understand that TET does not require an identity that is unique or cryptographically strong. The requirement is that from Bob's perspective it should be easier (cheaper, less complicated) to retain the current identity than to create a new one and it should be easier to have one identity then several ones. Such a requirement can be satisfied in many different ways. For example, Bob may face high entry barriers (complex sign-up process, verification, cost) before he can receive his identity. Similarly, he may be deterred by the high exit barrier where by dropping his identity he will lose certain valuable assets (e.g. his good reputation [Jordan2004]). As we want Bob to lock in his identity, we should also provide reasonable measures for him to defend this identity against attacks.

Let's consider as an example the recent (2005/2006) registration of .eu domain names to see how the identity should (and should not) be handled. The assumption is that such domain name should represent the stable identity and that its owner should have incentives to retain the name. The first two phases of the registration created significant entry barriers, both in terms of cost and in terms of complexity so that it can be reasonably expected that those who received .eu domain names during those phases will be inclined to retain them. They also offered quite reasonable legal support in case of domain disputes.

However, the last 'land rush' phase has been characterised by low cost and low complexity, creating less incentives for the prolonged use of those domain names. Unfortunately, there is no way to determine whether the given domain name has been assigned during the first two or during the last phase, so that Alice has no way of knowing whether 'bob.eu' is inclined to retain his name or not. Should there be a separate name space or a relevant indicator in the domain registry, the .eu domain name registration system could have been used as an example that satisfies TET requirements regarding identity.

## 11.4.2 Affinity

The requirement of affinity addresses the need to disclose Bob's cognitive and intentional capabilities. In the case of inter-personal relationships, it is closely related to the understanding of similarity (or the lack of it) between Alice and Bob. Understanding the extent of perceived intentional and cognitive abilities allows Alice to apply an appropriate strategy in her relationship with Bob so that she neither under-estimates nor over-estimates him, but calibrates her expectations accordingly. This is the support for Alice's ability to tell the man from the computer or from the proverbial Internet dog.

The concept of affinity as a prerequisite of trust is visible in the literature. Deutsch [Deutsch1962] and Kee and Knox [Kee1970] have discussed psychological similarity as an important element in the creation of trust. Similarly, Creed and Miles [Creed1996] listed perceived similarity as a requirement for trust. Further, there is a strong relationship between trust and empathy [Feng2004] whereas such empathy can be expected to be established only between subjects that are psychologically similar.

Alice faces an important task to understand who Bob is – not in terms of his identity (that we can conveniently assume for the moment to be guaranteed) but in terms of his intentions and cognition. Trust can be reasonably extended by Alice to Bob if he is able to be trustworthy – otherwise she may be better off with control as there is no point for Alice to extend the notion of trust and trustworthiness, e.g. on devices or systems without sufficient intentional capability. For example, the car should not be considered trustworthy, but only (hopefully) reliable.

The same applies for Bob's cognitive capabilities. If he is e.g. unable to reason or to take care of his future, there is little use in control elements that rely on future punishment. Specifically, if Bob is expected to work according to specific algorithm (e.g. Bob is a software agent), Alice cannot expect anything more from Bob than only following the algorithm. Any deviation from it (potentially even a beneficial one) can be considered a failure, decreasing confidence in Bob.

Empathy [Feng2004] seems to play a very important role in the creation of trust. By assuming emotional similarity between herself and Bob, Alice can expect cognitive and intentional similarity as well – so that she can assume that he is capable of acting in a trustworthy manner and that they both will be able to communicate and read each other's signals. Empathy is not enough to build confidence (Bob may be cheating or may be simply incompetent), but the perception of cognitive and emotional similarity seems to facilitate the usage of trust rather than control.

Affinity is Alice's subjective judgement – she may e.g. believe that men cannot be trusted while dogs and cars can, or that websites must not express too much of the learning capabilities. Potentially, Alice may have several pre-defined constructs – scenarios of how to deal with different agents. Specifically, when she comes to complex systems (such as a company, the Internet, etc.), Alice may feel that taking the intentional stance will provide her with a better explanation of the behaviour of such systems. In this case, Alice can also assume that such systems can be trustworthy.

The affinity requirement within TET is about providing only the assessment what (or who) Bob is, leaving the decision to Alice. As Alice's beliefs regarding trustworthiness-related capabilities are not known (i.e. dog – yes, cat – no?), the best that affinity can do is to provide information about Bob – describe Bob for Alice. However, this means that Bob may be required to reveal a lot about himself – more than he might have wanted. This is the trade-off that Bob is facing: to be considered trustworthy he should put himself behind his words, to reveal some facts about his person. It is known that the message that has been delivered without disguising one's identity is considered to have much stronger effect – the person who is open about himself is considered more trustworthy.

Bob may, however, be willing to create an improved image of himself – a public presentation that hides some dark secrets, highlights his better side and maybe even tweak some facts. He may look smarter, more compassionate or more competent. There is not a guarantee with the description of Bob. If Bob is persistent and smart, he may create and manage a consistent 'alter ego'. Interestingly, it may work for Alice – she may not want to know everything about Bob, but just enough to assess whether she should be confident about him within a given context. If Bob is a doctor (and consistently maintains his image as a reliable professional), Alice may not be concerned with regard to his driving skills.

The role of affinity is to properly adjust Alice's expectations regarding Bob. Affinity determines which policy Alice should apply while dealing with Bob (as well with others), what standards and expectations she should use while judging his behaviour. Affinity is a construct that allows us to discuss confidence across the wide variety of entities – from human to inanimate, within the same conceptual framework.

## 11.4.3 Honesty

When Alice wants to assess her confidence in Bob she is watching Bob over certain information channels. She may interact with him, face-to-face or through certain digital communication means. She may also ask others about their opinions regarding Bob – using others as her information channels. Whatever Alice chooses, she should be reasonably assured that the information she receives is true, i.e. that others do not intentionally lie to her and that the communication channel is free from distortions.

The term 'honesty' is used here to describe the behaviour where agents tell the truth and communication channels do not distort, damage or alter information. Honesty encompasses both 'technical honesty' – the technical ability to deliver signals and 'personal honesty' – the willingness to tell the truth. In the case of digital communication 'technical honesty' can be achieved significantly easier than the personal one. There are known technical methods (e.g. cryptography) that protect the communication channel from intrusion and messages from distortion. Further, digital communication does not suffer significantly from the noise problem. Therefore the main part of our consideration is focused on honesty of agents.

The prerequisite for any evidence-based decision is the qualified reliance on information, i.e. the ability to determine to what extent a particular piece of evidence is true. Staying away from the discussion of what truth is, we can simply state that TET should make telling the truth desirable while making cheating unwanted. It is probably impossible to expect total honesty, because of the nature of humans, potentially subjective meanings of truth and due to the inherent subjectivity of opinions.

It is hard to guarantee honesty and even harder to identify and eliminate those that are untruthful. Typically (e.g. [Fernandes2004], [Zhang2006]), honesty is equated with the majority agreement – one is considered honest if one's opinion is shared by the majority of others. Unfortunately, this scheme may also discourage whistle-blowers, early adopters or even agents with significantly different needs or tastes. It is also hard to determine changes in truthfulness if it appears in a middle of a longer relationship.

Honesty as such does not cause a particular relationship to be more trust-oriented and does not increase confidence. In fact, by learning certain facts about Bob, Alice may become less confident in him. However, honesty improves her chances to learn as much of him as it is possible, thus building a better foundation for her assessment.

Unfortunately, the ability to lie is intrinsically connected to the ability to tell the truth: one cannot be entirely eradicated without the other. Therefore TET can only discourage Bob (and other agents) from lying, but it cannot prevent it. TET may, for example, increase transparency (thus increasing Bob's chances of being caught lying), evict Bob if he is found not telling the truth, etc.

## 11.4.4 Decency

Decent behaviour is defined here as behaviour that conforms to set rules – specifically rules that are defined by Alice and that are expected to be followed by Bob. In this respect decency acts as a filter and amplifier (one of the principles of TET) that identifies evidences of conforming (and non-conforming) behaviour.

There are several systems that use the concept of appropriate behaviour to build confidence. Specifically, reputation-based systems are built on the assumption that one's decency can be established by a voting-style system where those who interacted can leave their feedback. Assuming the reasonable uniformity of requirements regarding decency (which imply the comparatively equal value of feedbacks), as well as the fulfilment of other requirements, such systems show great promise as architectural candidates for TET.

However, known reputation-based systems may fail to deliver confidence in the case of new entrants or where the past history does not give a clear indication of a particular behavioural trait. As they usually build on statistical analysis (e.g. [Kamvar2003]), they become more discriminating only when there is an accumulated history of the relationship. By going beyond the simple feedback rating, TET can possibly deliver richer information regarding confidence.

In general, expectations regarding decency are driven by the outcome of affinity: there may be different expectations regarding decent behaviour of a computer or a person, for example. In some cases decency will require just the contract to be fulfilled (specifically if Bob is a software agent) while in other cases Alice may expect evidence of the deep relationship (if Bob is a person).

### 11.4.5 Curiosity

The assessment of confidence works on the basis of evidence – Alice should be willing to interact with Bob (or with others) to determine whether she can be confident about them. Actually, the ability to engage in interaction (thus collect evidence) seems to be more important than the actual outcome of the interaction, even if only the calculative aspects of the relationship are considered. However, gathering evidence (specifically initially) is a process that consumes time and resources, thus increasing the overall barrier to entry. Further, Alice may neglect evidence-gathering, forcing herself out from the relationship with Bob because she does not have any experience with Bob.

Such problems can be solved if Alice is willing to experiment and gather fresh evidence. The role of curiosity is to encourage Alice's evidence-gathering by forming appropriate incentives. For example, Alice's curiosity may be dampened by the perceived high cost of failure. In such case the restitution mechanism may help by alleviating her worries and may encourage her to engage in some contacts with Bob.

Restitution seems to be an important driver when it comes to the acceptance and adoption of services [Lacohee2006], and several e-commerce sites have already implemented different restitution mechanisms with a visible impact on users' confidence. The money-back guarantee seems to be the single most important driver of experimentation [Bahmanziari2003]. Solutions such as clearly visible phases of commitment ('you do not pay anything until you press this button'), ability to cancel undelivered items or return unwanted ones are important elements of various restitution mechanisms.

Another approach may call for lowering the level of initially required commitment. The game theory [Ullmann-Margalit2001] brings interesting insight into the development of trust, by postulating that trust can emerge only if the 'soft' game of trust is available. Such a 'soft' game is structured similarly to the classical Prisoner's Dilemma, but does not have a severe penalty for defecting. The 'soft' game of trust facilitates safe experimentation as it inhibits judgement for long enough to establish the relationship.

Alternatively, Alice may be somehow compensated for her efforts to develop new relationships – not necessarily in monetary form. For example, her social reputation may depend on her entrepreneurship (e.g. the number of contacts she has made), so that her effort and her potential losses are offset by her improved social status. Such a reputation may influence also the level of her perceived honesty – in the expectation that the person who knows more people may deliver a better, more balanced opinion.

## 11.5 Tools

The outer shell of the TET concept consists of three abstract tools that should be available on any platform that supports TET. The purpose of identifying such tools is to structure the discussion about the potential functional features that should be delivered by the platform to facilitate TET. Those three tools therefore represent three clusters of functionality that should be delivered by the platform. It does not mean that the selection of tools is exhaustive or

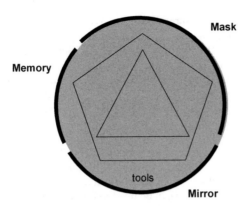

**Figure 11.4** TET tools

that the existence of those three tools will automatically make the platform TET-compliant. However, if the functionality represented by those tools is not present, then the fulfilment of TET requirements may be problematic.

In order to distance those conceptual tools from any possible technical implementations, their names were intentionally metaphorical. Those tools are metaphorically called: mask, mirror and memory (Figure 11.4).

There is no one-to-one relationship between requirements and tools, as there is no direct relationship between principles and requirements. Every tool must satisfy all requirements and only all the tools together deliver the functionality that supports TET. However, the concept of a mask relates closer to requirements of identity and affinity, the mirror addresses honesty and decency while the tool of memory is related to curiosity.

### 11.5.1 Mask

When Alice asks Bob 'Who are you', she may end up with some surprising answers. He may of course simply say that he is Bob, but he may also describe himself as a human, a medical doctor, a tired person, a chocolate lover, an unimpressed viewer, etc. All those responses answer the question and all form what is Bob's mask – the presentation of his identity that he elected to wear while dealing with Alice – or rather while dealing with Alice's mask.

From the perspective of TET it is irrelevant to what extent Bob's mask reflects the 'real Bob'. As long as Bob behaves according to his mask, TET will accept him in a way he would like to present himself, not looking beyond the mask. However, TET will also record Bob's effort associated with maintaining the mask and possible deviations from self-imposed standards. The mask is therefore not only liberating Bob but also restraining him – at a certain moment he will have to live up to his mask, even though it may not be in his nature. So, if Bob would like to be a dog, then TET lets him to be a dog, but then Bob will be confined to his canine rights and duties.

It is worth considering what happens if Alice's expectations regarding Bob's mask do not match Bob's actual behaviour, e.g. what if Bob presents himself as a compassionate person while he behaves as a hard-selling agent? We can expect that Alice will react with disappointment and distrust one she realises the difference. Surprisingly, Bob should not be smarter than his mask suggests. For example, personalised websites [Serino2005] are treated with suspicion, even though they deliver better service. The reason for this (apart from privacy woes) is potentially the breach of the original understanding of identity – the website is expected to behave in an automatic and repetitive manner, not with intelligence.

The mask is a tool that at the same time preserves privacy and enables communication. Information, once disclosed, can be copied and collected indefinitely so that there is little hope in technology to protect privacy. Bob, once disclosing that he is Bob, can reasonably expect that this information will always be present. It is only by creating multiple identities and by protecting contextual integrity [Nissenbaum2004] that Bob can expect certain privacy.

The mask can be thought of as the 'identity-as', the extended notion of identity that encompasses not only a unique and convenient label (e.g. name) attached to Bob, but actually delivers certain qualified classification schemes regarding different properties of Bob – his identity not as an individual but as a member of particular class (men, dogs, computers, etc.), qualified by the belief that Bob is what he pretends to be. This may also include expectations regarding Bob's intentional and cognitive capabilities. Therefore the mask carries both the identification and the recognition of potential affinity.

There are several technologies that can be used to implement the tool of the mask. Low-cost registration combined with certain feedback mechanisms may work quite well. For example, gossip can be used to determine the common opinion about Bob's mask, independent of the opinion about Bob's trustworthiness, etc. By talking with other members or by reviewing the collectively stored feedback information of the community Alice may come to the conclusion that Bob is what he pretends to be – or that he is hiding something or that his identity is in question.

There are also other possible uses for the mask. The central authority can issue a set of certificates stating who Bob is, possibly in small portions, so that Bob can to some extent control the disclosure of his personal information, e.g. as proposed by [Brands2000]. Notwithstanding the obvious advantages of such approach (strong cryptography, established technology, etc.), this particular implementation may be overly centralised and overly dependent on a trusted third party – the issuer of certificates.

## 11.5.2 Mirror

Being recognised as an equal (or superior) is a significant driver that shapes our social behaviour – in several good and bad ways. Peer pressure, group acceptance, informal culture – there are several names for the phenomenon that is used by the society to enforce certain behaviour upon us. They all are built on an expectation of the visibility of our actions and the immediacy of feedback. Alone, invisible and unsupervised, people tend to relinquish the social pressure and drift towards extremes.

This is exactly what they can do in a digital world. They are reasonably insulated from social pressure, invisible to locals and outside of the regular circle of supervision. They can do whatever they want. However, if they operate in a TET-enabled environment there is always an active tool that replicates a social pressure: a mirror. The mirror allows Bob to look at himself (and at others) from the perspective of the community he would like to belong to. The mirror collects his behaviour and tells him how he fares against the principles of the community.

The mirror is the complement of a mask: if we think of multiple identities as our masks, then what we need is a mirror, the mechanism where we can see the reflection of our actions. Such a mirror can be a simple one to look at oneself or it can reflect what society thinks of us (such as our reputation), can be instant or can retain the past (or project the future) but the mirror is always set to 'close the loop', to let us reflect on our own (and others?) behaviour.

Reputation is a simple communal mirror that reflects what the community thinks of certain actions of Bob. Reputation has its monetary value but it is even more important as an element of recognition. It was even mentioned in the Universal Declaration of Human Rights (1948) where Article 12 states that 'No one shall be subjected to arbitrary interference with his privacy, family, home or correspondence, nor to attacks upon his honour and reputation' [Rotenberg2000].

Alice can construct her own private mirror to view others her way: potentially attenuating their faults or maybe being more forgiving than the communal mirror. If she is particularly important to him, Bob can look at Alice's mirror to see how he fares in her eyes – and potentially alter his behaviour to live up to her expectations.

Contracts are examples of crude private mirrors. In the contract both parties express their expectations regarding the other party so that at any moment they can see whether they are compliant or non-compliant. Contracts are private mirrors, because the roles set in the contract apply only to parties that entered into such a contract. Contracts are crude because they are essentially tools of control, not of trust (at least in Western culture). Their primary role is to coerce all parties into the mutually agreed behaviour under the threat of punishment. However, contracts can be turned into trust-building tools [Sako1992] if they become instruments to express the willingness to trust.

The mirror also allows the possible future to be shown, constructing the 'what-if' propositions. Bob may see what he will look like if he supports Alice now and how if he betrays her. Such a forward-looking mirror is as much a control instrument as an educational tool. Bob may behave in an appropriate way because he is afraid of the consequences and eventually he will learn to behave appropriately. The mirror will shape the mask.

### 11.5.3 Memory

There are two important features of memory: one is to remember, while the other one is to forget. Both are necessary for the functioning of TET and both are necessary as elements of the public memory. We are not concerned here with private memories of agents – they are simply necessary from the operational perspective. What the memory tool represents is the shared communal memory – the one that can be influenced by all agents and used by them.

The case for remembering is an obvious one. For the benefit of both masks and mirrors, agents must remember what they and others have done or said. Facts and opinions are collected, stored and processed so that mirrors can show what others think about the mask and how the mask should perceive itself. Without collective memory the group is no more than a dispersed set of individuals.

The case for forgetting is a more complex one. TET should encourage Alice to be curious about Bob, to try collecting evidence and interact with Bob. The main inhibitor comes from the fact that Alice may be afraid of losing in failed interactions. Alice may be afraid of financial losses or of being psychologically hurt, etc. What she needs is quite often the reassurance that if things go sour she may withdraw with minimum loss. What she needs is a guarantee of restitution – a digital version of 'no questions asked, money-back guarantee'.

Between people the restitution is never perfect, as the perfect state requires the ability to 'roll back time', to nullify what has already happened with its consequences. Therefore, the restitution is often not in kind but in a form of monetary compensation – for the lost time, expectations, efforts, hopes, etc. However, TET brings restitution one step further: the collective memory can be altered to reflect the agreement that certain actions should have never happened. If Alice is dissatisfied with her interaction with Bob, she may request the whole interaction to be removed from the memory so that from the perspective of others it has never happened. It does not imply that curiosity and experimentations come free to Alice – she commits her time and resources and the restitution process may require even more resources.

Altering collective memory may sound a bit scary to some but this is what the technology and society are commonly doing. Databases operate on the basis of transactions where each transaction is in fact an experiment performed on the database. The set of operations within the borders of transaction are performed on the database in such way that it can be rolled back, nullified for as long as the transaction is not committed. The majority of real-life commercial transactions work this way as well, allowing both parties to withdraw from the transaction up to certain point, with no or little penalty. The withdrawal may not be entirely free for parties involved, considering time, effort or money already invested, but at least it restores the previous state.

Forgetting should be not also taken literally: it does not necessary imply the erasure of public records. However, it assumes the existence of digital equivalents of annulment, forgiving or absolving where records are updated to reflect that society should no longer be concerned with certain records from the past and that it should treat them as non-existent.

The memory tool can have several implementations, with the central registry the most promising candidate. Such a registry can serve as a source of all the facts that may contribute to the perception of confidence, the evaluation of mirrors (including e.g. reputation) and the verification of masks (including e.g. authentication).

However, one must be confident in the proper operation of the collective memory to accept facts that are delivered by it. Note that trust alone is not sufficient. The concept of the trusted third party running the registry requires implicit trust to be put in such a party by all participants – something that may be hard to achieve. Therefore the central registry should be designed in such a way that the necessary confidence should be attainable by control rather than by trust.

## 11.6 Example

Let us consider eBay's PowerSeller (described elsewhere in this book) as an example of a TET-related solution to see how it satisfies actual requirements and to explore its drawbacks and potential improvements. eBay has been created as a person-to-person trading system, but it was quickly noticed that some traders are significantly more active then others. Currently there is a reasonable number of traders who make their living on eBay, either by buying and selling their wares or by helping others (e.g. as trading assistants). Also several physical shops have their presence on eBay.

PowerSeller is an attribute of a trader that is assigned by eBay to those traders who have fulfilled certain set criteria: reasonable persistency, minimum turnover and acceptable behaviour. Feedback rating for PowerSellers is counted slightly differently than for regular traders, to take into account repeat business that seems to be characteristic of this group.

The system of PowerSellers is built on top of eBay's reputation-based scoring that stores all the feedback in the persistent, publicly available database (with some restrictions). On the monthly basis eBay filters this information and identifies PowerSellers. There is no process whereby the trader can express his desire to become PowerSeller or to sign up, it is eBay that makes the decision on the basis of published criteria. Interestingly, the trader also cannot opt out: if eBay detects that he fulfils criteria for becoming PowerSeller, he is automatically granted this status.

The status of PowerSeller is visible to all traders through a small icon located next to the trader's screen name. Depending on the volume of trading, the icon can take on one of several colours, signalling the tier level. Except for the volume, other requirements are identical for all tiers. Actual criteria for becoming a PowerSeller are reasonably easily available on eBay's website.

It has been noted elsewhere that PowerSeller status creates a perception of trust, by offering strong evidence of continuity (a seller remains within eBay and is willing to be bound by eBay rules and regulations), competence (high volume of transactions combined with high buyers' satisfaction) and motivation (assumed investment in building a reputation on eBay

Example                                                                      **201**

while upholding eBay's policy). It is not known whether the status of PowerSeller increases the actual profit margin.

Through PowerSeller, eBay is actually highlighting certain properties of the seller, properties that can be potentially found directly by the buyer, by browsing permanent records of transactions stored by eBay – at the expense of complexity. From the perspective of the model, eBay is filtering and selectively amplifying available evidence of trust that otherwise might go unnoticed – exactly what TET is supposed to do.

Considering TET, we can see that the concept of PowerSeller, together with other elements of eBay, may satisfy several of TET's requirements. As PowerSeller is an overlay on the eBay's system, we are not able to discuss PowerSeller without referring to underlying mechanisms provided by eBay. Whenever possible, difference between the eBay basic mechanism and PowerSeller have been made clear.

## 11.6.1 Identity

There is nothing specific about the identity of PowerSellers, as they re-use the identities of a regular eBay's trader. If the trader values the fact of being PowerSeller, then to a certain extent this creates an additional incentive to retain the identity, as becoming a PowerSeller takes time and effort. Unfortunately, PowerSeller status also creates a disincentive to retain identity if the seller's reputation becomes tattered, as it is potentially faster to re-build such a reputation under a new identity.

PowerSeller may potentially increase incentives to change one's identity prematurely if the trader may want to preserve the feeling of an amateur, the low-profile cottage seller and not indicate the volume of his trade. The reasoning behind this is very similar to the one that applies to a corner convenience store not willing to disclose its affiliation with a large chain: by being perceived as independent and local, the store attracts different kind of clientele and capitalises on a different loyalty. The simplest way to achieve this is to trade under several identities so that some of them will remain below the minimum PowerSeller volume.

If the large unscrupulous trader becomes a PowerSeller, he may also have incentives in splitting his personality to boost his own reputation. It has been mentioned that PowerSeller's reputation is calculated differently, not discounting the repeat trade. Therefore, by trading with itself (and providing positive feedback), the seller can easily attain the desired high status. The trader can actually bury some negative feedback in a heap of faked positive ones, creating an interesting opportunity to defraud others.

## 11.6.2 Affinity

The concept of PowerSeller definitely satisfies the requirement of affinity, as it identifies the specific group of traders. As their transaction volume grows, they are also more likely to be organisations rather than individuals, with all the advantages and disadvantages.

It is not known whether buyers perceive PowerSellers differently from ordinary ones – whether e.g. they believe that PowerSellers are in fact companies or shops, while 'regular' traders and individuals. If this is the case (and it is quite likely) then buyers may have different expectations regarding the relationship with those two groups. Potentially, PowerSellers may be required to provide more evidence of control while traditional traders can build on trust. This may in fact entirely circumvent the original intention of PowerSeller as an additional sign of trust.

## 11.6.3 Honesty

There is nothing special in PowerSeller that supports honesty. It is using the same feedback-based system that is characteristic to eBay. Honesty in this system is protected by its public character and by the ability to dispute (to certain extent) feedbacks. There is already an existing incentive to provide honest feedback as a low-complexity solution (honest feedbacks are unlikely to be disputed), with a certain slant towards overly positive feedbacks (that are even less likely to be disputed).

### 11.6.4 Decency

PowerSeller is effectively the 'decency filter', the seller profile, the mechanism that filters and amplifies signals that are related to the expected behaviour: presence, volume, opinion. By applying a different means of calculating reputation, PowerSeller actually upholds the concept that there may be different standards of decency, depending on the category of a trader.

The decision to create such a category is an interesting one, but its global nature (same rules apply to all traders) and its imposed rules (provided by eBay) are a bit problematic. Even though standards for decency are shared within society, eBay's traders are potentially too disperse in their opinion to benefit from only one profile (even with several tiers). It may be beneficial to allow traders to create their own decency standards that they can apply to sellers (and buyers). Potentially, the community may end up with a group of several accepted standards with each buyer and seller being able to apply standards that reflect their own expectations of decency.

### 11.6.5 Curiosity

The concept of PowerSeller does not encourage curiosity, but eBay is quite good in supporting it anyway. Curiosity and exploration are encouraged by relatively unfettered interface, e.g. the user does not have to log on to browse auctions, see and compare goods, review ratings, etc., a design decision that creates a feeling similar to a casual walk through the flea market. Similarly, joining the eBay community is relatively simple and is not associated with any special obligations.

Restitution is addressed by eBay through their payment protection programme and – in part – through the retaliatory power of negative feedbacks (where restitution is achieved by the restoration of a perception of justice, rather than through a transactional roll-back). Both mechanism are applicable to PowerSellers, even though the concept of PowerSellers does not bring anything new in this area. There may be a certain perception of the increased availability of restitution while dealing with PowerSellers (as compared to other traders), in the same way as dealing with large stores provides the comfort of no-questions money-back guarantee.

## 11.7 Another Example

In order to illustrate the concept in full, let's consider briefly yet another example, this time much simpler and relevant to the two-party relationship rather than to social and public space. Let us consider the relationship between the Internet service provider (e.g. broadband provider) Alice and individual user-customer Bob. For the sake of this discussion let's say that the service is somehow restricted: possibly the usage is metered and limited to some gigabytes per month or possibly Bob may lose the service if he visits inappropriate websites.

Such a relationship is asymmetrical: the provider has a significantly greater power and control compared to an individual user. By virtue of its position, Alice-provider can monitor all the traffic, filter it, impose limits and regulations, etc. It is important that the provider's cost of exit is relatively low: the provider is losing one of possibly thousands of accounts. Comparably, for Bob – the user, the cost of exit is relatively high, as the termination or a switchover of a service is a major disruption and an organisational and technical challenge. While Bob has no option but to rely on Alice (the blind reliance cannot be properly called 'trust'), Alice can actually trust Bob or control him. For Alice, in a long run, trust can be a cheaper option, specifically if Alice would like to extend her relationship with Bob.

Alice is willing to invest in building some Trust-Enhancing Technologies to make sure that Bob will behave properly. Potentially this is the option that is cheaper for her than controlling him and cheaper than losing him and then acquiring another paying customer. Bob, on the other hand, may be willing to use such tools on condition that they will also bind Alice. Therefore Alice-provider offers Bob a network-side monitoring tool that works as follows.

Bob can log in to Alice's website and can watch in near-real time all indicators of his behaviour that are essential for Alice – his usage profile, current bandwidth consumption, past performance, monthly averages, etc. together with the final assessment whether his behaviour is appropriate. Note that such indicators are not necessarily only the ones that come from the contract. Certainly, Bob will be informed (and contractually punished) if he abuses the service, but the purpose of this tool is to tell Bob what Alice considers 'normal'. For example, even if Bob may have unlimited broadband access, Alice may consider the continuous high level of traffic as a sign that Bob should not be trusted.

In a nutshell, Alice offers Bob a mirror that lets him view himself through her eyes, exactly the same way she can view him, and a collective memory that tells him how she had seen him before. Bob can experiment with such a tool: he may intentionally load a large file to see how the increased traffic is recorded or he may type the prohibited address to see if it has been blocked. By experimenting he gains assurance that such a mirror is genuine and that the memory works as expected.

This simple monitoring tool therefore provides a mask (log-in password), mirror (performance indicators) and memory (history of past encounters). Its honesty can be verified by exploration, and divergence from the preferred behaviour is clearly indicated. Even though the tool is operated by Alice, it can be used to develop a better assessment of confidence. For both of them: Alice and Bob alike, the tool provides a one-stop tool to assess Bob. However, the tool is also binding Alice – once the tool has been set up, she cannot deviate from the assessment provided by it. For Bob, the tool serves also as an early warning and educational tool: it allows him to correct his behaviour before Alice has to take any action.

Note that this tool works mainly to increase Alice's trust in Bob, not to increase Alice's control over Bob. Alice does not receive additional instruments of control beyond what she already has and Bob is not increasingly aware of such control (at least not significantly more aware). The tool provides a means for Bob to work towards behaviour that is approved by Alice. However, it also gives Bob some means to control Alice, providing a counter-balance to his over-reliance on her. From the perspective of a relationship this is beneficial: Bob's blind reliance on Alice can easily fail (for no reason at all) but Bob's confidence in Alice, supported by certain evidence, can withstand minor disruptions and disputes. Further, Alice's willingness to disclose her internal processing can serve as a sign of her goodwill.

## 11.8 Conclusion

Trust-Enhancing Technologies is the multi-layered proposition aimed at recreating in digital form the best social practices that should lead to improvements in the ability to assess the level of confidence – and eventually to build confidence. TET is built on three principles of amplification, transparency and incentives and defines five requirements that should be met by the platform: identity, affinity, honesty, decency and curiosity. Finally, TET defines three tools that should be available on the platform: mask, mirror and memory.

TET is designed to counter-balance the observed trend of disembodiment where participants in the digitally-mediated communication have increasing problems in the assessment of their confidence in others. Such a trend leads to the desire to replace trust-based relationships with control-based ones. By working on evidence that can be communicated and processed digitally, TET is not dependent on human factors such as emotions, affections or empathy, that can be inhibited by digital communication. Instead, TET leverages digital communication and its global nature to deliver substitutes and complements of such factors.

TET is also an educational proposition. By allowing all parties to properly assess their respective confidence, TET builds a platform where equilibrium can be established: and trust will be granted only to those who deserve it. This may facilitate the transition towards trust, driven either by the rationality of this behaviour or by associated emotional connotations. Such a transition may affect not only human participants but also software agents that may alter their behaviour (or be re-programmed) to benefit from the equilibrium. By facilitating the assessment of confidence, TET will eventually create platforms where trust will become the best choice.

# Bibliography

[Adams2002] Carlisle Adams and Steve Lloyd: *Understanding PKI: Concepts, Standards, and Deployment Considerations*. Sams Publishing. 2002.

[Bacharach2001] Michael Bacharach and Diego Gambetta: Trust as Type Detection. In Cristiano Castelfranchi and Yao-Hua Tan (eds): *Trust and Deception in Virtual Societies*. Dordrecht: Kluwer Academic Publishers. pp. 1–26. 2001.

[Bahmanziari2003] Tammy Bahmanziari and J. Michael Pearson: Is Trust Important in Technology Adoption? A Policy Capturing Approach. *The J. of Computer Information Systems*. 43(4): 46–54. 2003.

[Brands2000] Stefan Brands: *Rethinking Public Key Infrastructures and Digital Certificates: Building in Privacy*. Cambridge, MA: MIT Press. 2000.

[Creed1996] W.E. Douglas Creed and Raymond E. Miles: Trust in Organizations: A Conceptual Framework Linking Organizational Forms, Managerial Philosophies, and the Opportunity Cost of Controls. In: Roderick M. Kramer and Tom R. Tyler: *Trust in Organizations*. London: Sage Publications. 1996.

[Dennett1989] D.C. Dennett: *The Intentional Stance*. Cambridge, MA: MIT Press. 1989.

[Deutsch1962] Morton Deutsch: *Cooperation and Trust: Some Theoretical Notes*. Nebraska Symposium on Motivation. University of Nebraska Press, pp. 275–318. 1962.

[Egger2000] Florian N. Egger: From Interactions to Transactions: Designing the Trust Experience for Business-to-Consumer Electronic Commerce. PhD thesis, Eindhoven University of Technology, The Netherlands. ISBN 90-386-1778-X. 2000.

[Feng2004] Jinjuan Feng et al.: Empathy and Online Interpersonal Trust: A Fragile Relationship. *Behaviour and Information Technology*, 23(2): 97–106. 2004.

[Fernandes2004] Alberto Fernandes et al.: Pinocchio: Incentives for Honest Participation in Distributed Trust Management. In C.D. Jensen et al. (eds): *iTrust 2004*, Berlin: Springer. LNCS 2995, pp. 63–77. 2004.

[Fukuyama1996] Francis Fukuyama: *Trust: The Social Virtues and the Creation of Prosperity*. New York: Touchstone Books. 1996.

[Giddens1988] Anthony Giddens: *The Consequences of Modernity*. Cambridge: Polity Press. 1988.

[Jordan2004] Ken Jordan, Jan Hauser, and Steven Foster: The Augmented Social Network: Building Identity and Trust into the Next-Generation Internet. Available at: http://www.firstmonday.dk/ issues/ issue8_8/ jordan/. 2004.

[Kamvar2003] Sepander D. Kamvar, Mario T. Schlosser and Hector Garcia-Molina: The Eigentrust Algorithm for Reputation Management in P2P Networks. In: *WWW2003*, 20–24 May. Budapest, Hungary. 2003.

[Kee1970] H.W. Kee and R.E. Knox: Conceptual and Methodological Considerations in the Study of Trust and Suspicion, *Conflict Resolution* 14(3): 357–366. 1970.

[Lacohee2006] Hazel Lacohee, Stephen Crane, and Andy Phippen: *Trustguide*. Available at: http://www.trustguide.org. 2006.

[Nissenbaum2004] Helen Nissenbaum: Privacy as Contextual Integrity. *Washington Law Review*, 17: 101–139, 2004. Available at: http:// crypto.stanford.edu/portia/papers/ RevnissenbaumDTP31.pdf. 2004.

[Panteli2001] N. Panteli and P. Dawson: Video Conferencing Meetings: Changing Patterns of Business Communication. *New Technology, Work and Employment*, 16(2): 88–92. 2001.

[Rotenberg2000] Marc Rotenberg: Protecting Human Dignity in the Digital Age. In *Infoethics2000: Ethical, Legal and Societal Challenges of Cyberspace. Third Int. Congress.* UNESCO, Paris, France, 13-15 November. Available at: http:// webworld.unesco.org/ infoethics2000/documents/study_rotenberg.rtf. 2000.

[Sako1992] Mari Sako: *Price, Quality and Trust: Inter-firm Relations in Britain and Japan* Cambridge: Cambridge University Press. 1992.

[Serino2005] Catharina M. Serino et al.: Making it Personal: How Personalization Affects Trust over Time. In *Proc. of the 38th Hawaii Int. Conf. on System Sciences*, 2005.

[Turkle1997] Sherry Turkle: *Life on the Screen: Identity in the Age of the Internet*. New York: Simon & Schuster. 1997.

[Ullmann-Margalit2001] Edna Ullmann-Margalit: Trust out of Distrust. Available at: www.law.nyu.edu/ clppt/program2001/readings/ullman_margalit/Trust%20out%20of% 20Distrust.pdf. 2001.

[Zhang2006] Jie Zhang and Robin Cohen: A Personalized Approach to Address Unfair Ratings in Multiagent Reputation Systems. In: *Proc. of Fifth Int. Conf. on Autonomous Agents and Multiagent Systems AAMAS-06*; Rino Falcone et al. (eds): *WS8: Ninth Int. Workshop on Trust in Agent Societies (TRUST)*, Hakodate, Japan. 2006.

# Part III
## Confidence in a Convergent World

# 12

# Convergence and Confidence

The more elaborate our means of communication, the less we communicate.

(Joseph Priestley)

## 12.1 Introduction

The majority of our social interactions require a certain amount of confidence, specifically a certain amount of trust. There are some obvious examples such as a patient's trust in the doctor's ability to perform major surgery. There are, however, several smaller examples: confidence in the postman to deliver the mail, trusting a solicitor to arrange legal matters properly, confidence in an operation of the police, etc.

Commerce is one activity that permeates all society – we are almost continuously both buyers and sellers on several different occasions, buying, selling and exchanging goods and services. Not surprisingly confidence is an essential element of commerce (e.g. [Bohle2000]) – and is an essential element of e-commerce as well.

The importance of confidence (specifically trust) can be explained by the fact that every commercial relationship includes the 'trust gap', the period of time where one party has already fulfilled its part of the transaction while the other still can withdraw from it. Salary is paid once a month, requiring trust in the employer's capability to pay. Food is consumed before it is being paid for, on the basis of trust the restaurant has in its patrons. Homebuyers (specifically in the UK) experience the very long 'trust gap' as the seller can withdraw from the transaction up to the last moment.

People are very experienced in building and maintaining confidence – as long they can deal with other people face-to-face. People are continuously seeking evidences of controllability and trustworthiness to the extent that [Habermas1986] every statement (both verbal and non-verbal) is thought to be an evidence of trust (or distrust), along with its actual content and embedded relationship.

*Trust, Complexity and Control: Confidence in a Convergent World*   Piotr Cofta
© 2007 John Wiley & Sons, Ltd

Digital communication has connected people but at the same time it has created barriers to traditional forms of evidence that can build trust. E-mail or short message is quick and timely but it leave no space for hesitation, wink or casting doubt. There is also a reasonable doubt about the actual origin of the message, its integrity, about the ability to properly interpret the cultural or emotional background of both sender and recipient – certainly not the perfect vehicle to convey confidence and trust.

Therefore, the main impact of digital communication on commerce may not be the technical facilitation of e-commerce but in re-defining and recreating the social trust that is essential for the commerce to flourish. This chapter explores the impact that the convergence in communication may have on confidence in e-commerce. We start from certain clarifications regarding the convergence, to determine main impact areas. From there, the complexity-based model of confidence delivers the framework that can be used to analyse what kind of impact convergence may have on confidence. An early version of this chapter was presented in [Cofta2005].

## 12.2 Convergence

Convergence in communication has gained significant visibility. Convergence itself is a word that encompasses several activities, so that it is necessary to narrow the scope of its meaning for the purposes of this discussion. Convergence can be seen here as a trend in commu-nication that enables the creation of all-encompassing communication from separated and non-integrated technical links. Convergence is not only about making a fixed connection mobile or about providing continuous coverage and seamless handover. Convergence can be perceived through its four quadrants (Figure 12.1.) [Mallett2005], listed briefly below:

- Fixed and mobile – seamlessly integrated fixed and mobile communication
- Media and network – network-centric media distribution
- Communication and IT – continuous access to the most appropriate information
- Voice and data – towards true multi-media experience.

Convergence – in general – is about making different things become similar, letting them merge together, move in the same direction, become one. This might imply that the land of convergence might be a land of plenty, but also the land of boredom and monotony. Fortunately, this is not the case. Convergence drives several important changes to the way

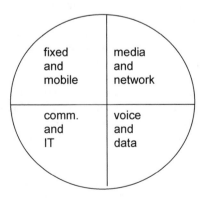

**Figure 12.1**   Four quadrants of convergence

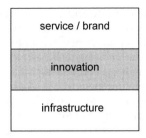

**Figure 12.2** Three layers of convergence

technology is deployed and perceived and may significantly influence the relationship between society and technology, but it does so in many different, often creative ways. The following is the discussion of some of the most important trends.

In a process that will be frequently referred to throughout this chapter, convergence splits the market into three layers (Figure 12.2). On the bottom there is a layer of infrastructure providers, owners and operators of the actual communication infrastructure. Their product is the wholesale commoditised and convergent communication. The topmost layer is occupied by service providers that deliver customer solutions by packaging and enriching infrastructure offerings. In the middle there is a growing innovation layer where creative players add value to the infrastructure offering before it is packaged by service providers, by delivering new technologies and services that have a wide appeal.

It is expected that the innovation layer will be the most interesting market area associated with convergence. While infrastructure providers will sweat their assets and while service providers will engage in a brand war, the middle layer will provide peace and flexibility to develop exciting solutions.

## 12.2.1 Pervasiveness of Technology

The standardisation and unification of technology that are caused by convergence come at enormous speed. Whatever is considered to be the best standard replaces obsolete technologies, thus creating the unified and monopolistic (from the perspective of technology) platform. The relative triumph of IP technology (used by the Internet) over solutions such as ISDN, ATM, etc. is an example of technical unification. BT's 21CN project to replace the core telecommunication network with its significantly improved IP-based equivalent is yet another example of such unification.

The important driver behind unification is savings. One technology means less maintenance cost, more standardised equipment, etc. that certainly drives the economy of scale. Even though this may lead to price wars, the leverage of the economy of scale offered by a unified technology potentially offsets those concerns. Unification brings some downsides as well. A single technology is more susceptible to successful large attacks and it reduces survival opportunities in the case of a major fault. Similar to large agricultural monocultures, the single well-adjusted parasite can damage the whole system.

However, it is interesting to note that current winning technologies share certain common characteristics. They are generally non-monopolistic, built on open standards and devoid of central authority that can enforce or guarantee service delivery or quality. The acceptable level of service can be reasonably assured (even though it is not always guaranteed) and any additional quality of service is the chargeable option, and one that is quite hard to deliver consistently. The user is increasingly facing uncertainty and it is the user's device that is supposed to somehow manage irregularities in service quality (e.g. Internet packet delays).

The main outcome of unification and openness is the pervasiveness of technology. As technology becomes cheaper and more compatible, it increases its coverage and ventures into all possible application areas. In this process new technologies quite rapidly lose their shine. However, they gain in coverage, proliferating well beyond their original application area.

In this process, convergent technology becomes a part what can be called 'an augmented nature', the part of our everyday environment that is so pervasive that it is taken for granted. As the technology defines its alternative geography (e.g. in a form of network coverage or WiFi hot-spots), people increasingly learn to rely on its availability. Such familiarity breeds confidence, similar to everyday confidence in the sun rising every morning and become part of our ontological security. Trust in technology gradually emerges, not from any rational assessment but more out of convenience, sustained by continuous fulfilment of expectations. Occasional breaks, errors and non-delivery are treated much like atmospheric phenomena – with annoyance but without anger.

## 12.2.2 Brand Virtualisation

At a certain level of technical abstraction, the underlying mix of technologies can be conveniently presented as a unified, standard technological solution [Cofta2006]. Converging fixed and mobile telephony means that Alice is no longer concerned whether the call is routed along fixed lines or over radio waves. In fact, she is no longer concerned whether the call is routed in a traditional sense at all, as for her the voice call is no different than any other data streaming application.

However, at the same time this standard technical common ground is increasingly re-packaged by an increasing number of brands. They are quite often virtual in a sense that they do not have to control (or own) any of the underlying technologies or technical infrastructure. Mobile virtual network operators (MVNO) demonstrate that the brand and product packaging can be more important (and potentially more valuable) than the network infrastructure itself. Brands effectively become stickers on the product mix, freely selected from the 'basket' offered by the underlying technology.

This separation of brands can have a significant impact on user confidence, as the brand is no longer a guarantee of market persistence. It is also not a sign of a significant investment or technical expertise, as both can be acquired from infrastructure providers. The brand also cannot guarantee the quality of offering across all layers of virtualisation, down to the last mile of copper wire. The actual product or service, its quality, durability, etc. are all negotiable, and can appear (and disappear) at the flip of the digital switch. Considering that the only real asset of the brand is the brand itself (and more specifically its reputation), it is a particularly tough call. The brand-based market [Tadelis1999] tends to continuously generate new brands and consume brands of good reputation, thus preventing long-term differentiation.

The space between increasingly standard infrastructure and increasingly separated brands is an area of opportunity: the innovation layer [Mallett2005]. This area, free from constraints of 'resource sweating' and liberated from continuous needs of technical expertise, can deliver novel solutions that may allow brands to flourish. Increasingly, convergence is driven by this layer, coming from a purely technical exercise into the market one. Triple or quadruple play (fixed line, mobile phone, broadband and television) is a middle layer innovation, owing as much to technical as to business convergence, but other (potentially better) propositions enter this area.

There is a certain parallelism between brand virtualisation and the proliferation of user's identities. Both are created on top of an infrastructure, to repackage certain packages of service, content, delivery, etc. Both put significant stress on reputation. Finally, both can easily disappear if they are no longer needed. It seems that both sides (service providers and individuals) assume the same strategy, but that this strategy may not be beneficial to confidence.

## 12.2.3 Cross-pollination of Technologies

Technologies that converge have usually developed in separate silos, creating solutions that were specific to their respective environments and optimised for the purpose of a given technology. As the convergent technology evolves, it has the opportunity to intercept and integrate mature solutions from several technologies in a way that may suit the needs of the convergence (rather than the need of a particular technical silo). Such cross-pollination of ideas and solutions is beneficial for the leading technology as well as for the one that has been marginalised, as even the marginalised one can provide a valuable contribution.

There are several areas of technical tension that can be resolved for the mutual benefit if technology can be used outside its original scope. For example, SIM (Subscriber Identity Module) has been developed for the purpose of user identification and authentication in GSM networks. SIM is probably one of the most widely deployed smart cards in the world (at least at the time of writing) – every GSM phone has at least one of them, and some users carry hundreds. The popularity of this technology has been noticed by solutions that originated from the Internet technology and has been adopted in a form of specific authentication proto-cols (EAP-SIM, GAA). SIM itself evolved under pressure from the multimedia applications into ISIM. Such cross-pollination positions SIM strongly to become a primary authentication device in the future.

In a similar vein, trusted computing, originally developed for personal computers and servers, is currently scaled down to be implemented on mobile devices or even particular components (such as hard drives). Similarly, payment schemes 'invade' mobile communica-tion, more efficient video compression algorithms deliver new types of content across new platforms, smart cards migrate into memory cards while SIM increases its on-board memory, EMV migrates to authentication, etc. Contrary to new developments, such cross-pollination works for established, tested and approved technologies so that the adoption time can be very short and the benefits can be enormous.

# 12.3 Convergence and Confidence

In the discussion below we will look at the impact of convergence on confidence, in the expec-tation that confidence will drive the further uptake of e-commerce as well as the adoption of other, more complicated services that can be built on the basis of convergent communication networks. Following the complexity-based model of confidence, we will examine the impact of convergence on enablers (that allow Alice to gather evidence about Bob) as well as on evidence of trust and control (in expectation that this may not only improve Alice's ability to assess Bob but may also actually improve Bob's behaviour). We will look for two types of impact: one that influences the flow of evidence and another that may change the behaviour of Bob so that Alice can be more (hopefully not less) confident in him.

An overview of the analysis is presented in Table 12.1. Each construct from the model of confidence is represented as one row. Each column in this table represents the partic-ular impact area of the convergence, as discussed above: pervasiveness and standardisation, virtualisation of brands and cross-pollination of technologies. For each row, the table lists what elements of convergence may have impact on the confidence. More detailed discussion follows.

## 12.3.1 Enablers – Similarity

The construct of similarity states that if Alice finds Bob similar to herself, she is not only more likely to be confident about him, but – first and foremost – she can develop a better model of how Bob may behave, so that she may consider trusting him. Conversely, if Bob is found

**Table 12.1**   Impact of convergence on confidence

|              | Pervasiveness | Virtualisation | Cross-pollination |
|--------------|---------------|----------------|-------------------|
| *Enablers* | | | |
| Similarity | | Personal brands Personalisation | |
| Honesty | | | Technical honesty Trusted third parties |
| Identity | Portability and single sign on | Privacy and contextual integrity | Reliable identity |
| Availability | Rich and pervasive experience | | Innovation and excitement |
| *Trust* | | | |
| Continuity | Standardisation of expectations | Advantage of core players | |
| Competence | Level playing field; best effort | Perceived competence as differentiator | Unified technical competence across user experience |
| Motivation | Transparency and understanding | Symmetric reciprocity of behaviour Loyalty/disloyalty | Intimate trusted advisor |
| *Control* | | | |
| Knowledge | Increased experience | Reputation as differentiator | Portability of expectations and knowledge |
| Influence | Transparency; holistic approach | Obscurity | Secure and trusted platforms |
| Assurance | Social acceptance | | Managed services |

to be different, Alice may not understand him, may not trust him and may have problems developing confidence in him.

Personalisation is the most relevant aspect of convergence that is related to similarity. It is a feature of one agent (e.g. Bob) that allows him to change his behaviour depending on an agent he is interacting with – changing, e.g. the subject of conversation, presentation, his look, etc. For example, Bob may behave differently depending whether it is Alice or Carol that is dealing with him. While certain personalisation is characteristic of personal contacts (as it identifies intimate knowledge of Alice's needs), there is a perceived difference between a casual conformity and the complete change of Bob's personality.

Convergence promises to bring digitally-maintained personalisation to the whole new level. As Alice is freely moving between different communication channels, different media and various services, it is increasingly possible to build tracing solutions that may monitor her behaviour and develop a model of her habits. Such solutions can be subsequently used, e.g. to deliver always exactly what she wants, how she wants it and where she wants it.

The integrated and inter-connected nature of convergent networks makes it possible and feasible to record and process the whole of Alice's communication, no longer constrained to individual communication silos. What comes out of this can be called 'personal brand', 'market of one' or 'micro-targeting', but the outcome is the same: Bob is learning about Alice so that his offerings and interactions are perceived by Alice as more and more intelligent, predictive and friendly – not only human-like but almost super-human-like, even though they are in fact the automated responses of a clever software.

Certainly, personalisation has some advantages. Even though Alice may rationally know that she is not interacting with a human being, emotionally she may cling to her anthropomorphism to develop the more intimate relationship with the service provider's automated advisors. Alice is also more likely to follow advice (and adverts) that suit her taste, satisfy her needs and match her current activity. Alice is also more likely to trust (rather than control) an advisor that seems to be intelligent, human-like and empathetic.

However, personalisation is also a threat, as Alice is more likely to reject an advisor that did not live up to her expectations, e.g. has been thought to be human but is overly automatic. There is no space for the acceptance between machine-like and human-like behaviour: it is either automated machine or perfectly human entity that gains acceptance. There is also no space for super-human abilities where Alice is surprised and scared of Bob's knowledge of her.

The impact of convergence on similarity is therefore mixed. Assuming that service providers will pursue short-term benefits of emphatically targeted marketing, they may end up alienating their customers. Even assuming that privacy is not an issue (and certainly privacy is an issue), imperfections in implementations will eventually show up, frustrating customers. However, if service providers are honest about employing 'clever automaton' rather than humans, and if they still provide communication channels to genuine humans (possibly not to scripted call centres), convergence may benefit both.

It is necessary to note another question, currently of lesser but of a growing importance. Service providers are trying to solve the reverse yet similar question: whether Alice is human. As Alice can deploy her robots, her automated bidding and trading agents, her search engines etc., she is increasingly perceived to be non-human and thus she is no longer responsive to emotions and empathy.

The proliferation of convergent technology means that increasingly agents will talk to other agents. Prohibiting agents (as it is currently policed by some trading sites) is not a viable long-term option, at least because the convergent technology blurs the distinction between Alice and her intelligent agent. Separate discussion is needed on how agents can establish confidence in other agents in cases when the similarity cannot be easily assumed. Some elements of this discussion are included in this book.

## 12.3.2 Enablers – Honesty

Honesty helps with confidence in several ways: Alice may use honest agents to solicit their opinion on Bob so that her knowledge increases; Alice may also want to collect opinions from honest sources regarding control instruments that she can use, Bob's beliefs, etc. Finally, Alice may also want to internalise her relationship with Bob (and eventually trust him) if she believes that Bob is honest.

It seems that when it comes to honesty, convergence brings both opportunities and challenges for Alice. The proliferation of technology implies that Alice has access to information from a much wider range of agents – both human and technical. Such agents may help her in developing a better understanding of Bob, specifically if her own interaction with Bob is limited or has not yet started. The 'small world' phenomenon [Milgram1967] suggests that there is always someone who may indirectly tell something about Bob. Social networks, referral sites, reputation-based trading sites, etc. are using honest contributions (feedback, opinion, invitation, etc.) to build confidence between remote parties that have never interacted with each other.

Unfortunately the easiness with which Alice can acquire an opinion about a stranger is also a threat and can turn against her – at least sometimes. Because opinions are so pervasive and so easy to obtain, she may tempted to interact with more strangers than is justified, thus exposing herself to additional risk. The gain from widely available opinions and reputations can be offset by the unnecessary risk that she feels she can take.

Note that reputation-based systems are not carved in stone. They are inherently vulnerable to manipulation if they allow for unconstrained access [Gavish2006]. Positive reputation ratings that Bob has from strangers are of no value to Alice, because she knows nothing about the honesty of those strangers. Digitally colluding agents remove the perception of safety in numbers.

The situation is not symmetric: while individual Alice faces additional risk, large service providers directly benefit from convergence. If they have access to multi-channel facilities that can be used to assess whether Alice is honest, they can develop their detailed profile of Alice, and then diversify to minimise individual risk. If not, they can disclaim responsibility for any loss that comes from dishonesty or they can accept responsibility (and losses) only to prevent Alice from switching to another service provider.

The solution is not in removing automated agents from the equation: they have their role to play as they free humans from tedious and repetitive tasks. The key is to make those agents traceable back to their creators. There is nothing wrong in thousands of agents if those who interact with them know that they execute tasks assigned by the same person, and it is the honesty of this person that really counts.

There is therefore an important role to play in the convergent world: to guarantee the link between physical and digital, between Bob and his agents. Such role can be potentially played by convergent trusted third parties (TTP) and there are plenty of existing technologies to choose from: PKI, SIM, EMV. The process of convergence will eventually pick one of those technologies and will use it (with associated institutional and business structures) throughout all the converging sectors. Some interesting aspects related to such future are discussed later in this book.

## 12.3.3 Enablers – Identity

Without a guarantee of a reasonably constant identity, Alice is unable to gather and process historical evidence about Bob; without Bob's common identity known to other agents, Alice is unable to relate evidence about Bob that come from such agents. Note that Alice may not be concerned with who Bob 'really' is (or who is the person behind Bob), but she is definitely concerned with her long-term ability to identify Bob whenever she encounters him.

Unfortunately, convergence made identity particularly problematic. It is not only that 'on the Internet nobody knows you're a dog', but it is also that everybody can pretend to be whoever he wants and can switch identities as desired. Separate user profiles and log-on names for different services may be a source of usability problems, but the primary concern is that they permit the creation of disparate identities. Even though Alice's behaviour can be traced across a variety of communication channels, it becomes gradually disassociated from Alice.

Another aspect of identity that becomes important in convergent systems is the identity of agents. The question of identity has usually been associated with humans (or possibly with an organisation or a group of people). As devices and software agents proliferate and the convergent structure becomes pervasive, the identity of an agent may become more important than the identity of a person. However, such an identity is even weaker than the identity of a person, as it is missing the important link with one's physical body. Methods that have been developed to demonstrate one's identity (authentication) may no longer be relevant if the entity is question is a software agent.

The pervasiveness of technology creates the need for a usable and portable identity, the one that can easily be transferred between different communication channels. It is an expectation (not always fulfilled) that the single identity should be sufficient to select, acquire and consume a desired service or content. Some Internet merchants understand this desire quite well and offer not only an integrated identity to their customers, but also – through affinity schemes – extend such identity to cooperating partners. Even though it works only within an affinity circle, the convenience offsets limitations.

Further, Liberty framework with its concept of a federated identity allows several service providers to link identities and share the authentication process, thus overcoming the restrictions of close relationship between them. Should this trend continue, we may eventually end up with only one identity (or a small set) that works for the majority of service providers. However, another aspect of convergence seems to counter-act the expectation of the single identity.

The virtualisation of brands can potentially generate a significant number of new identities, further proliferating the number of passwords, cards or tokens that are needed in a supposingly convergent networks. On the provider side, the low cost of establishing a new operation makes it possible to experiment with several products, each potentially deployed under a new brand identity and each potentially requiring the user to assume a new identity. For example, as mobile operators differentiate on tariffs and develop virtual operation, cost-savvy customers can carry a handful of pre-paid SIM cards to be able to select the most optimum one. Hence the surprisingly persistent identity of SIM becomes yet another short-lived token, used among several others.

The need for multiple identities is further fuelled by the desire for privacy, the expectation that information that is considered private will not be used in a manner that is unintended (contextual integrity [Barth2006]). As identity theft can be damaging to the business and traumatic to individual, both protect their identity assets by multiplying them (preferably at low cost), so that the loss, even if painful, is limited to the single identity. While the legal protection has been paid usually only lip service, and while privacy-enforcing technologies are too often considered to be an unnecessary burden, a certain privacy can be achieved by maintaining alternative identities. Such identities, separated and used for different purposes, can be thrown away if abused. Virtual brand is matched by virtual identity, as individuals learn to proliferate and manage the growing number of their identities. Some service providers may benefit from this: for example, the one-time disposable credit card number prevents identity theft and is an interesting offer. However, in general, the way privacy is handled today actually prevents the creation of confidence.

It does not seem that the tension between the proliferation of identities and the consolidation of them can be solved any time soon, unless core infrastructure providers commit themselves to delivering certain solutions in this area. Convergence may deliver solutions to the growing pains of identity, its management and privacy. The cross-pollination of technology may, for example, revive the concept of SIM as a source of identity well beyond its originally intended scope. Standards such as ISIM, GAA, EAP-SIM, etc. allow the gap to be bridged between the managed mobile network and the unmanaged Internet while solutions such as software-only SIM make SIM-based identities available to the wide range of convergent devices, balancing security requirements with deployment cost. Elsewhere in this book, the concept of privacy-preserving authentication demonstrates how disparate identities can be consolidated without the loss of privacy.

## 12.3.4 Enablers – Availability

The assessment of confidence thrives on evidence. The more evidence is available, the less uncertainty Alice faces while she tries to determine whether she can be confident about Bob. In the absence of evidence or when evidence is old or irrelevant, Alice is no longer in a position to judge Bob. Availability of evidence has two sides: Alice must be able to find evidence (i.e. Bob must produce it) but Alice must be also willing to search for it. Finding evidence may consume time and resources and may not be always free from problems.

It seems that convergence can help out on both counts: Bob's production and Alice's willingness to search. First, the universal access that is made possible by the proliferation of technology increases in itself the flow of evidence and reduces the risk of such evidence being old or irrelevant. The variety of communication means combined with the omnipresence of the infrastructure and reasonable pricing has already significantly improved the flow of evidence. Probably this is the most unanimously positive impact of convergence on the assessment of confidence.

For example, mobile phones have significantly increased the intensity of our communication (thus the amount of evidence that are at our disposal) and made it more relevant to current events and location. One may complain that 'texting' or mobile calls are more casual, shallow and do not lend themselves to deep thoughts characteristic of letters. However, in the majority of cases, the role of this casual communication is not to convey thoughts but to frequently exchange simple signals that build up our confidence.

Further, convergence popularises new interaction styles, incorporating location, presence, diaries, video or avatars (all of them in several technical implementations). Such multi-channel, multi-sensory communication significantly increases the ability to gather information, as it delivers evidence that is both rich and timely. Even though the future adoption of particular communication forms is not always clear, it is apparent that the rich communication content generated by individuals is increasingly complemented by the growing convergence and standardisation of technologies.

Since 'to google' became a verb (see e.g. *Oxford English Dictionary*), finding and correlating such evidences is becoming reasonably easy, and convergence brings the increased amount of information sources into the sphere where they can be searched and organised. Certainly, there may sometimes be doubts about the validity of certain information sources, but it is its abundance and availability that matter here. It is actually hard to stay out of search engines, data integrators, central registers, etc, to the extent that some people take it as a form of modern artistic challenge – not to be found on any search engine despite being professionally active.

While evidence is abundant and easily reachable, Alice's willingness to gather evidence is also positively influenced by convergence. In order to try new contacts, information sources or evidence, Alice should be able to overcome the cost associated with it – in terms of both resources and emotions. While search engines and inexpensive communication contribute to lowering her cost, better regulations offer to offset such cost even further. It is not only that Alice can easily find the service or a person and check his reputation and offering, but it is also that she can engage in several forms of interactions reasonably safely. The availability of certain forms of restitution [Lacohee2006] makes Alice even more at ease with evidence gathering.

Convergence facilitates also the excitement that allows Alice to offset her emotional barriers and some of her potential emotional losses. Convergence drives innovation into the communication, through the 'innovation layer' between commoditised infrastructure and customer-facing branded service providers. The potential innovation area is quite large and barely explored, so that Alice may hopefully expect an uninterrupted stream of innovations to sustain her excitement.

## 12.3.5 Trust – Continuity

Continuity, sometimes called 'the shadow of the future' is Alice's reasonable expectation that Bob will live up to the norms and standards that are acceptable for her. This implies that such norms and standards should be defined and recognisable and that there is a visible commitment from Bob to adhere to them. The communication of continuity is usually quite hard, even in direct communication between people and is supported by recognised common beliefs, shared institutional membership, etc.

The main benefit of convergence with regard to continuity may be in the creation of actual common expectations and norms. This is facilitated by the proliferation and standardisation of convergent technologies. In fact, users of convergent technologies will eventually share the same set of standard technologies whether they like it or not. By standardising technology and expectations, convergence may decrease the ability to differentiate, but it also decreases psychological barriers to use the technology. Alice gradually learns what to reasonably expect from the device, from the network and from Bob. In this process, Alice also learns how to identify whether Bob is bound by the same set of norms.

However, technology itself is neither the most innovative nor the most exciting place from the perspective of continuity, as it gradually binds both Alice and Bob regardless of their intentions. Increasingly, not the infrastructure technology itself but the 'innovation layer' [Mallett2005] will become the place where continuity will be demonstrated and expected. We may expect tools that are designed specifically for the purpose of communicating continuity. The existing demonstrated shared membership of on-line groups is the harbinger of them.

Continuity can also potentially suffer from the introduction of convergence, as the proliferation of virtual brands may not allow us to build long-term expectations. An understanding that brand has little incentive to stay unchanged may lead to short-term exploitation rather than to long-term commitment. In light of this, the ownership of the infrastructure (server farm, network, bank branch, high street shop) may become a perceived value, a marketing asset, as it signals long-term commitment.

## 12.3.6 Trust – Competence

There is no point in Alice trusting someone who is not competent enough to support her. Even if Bob is eager and willing, Alice will not benefit from this if he does not have skills, resources or abilities to help her. The perception of competence is a prerequisite of trust. When Alice is dealing with Bob with the aid of a convergent service, Alice's perception of Bob's competence is actually a mix of a competence of Bob himself and of his service provider. From her perspective, if Bob's website is slow, it does not matter whether it is because his design skills are sluggish or whether he has chosen an inefficient provider.

Convergence influences competence in several ways, of which not all are potentially beneficial but all are important. Surprisingly, first we can see the overall decrease in expectations regarding technical competence. Technologies that are currently introduced by the convergence put less stress on the overall guaranteed quality while they increasingly employ the 'best effort' approach. Very little is technically guaranteed: we share bandwidth, spectrum and name space. Even though 'WWW' is no longer the 'world-wide wait', response time and quality are unpredictable. The upside is the low cost and simplicity of an infrastructure. The downside is the potential degradation in quality. Certainly, technologies such as MPLS (Multi-Protocol Label Switching) can partially offset the problem, but it is more interesting that Alice gradually accepts this situation, thus somehow lowering her everyday expectations.

At the same time, basic technical competence becomes an expected commonality. Gradually everybody is using comparative technology in comparative settings: either because technology is being built from standard components or because it is leased from large infrastructure providers. Even though there is less expectation for technical excellence, there is more expectation for technical competence. Alice has gradually learned what to expect from the average Bob. Short outages, some jitter and delays are increasingly acceptable while severe technical problems are increasingly not. From the perspective of confidence the shared expectation regarding technical competence is being developed.

Such expectation can be used as a starting point for a competence-based differentiation. Once Alice understands what is a reasonably expected reference point, she is also in a better position to understand signs of technical competence and excellence – if communicated clearly. Therefore convergence may benefit Bob in demonstrating his competence – if he knows how to do it. This situation may be potentially advantageous to large infrastructure operators, as they are in the best position to deliver offerings that go beyond what is mediocre. However, such a strategy will positively influence confidence only if Alice is able to clearly identify evidence of such competence – something that is not easy for the technical competence alone.

For example, Bob may want to communicate his competence through design. It is known that aesthetics is often considered a sign of competence (e.g. [Karvonen2000]), and Alice is willing to trust Bob more if he runs the site or the service that is pleasing in its design. Such an assumption is partly justified on the ground of the amount of work that has to be spent on developing such a site, but it is also partly justified on the basis of perceived similarity between Alice and Bob.

The value of aesthetics may not decrease, but expectations may shift: convergence also affects the design part and makes sites of similar functionality look similar (see e.g. the proliferation of websites that use content management to lay out their content). This, combined with the availability of tools and templates, sets Alice's expectations high and raises the bar for demonstrating exceptional competence. We may therefore see a shift from the demonstration of purely technical competence towards the competence that encompasses several aspects of the interaction.

There is also an unexpected challenge for the competence from the cross-technology convergence. For as long as Alice has been aware of the variety of technologies she is dealing with, she could have tuned her expectations accordingly. However, the unification delivered by convergence means that Alice is increasingly unaware of technical details, and her expectations regarding technical competence may no longer be reasonable. If, for example, Alice is enjoying a video conference over WiFi and then seamlessly drops to 3G, the quality of the image may decrease. As she is unaware of the change-over, she may perceive it as a lack of competence – either Bob's or their service provider or both. Interestingly, shielding her from technical details may actually decrease her trust.

## 12.3.7 Trust – Motivation

Motivation demonstrates that Bob is willing to help Alice, not because he is forced to but because this is his intention. The encapsulation of Alice's interest in Bob's interest [Hardin2002] is typical evidence of motivation when Bob can be believed to help Alice to help himself.

Convergence, through the pervasiveness of technology, creates an environment where motives can potentially be better understood and interpreted, but only if care is taken to explain them. Specifically the 'innovation layer' between technology and brands is important for motivation. New business models, new pricing schemes, or unorthodox usage of technology can be interpreted as evidence of motivation – or the lack of it. Shared understanding of how and why parties engage in convergent communication is essential to foster a better understanding of other's motivation.

The volatility of brands combined with the relative stability of technology creates an environment that may increase the symmetry of reciprocal behaviour, thus allowing one party to understand (sometimes through trial and error) the motivation of the other party. While it has always been reasonably easy for the large party to monitor and experiment with its relationship with smaller parties (e.g. large operator and its customers), the situation has not been symmetric. The exit cost for the smaller party has been too high to reciprocate mishandling so that the smaller party has been potentially subject to abuse while the larger party was not able to calibrate its message and listen to the smaller one. With virtual brands and almost-zero switchover cost, both parties are in the same position regarding signalling and both can better learn about the other's motivation.

Convergent technology may also facilitate the automation of the process of the assessment of motivation (and potentially other evidence of confidence). The concept of the Intimate Trust Advisor (ITA) [Cofta2003] can be realised only in an environment that is technology-intensive yet reasonably standardised, so that evidence can easily be quantified and assessed.

### 12.3.8 Control – Knowledge

Knowing Bob is a strong driver may potentially boost Alice's confidence, as she can reasonably expect that Bob's behaviour will follow the established pattern. Even though she may not be sure to what extent Bob's behaviour can be attributed to his trustworthiness and to what extent to the presence of external circumstances, knowledge allows her to plan her actions.

It has already been stated that convergence facilitates the growth of the amount of information about Bob that is available to Alice. Such growth can be the result of planned action (e.g. tracing Bob's shopping behaviour) or the availability of new sources (e.g. reputation in on-line communities): both facilitated by convergence. However, it is even more important that the actual interpretation of different bits of knowledge may also converge, i.e. that potentially there will be only one dominant digital reputation system, that will provide the complete digital profile of Bob. Such convergence has not happened yet, mostly because there are discrepancies regarding the interpretation of the reputation value. However, in line with technology standardisation trends, reputation ratings will eventually reach the level of *de facto* standardisation.

Considering that virtual brands have effectively nothing but a reputation to compete on, the importance of such reputation is paramount. Knowledge of a brand's past and present reputation may be the only differentiating factor so that brands may spend significant resources in building and defending reputation ratings. Unfortunately, this also means that brands will have strong incentives to discredit competing brands as well as to buy well-performing ones. Such a war on brands may negatively impact on Alice's ability to build knowledge and may result in the decrease in her confidence about the whole system.

Portability of technology facilitates the portability of knowledge. Within convergent systems Alice can effectively re-use and leverage her knowledge gathered through early non-convergent technologies. This will improve her confidence in technology itself, but it will also allow her to more easily consolidate her knowledge across different convergent communication means, resulting in a shorter learning curve, better assessment and greater convenience.

### 12.3.9 Control – Influence

Influence is the most obvious component of control and it exists in several flavours: from coercion to persuasion. Alice, if she wants Bob to behave in particular way, can directly or indirectly apply certain pressure (psychological, physical, etc.) to make sure he does what is good for her.

The convergence changes the balance of power between Alice and Bob, but it does so in several different ways. First, the pervasiveness of convergent technology increases Bob's visibility thus driving greater transparency into the relationship. This itself may be beneficial for confidence, as it decreases the 'trust gap'. In addition, this also increases Alice's ability to influence Bob because she is better informed about the actual stage of their relationship.

At the same time, as Alice can 'see' Bob through several different channels, media and activities, she can also possibly influence him using contexts that are different from the one associated with transaction. Bob is increasingly reachable, through the convergent communication – e.g. even now Alice can offer him an individual showing of a premium TV programme as an incentive to call. Therefore convergence delivers Alice more instruments to influence Bob.

However, the virtualisation of brands may work here against this trend, increasing opaqueness and decreasing Alice's ability to influence Bob. Regardless whether Bob has deployed his own virtual brand or whether he is using a virtual service provider, his ability to switch or drop virtual brands means that Alice has less control and less influence over him. Bob can hide his movements (e.g. by using several phones from several operators) and can slot his activity into several compartments in a way that prevents Alice from inter-linking them. From Alice's perspective, Bob's use of several virtual brands may be considered a sign of unwillingness to be controlled and such a sign may lead to her lower confidence in Bob.

Alice's confidence in Bob depends on her confidence in the device that Bob is using. The convergent device is not only increasingly open, but it can be also increasingly secure. Openness and security are not usually positively correlated, as the open device (such as a general-purpose laptop or PDA) must accept modifications to its software and hardware that introduce new vulnerabilities. However, cross-pollination of technologies allows various technical solutions (such as Trusted Computing Platform or smart cards) to be used across a wide range of convergent devices. Note that neither TCP nor smart card alone can make Alice confident in Bob, but they both are instruments that can be used to make sure that Bob's device is not acting against Alice.

## 12.3.10 Control – Assurance

In the case of assurance, Alice is influencing Bob's thinking about his future by instruments that can apply reward or punishment, depending whether Bob has supported her or not. The most typical example can be the use of a legal system to enforce the contract or the bonus system that promotes more effective work.

Convergence brings two important factors to assurance: social acceptance and managed services. Social acceptance is driven by the pervasiveness of technology. As technology became pervasive, social behaviours associated with it gradually became social norms so that deviations from such norms are increasingly unacceptable. This is self-perpetuating to a certain extent: the formation of norms related to technology increases confidence in such technology (as it is perceived to be used by everybody else), and this leads to further reinforcement of norms. Convergence, by accelerating the adoption speed (e.g. by lowering cost) and by increasing exposure (e.g. by increasing availability) accelerates this process.

From the perspective of confidence, it is not only the adoption of convergent communication itself. It is also that Bob will increasingly find himself bound by social rules of such communication, so that the communication exists within enforceable social context. For example, e-mail, due to its technical nature and limited availability, has been considered inappropriate for messages that were legally binding, so that e-mail-based contracts or promissory notes were not admissible. However, as the technology gained popularity, it became socially acceptable, legally admissible and technically feasible to enforce such messages. The popularity drove adoption and this facilitated social control.

There is a tension within convergent systems between managed and unmanaged ('free') services and infrastructure. It can manifest itself as a conflict of interest between managed and unmanaged access networks (e.g. GSM and WiFi), DRM-protected and open content, paid-for and freely accessible applications, etc. It seems that unmanaged solutions are gradually gaining the upper hand and that managed solutions may refer to legal instruments to keep them at bay. However, managed solutions have certain value to offer from the perspective of confidence: assurance.

Unmanaged access or content is free to use, but it usually does not deliver any means to build assurance: agreements are not enforceable, support is based on voluntary contribution, there is no process to resolve conflicts, etc. If Bob delivers non-managed access, he cannot be held liable if something goes wrong. If Bob uses a non-managed network to deliver his service, it is unlikely that he can guarantee continuous access. Certainly, those woes are offset by the benefit of a low price, but they can undermine Alice's ability to develop confidence in Bob, if Alice wants to build on the basis of assurance.

## 12.4 Conclusion

Convergence means many things to many people. Here, convergence is discussed as a trend that consolidates various areas of communication to influence and facilitate different forms of activities such as knowledge building, trading, consumption of content, etc. Three most important aspects of the convergence have been identified: (1) its uniformity and pervasiveness; (2) virtualisation of brands on top of standardised infrastructure and; (3) cross-pollination of technologies.

It seems that convergence will have a significant impact on different aspects of confidence, both in terms of Alice's ability to determine whether she should be confident about Bob and Bob's ability to behave in a confidence-building manner. However, the impact may be of a mixed value. While convergence definitely re-defines the technical landscape of a relationship between Alice and Bob, we may not expect its immediate impact on their behaviour. They both face a challenge of tuning their signals and instruments to capabilities of convergent communication.

The pervasiveness of communication and increased exposure to convergent solutions can be seen generally as a positive trend, calibrating expectations, creating norms and facilitating evidence gathering. Transparency, rich communication and accessibility abound. Despite certain problems (specifically in the area of identity), convergence looks promising.

Similarly, the cross-pollination of technologies brings significant value to the creation of confidence by introducing a mix of matured technologies that can improve confidence, deliver innovations, drive excitement and otherwise develop the area to the benefit of communicating confidence.

The virtualisation of brands is a benefit of a mixed value: it highlights the importance of reputation and confidence but it also offers opportunities to contradict the process of confidence building by offering near-anonymity, discontinuity and mediocre quality. Even though the virtualisation delivers several important values from the business perspective, its potential negative impact on confidence must be better understood. Specifically, there may be a new role for infrastructure providers that offers deep differentiation through offerings that boost confidence.

It does not seem that convergence itself can influence Bob to become more trustworthy if he does not want to, it is only Alice who receives potentially better tools to assess Bob. However, the possibility of positive impact of technology on Bob cannot be rejected – society shapes the technology but it is also shaped by the technology. An example of a technical environment that may drive Bob towards better behaviour seems to be possible and is discussed in this book as Trust-Enhancing Technologies.

## Bibliography

[Barth2006] Adam Barth et al.: Privacy and Contextual Integrity: Framework and Applications. In *Proc. of the 27th IEEE Symposium on Security and Privacy, IEEE Computer Society*. Available at: http://www.adambarth.org/papers/barth-datta-mitchell-nissenbaum-2006.pdf.2006.

[Bohle2000] K. Bohle et al., Electronic Payment Systems: Strategic and Technical Issues, Background Paper Nr. 1 of the EPSO, Institute for Prospective Technological Studies, December. 2000.

[Cofta2003] Piotr Cofta and Stephen Crane: Towards the Intimate Trust Advisor. In *Proc. of the First Int. Conf., iTrust, 2003*, Berlin: Springer. Lect. Notes in Comp. Sci. 2692. 2003.

[Cofta2005] Piotr Cofta: Impact of Convergence on Trust in e-Commerce. In *Proc. of Networking and Electronic Commerce Research Conf. NAEC2005*. 2005.

[Cofta2006] Piotr Cofta: Trust in a Virtual World: The Value of Mobile Experience. Presentation for Joint Security and Dependability/Mobile and Wireless Workshop 11–12 May. Brussels, 2006.

[Gavish2006] Bezalel Gavish: How Can We Reduce Fraudulent Activities on Auction Sites? Keynote speech on Networking and Electronic Commerce Research Conf. (NAEC). 2006.

[Habermas1986] Jürgen Habermas: *The Theory of Communicative Action*. Vol. 1. *Reason and the Rationalization of Society*. Cambridge: Polity Press. 1986.

[Hardin2002] Russel Hardin: *Trust and Trustworthiness*. New York: Russell Sage Foundation. 2002.

[Karvonen2000] Kristiina Karvonen: The Beauty of Simplicity. In *Proc. of the ACM Conf. on Universal Usability*. 2000.

[Lacohee2006] Hazel Lacohee, Stephen Crane, and Andy Phippen: *Trustguide*. Available at: http://www.Trustguide.org.2006.

[Mallett2005] Colin Mallett, Richard Dennis and K. Jit Singh: 2005: An Interesting Year for Convergence?. *Hong Kong Mobility Roundtable*, 2–3 June. 2005.

[Milgram1967] S. Milgram: The Small World Problem. *Psychology Today* 1(61). 1967.

[Tadelis1999] Steven Tadelis: What's in a Name: Reputation as a Tradeable Asset. *American Economic Review*, 89(3): 548–563. 1999.

# 13

# Confidence Creation in e-Business

Caveat emptor

((Let the buyer beware) – Latin)

## 13.1 Introduction

The use of digital communication in business has a surprisingly long history, as it can be traced to the early days of telegraph. Traders, buyers and sellers used to rely on Morse code to investigate, set prices, make and implement trading decisions. More recently, the mobile phone, before becoming the omni present accessory, has been well adopted by the business community as an instrument of convenience and necessity.

The development of the Internet facilitated new types of commercial activities. Rapid communication that was usually confined to the domain of a few well-connected investors, became an easily accessible activity of millions. Banks, investment brokers, stock exchanges – they all went to the Internet, decreasing their cost and increasing their customer base as well as their revenue. In this rapid and massive process the perception of confidence, an essential element of every transaction, has been re-defined. New types of services, available mostly through digital media, face an important task to develop confidence-building methods in a new, digital environment.

From the very onset, all types of e-commerce demonstrated significant interest in confidence. There are several models that cover the 'decision to trust', the mental process that enables the user to proceed with transaction despite certain obstacles. The popular interdisciplinary McKnight's model ([McKnight2001] with later changes in [McKnight2003]) has been specifically developed on the basis of e-commerce. Tan [Tan2000] provides a simple model of trust creation for the purpose of e-commerce transactions. Egger [Egger2000] has developed quite a detailed list of trust-building elements for websites.

We will demonstrate here that eBay's activities related to confidence-building go beyond the well-known reputation system and form the specific pattern, following the framework defined by the complexity-based models of confidence. Other Internet-based solutions that rely on consumer confidence (e.g. in e-commerce, e-health, e-government, etc.) may benefit from understanding such a framework, as it provides an efficient method to determine areas of activities that should be addressed in order to achieve a higher level of confidence.

We will not consider eBay's involvement with enablers that should facilitate the assessment of confidence. eBay's practices in this area are not much different from those used by other sites (even though they are quite comprehensive), so that they do not seem to explain the success of eBay. We also do not concentrate here on components that are common to several Internet-based companies or that are quite obvious. Elements such as personalisation (e.g. the user home page), facilitation of social interaction (e.g. user groups), technical support (e.g. FAQ) or real-world conferences are not mentioned here, even if they also contribute to the overall perception of confidence. We discuss here elements that differentiate eBay from other companies and that are likely to contribute to its success.

An early version of this chapter was written during the first half of 2006 and was presented as [Cofta2006].

## 13.2 eBay

eBay (www.ebay.com) is the largest e-commerce site, dedicated to the global person-to-person trading. The commercial success of eBay can be attributed to its ability to build and maintain trust and confidence in electronic transactions between relative strangers. eBay is primarily known for mastering the feedback-based reputation system where after each transaction both parties rate the service they have received. Such feedback, stored in a central database, is used to calculate the overall reputation rating for each user. Such rating is publicly available (even though feedback details may be kept private) and is known to influence buyers' decisions and the profit margin of the seller [Resnick2006].

The phenomenal growth of eBay created significant interest in its operation, including analysis of its business strategy (e.g. [Chen2005]), and of course its famous reputation system. This unique position of eBay cannot only be attributed to its relatively early market entry (it was not the first online trading site) or to its flawless operation (it made some errors). There is an understanding that even though eBay's technology can be easily imitated, eBay's phenomenon cannot be easily matched because of the high level of confidence it has managed to create and maintain while operating over the Internet.

The unique position of eBay should therefore be attributed to its ability to create and sustain a high level of confidence ('trust') among its users. Certainly confidence (and trust) are prerequisites of any commercial relationship and in the longer run both merchant and customer tend to select tools and institutions that offer the highest confidence and are willing to pay a premium for it. Specifically, eBay managed to create not only an environment where two relative strangers can trade with reasonable confidence, but also where such a reputation (and consequently confidence) are one of the primary attributes of a trader.

There is a significant amount of research that discusses reputation-based trust, specifically in the form that is present on eBay – the centralised reputation system, from Resnick's original investigations [Resnick2001] to propositions of a more formal approach [Josang2002] or the wide exploration of eBay's phenomenon [Bunnell2000] and efficiency of its reputation system [Bolton2003]. A recent book [Chang2006] discusses the meaning and the creation of trust in web-based systems, with special focus on opportunities brought by web services.

However, the operational structure of eBay goes well beyond the reputations system and consists of several components that only together form the overall perception of the company and the way it handles the issue of confidence. As different types of user may come into contact with different components of eBay (depending on the profile of their activity), they may view eBay from different perspectives. However, as all groups of users benefit from confidence, all components of eBay should contribute to the perception of confidence.

## 13.3 eBay and the Model of Confidence

Even though eBay relies on confidence between its users, this does not imply that eBay is somehow 'creating' confidence. The complexity-based model of confidence provides the framework that will be used to structure the analysis of eBay's operational components. Not going into details, it is worth mentioning that the model assumes that confidence can be assessed through trust or control, on the basis of evidence. Control instruments require further confidence in their operation. Such creation of confidence can be conceptually divided into four categories of entities: world, domain, agent and transaction (see Figure 13.1).

While confidence cannot be automatically created, eBay can deliver tools that facilitate the proper assessment of confidence, in the expectation that the economy (e.g. price differentiation) will drive users into the appropriate behaviour. eBay can also deliver tools that – if such tools are later elected by users – can increase control of the transaction, facilitating the development of initial contacts. By combining those two: better assessment and access to tools, eBay can expect that eventually users will move from control to trust in their relationships.

Therefore eBay is interested in both: better assessment and the availability of tools, expecting that this will lock its users into their relationships. As such relationships can be maintained conveniently only through eBay, this eventually locks all of them into the relationship with eBay, which is perceived as the most confident place to do business. This self-perpetuating scheme cleverly replicates some of the existing social tools and combines them with new interaction forms that are facilitated by the digital communication.

Figure 13.2 provides a quick overview of eBay's tools, components and initiatives that address different aspects of the confidence assessment and building. The rest of the chapter will be devoted mostly to more detailed analysis of those components, demonstrating how and why they influence confidence.

The following analysis of eBay's different components is structured according to the model. For every construct of the model, the analysis demonstrates how eBay has been able to maintain evidence that drives the perception of confidence – either by enhancing evidence that is already present or by developing tools that can deliver such evidence. Some of those components are well known (like the reputation system) while others (like the use of additional communication channels) have only recently been (written in mid-2006) announced and are not yet fully integrated into the fabric of eBay.

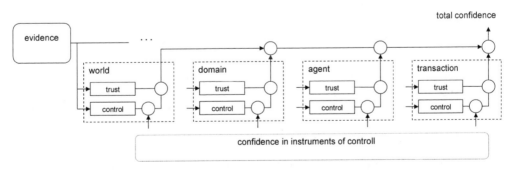

**Figure 13.1** The complexity-based model of confidence

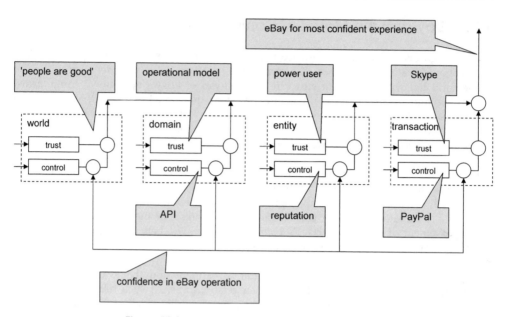

**Figure 13.2**  eBay confidence-building framework

### 13.3.1 Confidence in the World – People Are Good

Users' general confidence in the world ('propensity to trust') is not directly addressed by any mechanism available within eBay, as it exists only beyond the experience provided directly by eBay. However, eBay apparently understands the fundamental role of the ability to be confident in the world (specifically to trust the world) as the enabler and the prerequisite to any form of cooperation, including e-commerce. Research [Marsh1994] has demonstrated that the 'optimistic' default attitude leads in general to better results than the 'pessimistic' one, and that if they are not optimistic, people tend not to cooperate at all.

Therefore eBay is active in the area of promoting general confidence in the world. Not being able to provide mechanisms within eBay, it resorts to actions that are external to the trading activity. First, eBay defines itself as the community-building company, exemplified by the activity of eBay Foundation. Even through the Foundation is relatively small (and does not attract much publicity), its existence itself means that eBay would like to be associated with confidence-building in general.

Further, and more visibly, eBay attempts to reinforce the perception of the world as a relatively friendly place through its advertising activity. Its 2004 campaign [Howard2004] 'People are good' portrayed friendly strangers coming to help when in need, by capturing potential everyday experience (refreshingly different from several campaigns that depict rather selfish individuals). Neither activity has been particularly visible (the campaign had a limited scope), but the both demonstrate certain understanding of this area of confidence-building. We can only hope that eBay will continue to improve our perception of the world.

### 13.3.2 Trust in Trading Domain – Operational Model

Considering the domain of personal Internet trading (i.e. person-to-person) in general, it is necessary to realise that from an eBay perspective there is no great difference between the whole domain and eBay itself – eBay defines the world of personal trading. This can be attributed to eBay's market share and to eBay's ability to create the complete trading experience within the domain of eBay. Therefore, from eBay's perspective, trust in the trading domain can be equivalent to trust in eBay.

The creation of trust is based on the availability of positive evidence of three kinds: continuity, competence and motivation. The concept of auctions, the core idea of eBay has been known for as long as the trading itself. eBay automated the process, added the electronic front-end and created the global market but it did not modify the concept itself. Various types of auctions that are supported by eBay are the electronic versions of well-known business practices, including their well-established names. This simple move incorporated eBay into the overall fabric of business tradition and created the strong perception of *continuity*, by effectively re-using existing evidence of trust in the new area. The value of this move can be seen clearly in the light of new operational and business models that were tried extensively during the dot-com boom. Such models (even if some might eventually be viable) alienated the user and suggested the ephemeral relationship that could not support trust.

The *competence* of Internet-based companies is frequently judged by its technical and a esthetical quality [Karvonen2000], thus equalling their ability to perform technical tasks (such as running the website properly) with their ability to fulfil business expectations (such as delivering goods). eBay seems to understand it well as it is developing appropriate evidence. eBay is addressing competence through continuous innovation and refinement of its web portal, as well as its continuous technical development. Looking beyond the Internet image, the size of eBay, combined with its relative longevity, serves as another proof of its *competence*, as incompetent companies have usually been unable to survive the dot-com crash. Further signs of competence can be drawn from eBay's ability to withstand attacks (eBay has received more than its fair share of denial of service attacks, frequent phishing and spam attacks with no visible impact on its operation) as well as its visible section dedicated to security.

The business model of eBay is well understood: it generates its revenue by charging certain fees on placement, transaction and additional services. The structure of those fees as well as its level is reasonably stable and can be obtained on request. Again, as simple as it is, the business model can be easily understood and creates the perception that eBay is *motivated* to work for the benefit of its users – for purely financial reasons. No additional justification (e.g. market experiment, goodwill operation, etc.) is needed. Such evidence is well understood among eBay's users who are themselves interested and skilled in trading.

### 13.3.3 Control over the Trading Domain – API

The perception of control over the trading domain (which in this case is represented by the control over the trading capability of eBay) is strongly influenced by the availability of APIs (Application Programming Interface) [Mueller2004]. Such APIs allow the limited 'back-end' rapid access to the trading mechanism of eBay and its integration with the user's computer infrastructure. Similarly to the previous context, the dominant nature of eBay equals the concept of trading domain with eBay itself: controlling eBay means controlling the trade.

The selection of functions available through API is aimed generally at sellers who can place items, initiate auctions, query results, etc. Buyers can make use of some functions, but the most interesting one – the ability to bid on the auction – is not present (apparently to discourage automatic bidding). Implications of this decision go beyond the analysis presented here, but it is worth noting that it impairs the personal nature of relationships (and negatively affects e.g. similarity), by allowing the seller side to be automated while requesting personal operation by the buyer. This short analysis will be performed from the seller position.

The APIs do not provide any functionality that cannot be otherwise accessed through the regular browser-based interface. Therefore, it is the ability to access such functionality faster and in an automated manner as well as the ability to integrate it with their infrastructure that creates the needed perception of control.

Access to API is not particularly restricted (every seller can use them, provided that they fulfil certain minimum requirements), but it requires significant investment from the seller side and is probably not widely known to the casual user (specifically to the casual seller). Buyers are usually not even aware that APIs are available, unless they are interested in being sellers.

The seller sees the heightened *influence* over the trading process as he can create actions as well as respond to events much faster and (assuming integration) in a manner that is automatic and consistent with current business practices.

The API increases also the *knowledge* that sellers have about all the related actions and about other users. Things that are not practically possible through the manual interaction are becoming increasingly available through APIs. For example, the seller can easily have complete information about all its bidders/buyers (including their bidding history), so that the seller can e.g. alter further submissions depending on the profile of bidders.

Finally, the real-time nature of API connection means that the seller can introduce the automated *assurance* schemes that can raise alarms or take automatic corrective actions in case of suspected problems. For example, the API allows automation of the processing of non-paying bidders, but it can be also used to detect (and recover from) situations such as infrastructure failure, organised bidding, etc.

Note that there is an interesting relationship between the usage of APIs (which fall into the domain of control) and trust in users, as discussed later in the context of PowerSellers. The existence of API allows sellers, e.g. to develop attractive applications. The existence of such applications, that signals the investment by the seller, may potentially become evidences of trust in such a user (continuity, competence and motivation). Certainly, the user who has invested in integration can be expected to continue its operation, has demonstrated competence and is willing to observe eBay's rules. This is one of the examples (discussed later) where eBay seems to be able to create a self-propelling system which will use one component to further leverage another component.

## 13.3.4 Trust in Agents – PowerSeller

eBay has introduced the special category of PowerSellers. In order to achieve the status of the PowerSeller (and the associated recognition, visible e.g. through a special icon next to the seller's name), the user must (among other requirements) maintain the high level of positive rating (currently 98 per cent) and an excellent sales track performance rate.

The concept of the PowerSeller may look like an additional recommendation from eBay and as such should be classified as an instrument of control. Closer inspection of the PowerSeller program indicates that it is actually not the recommendation. eBay takes a neutral stance and never claims that it has a particularly high confidence in PowerSellers or that it wants buyers to have such confidence.

What eBay is doing is to highlight certain properties of sellers, properties that can be potentially found directly by the buyer, by browsing permanent records of transactions stored by eBay – at the expense of time and complexity. From the perspective of the model, eBay is consolidating and selectively amplifying available evidence of trust that otherwise might go unnoticed or overlooked (it is e.g. unlikely that the average buyer will really browse permanent records to build his own reputation rating).

From the trust-building perspective, PowerSeller amplifies the strong perception of *continuity* within eBay. Requirements ranging from the minimum presence of 90 days to the minimum transaction volume of £750 (UK site) create the impression of a seller that has invested some of his time and is therefore likely to remain on eBay and will be willing to be bound by eBay rules and regulations. Further, the PowerSeller's reputation rating differs from the regular one to favour regular trading relationship – again the strong sign of expected continuity.

Similarly, the minimum required service standards amplify the perception of *competence*. The volume of trading required from the PowerSeller (from 100 items per month to 5000 items for the top tier PowerSeller) means that the seller is certainly competent in selling and that the seller spends significant amount of time on activities related to eBay to be able to proceed smoothly with the transaction.

The *motivation* aspect is amplified by the fact that the PowerSeller must uphold eBay community values and must not violate any policies (within the forgiveness period of 60 days). Even though the verification of the adherence to those values is again performed by eBay, this is the best evidence of moral standing of the seller that is available on the Internet, specifically that it comes at no expense to the buyer.

## 13.3.5 Control over Agents – Reputation

eBay is widely known for its reputation rating mechanism (see e.g. [Dellarocas2003] for an overview of research in reputation mechanisms) that collects, processes and then globally presents feedback from all transacting parties. The reputation mechanism concentrates entirely on confidence in trading agents. It is implicitly assumed that all participants are confident in the proper operation of the rating mechanism itself, as the reputation-based system can be classified as a level-removed interaction.

The rating model does not address the current transaction, leaving the assessment of confidence in this context to individual traders. Buyers and sellers are not bound by the rating and they can proceed (or withdraw) regardless of it. It is known, however, that decisions to proceed are heavily influenced by the reputation of the party.

The reputation system of eBay is classified here as an instrument of control, even though the reputation itself is usually associated with the notion of trust. This follows e.g. Lewicki's observation [Lewicki1995] that the reputation is a hostage of a system – clearly indicating its control aspect. This observation is specific to the way the reputation mechanism is used by eBay (and in similar social contexts). It does not preclude the usage of reputation to build trust in other contexts and applications. We will therefore be looking at the reputation mechanism from the perspective of an instrument of control, i.e. we will be checking whether its existence increases confidence by delivering additional control.

Reputation is the instrument of *influence* one party has over another one. In the case of eBay, reputation is an instrument of control the buyer has over the seller. The seller's reputation influences the buyer's decision, with buyers more likely to trade with sellers of good reputation. There is a premium on price that sellers of a good reputation enjoy. There is no symmetric influence of the seller's decision as the seller is bound to sell to the winning bidder regardless of its reputation.

The *knowledge* about the agent brought by its reputation translates into expectations regarding the transaction. Reputation, in fact, is an expectation of one's future behaviour on a basis of past ones, being an encapsulation of knowledge and extrapolation into the future. Highly reputed sellers (and buyers) are expected to conduct transactions smoothly and rapidly resolve potential problems.

Finally, reputation is the ultimate *assurance* tool. If the transaction falls through, the reputation is the best (and sometimes the only) retaliation tool that is within the easy reach of both parties. Dispute resolution on eBay prevents this tool from being abused, but its value must not be under-estimated. Losing reputation is one of the worst things that may happen to the seller, specifically that there are so many sellers with a stellar rating.

As it has been mentioned before, the reputation tool conveys also certain evidence of trust. Specifically, the large response base contributes to the perception of *continuity* ('everybody is doing it') while the high rating raises expectations regarding both *competence* and *motivation* ('everybody believes he must be good'). Note, however, that eBay has been able to develop a much better tool to convey evidence of trust (e.g. Power Sellers) and trust conveyed by the reputation mechanism can be sometimes overshadowed by such a tool. Still, reputation is the only tool that delivers evidence of trust regarding buyers.

## 13.3.6 Trust in Transactions – Skype

The formation of trust during the transaction is one of the key contributors to the formation of trust in agents, as it is the transaction time that delivers the majority of trust-related evidence. At the same time, the relatively short period of transaction does not lend itself to the formation of a complex relationship, but is usually sufficient to establish a sufficient yet shallow perception of trust, on the basis of limited but important evidence.

Unfortunately all the above is true if the transaction is conducted face-to-face, where trust can be built through evidence of behaviour, appropriate dress code, body movement, intonation, facial expressions, etc. Digital media are particularly non-conducive to evidence of trust, as they restrict communication to a very limited spectrum. Trust is notoriously hard develop through digital communication means.

We can speculate that one of the reasons behind the acquisition of Skype is eBay's belief that more personal communication (voice, video) will introduce (re-introduce) the perception of face-to-face contacts and will therefore reinstate the missing channel that can be used to deliver evidence of trust during the transaction time.

Speculating further how eBay may use the voice (and potentially the video) communication, it is worth looking again at the construct of PowerSellers. Similar to its strategy there, eBay is unlikely to position itself as a creator of evidence of trust, but it will restrict its role to the enabler, amplifier (and to certain extent certifier) of such evidences. Evidence itself will be always generated by the other party. The simplest form of it is to enable unconstrained phone communication between parties (specifically that Skype calls are free).

Note that eBay is facing here an interesting dilemma. On the one hand, it can attribute its phenomenal success to the fact that digital media has unified its traders, distancing their digital identity from the actual one. This pseudonymity decreased the potential cultural tension while the asynchronous nature of auctions enabled trading across time zones. The synchronous communication (such as Skype) may be more conducive to evidence of trust but it may also introduce elements of culture differentiation, language perception and may alienate traders from distant time zones and cultures, thus potentially decreasing trust.

## 13.3.7 Control over Transactions – PayPal

Commercial transactions are non-atomic by their very nature, i.e. there is always a 'trust gap' where one party has a clear advantage over the other one – whether there are goods shipped on credit or there is payment advanced. The ability to control transaction attempts to reduce the gap to a manageable size, using different tools.

eBay does not provide the goods delivery service (not yet, at least), so that its effort has concentrated on the control of the financial side of the transaction, protecting the position of the buyer. The current selection of instruments leverages eBay's integration of PayPal, the lightweight worldwide money transfer system.

eBay in principle does not mandate any particular payment scheme, thus allowing the buyer to select the instrument he is most comfortable with (provided that the seller accepts it). This addresses the need for the *influence* the buyer may want over the payment process.

At the same time PayPal is so deeply integrated with eBay trading process that it seems to guarantee the best *influence* – becoming the default choice for many.

eBay does not provide tools to monitor the transaction, but it encourages both parties to communicate and discuss in case of problems or even just to verify the progress of the transaction. From this perspective, the *knowledge* of the identity of the other party is the essential foundation of such instrument of control (not to mention that this positively influences one of the key enablers). PayPal provides the thorough verification of the seller's (and buyer's) identity, delivering a notion of a strong identity – stronger than in the case of an average user. Note that eBay resorts to more organised communication in case of dispute and it is reasonably easy for eBay to standardise the communication regarding the whole transaction, introducing another instrument of control over the transaction.

The *assurance* is addressed by the multi-level structure of control and this is apparently one of the current focal points of eBay. The buyer is protected by PayPal's and eBay's buyer protection program (depending on the payment instrument, with preferences for PayPal) and is indemnified against dishonest sellers up to a certain amount. If this is not sufficient, the dispute process can be supported by eBay's Trust and Safety team.

### 13.3.8 Fundamental: Confidence in eBay Technology

The confidence in control elements of the model (i.e. the use of API, the reputation system and the functionality of PayPal) depends on the confidence in the proper technical operation of eBay. The recursive nature of the complexity-based model says that such confidence can be gained through the process identical to the one described earlier, just focused on the operational aspect only.

Without going into detailed discussion, eBay should deliver six categories of evidence regarding its own technical operation: (1) technical continuity (ability to withstand overloads and attacks); (2) competence (smooth handling); (3) motivation (impartiality in information handling); (4) influence (acceptance of technical feedback from users); (5) knowledge (technical transparency); (6) and assurance (recovery from major crisis).

Interestingly, for the moment it seems that neither eBay's lax approach to privacy [Miller2005] nor its support for censorship has attracted much the public attention [Leyden2006]. It has even survived the open revolt of its own users [Haines2004], not to mention phishing, denial-of-service attacks and domain hijacking.

It is the understanding that currently eBay asks its users for a lot of 'blind trust' without delivering sufficient supporting evidence. Even though the everyday eBay experience is very positive (justifying such trust to certain extent), even a single incident may have a disproportionate damaging effect.

## 13.4 Potential Development of eBay*

The logic of the model suggests certain development areas that eBay may take into consideration in the future, on the assumption that eBay's goal will remain unchanged, i.e. eBay will concentrate on being the trading place that allows the best assessment of confidence and that generates the highest confidence.

---

* The first version of this chapter was written in January 2006. More than a year has passed between then and when this chapter was incorporated into the book. *This time has already provided evidence that eBay is continuously working on increasing confidence.* The original text below that was written in 2006 has been left mostly unchanged (apart from minor editorial improvements), *and most recent comments have been added in italic.*

Note that the author is not affiliated with eBay and does not have access to any insider or confidential information. The following are the speculations that are driven by the logic of the underlying model discussed earlier,

1. Expand individual areas of the model. The simplest direction is to expand concepts that have been already introduced beyond its original scope, to different components, to enhance existing instruments. For example, the concept of PowerSellers can be brought forward to create new categories of 'lifetime members', 'professional members', 'certified members', etc., where each subsequent category of membership will be to certain extent certified by eBay – on the basis of available evidence. Alternatively, the eBay business model may change towards the 'cooperative' one where particularly trusted users will be granted governing rights over parts of the market, including the revenue share. Other suggestions have already been included in this chapter.

   *There are indeed five levels of PowerSellers – from bronze to titanium. However, the qualification depends only on the value of monthly sales, with identical requirements regarding behaviour. So, titanium seller is no better than bronze – it is only bigger. From the perspective of trust creation it is doubtful whether those categories bring significant value to the buyer (actually, they are not even immediately visible to the buyer).*

2. Focus on areas of low coverage. Some areas are only scantily covered and definitely work should be done to improve the coverage. For example, the transactional trust has only recently been addressed (by Skype). Other means of developing transactional trust (such as instant messaging, or potential photography of a trading partner, not to mention video) can be considered. Similarly, the seller's portion of transactional control may require improvements.

   *Since then, Skype has been integrated into the fabric of eBay interaction so that it is possible to make easy (and free) Skype calls between traders. It is too early to say how such a service will be accepted, specifically that eBay is developing Skype in several directions (instant messaging, video conferencing, ringing tones, Skype for developers, etc.).*

3. Cross-link areas. It has been already noted that trust generated by PowerSellers links with the control provided by APIs. More cross-linking can be expected that will develop eBay into the fully integrated set of instruments, eventually leading to the perception of eBay as the universe of its own, that does not need any evidence apart from those that are generated within eBay.

   *The current strategy is indeed to cross-sell between three core businesses: eBay auctions and shopping, PayPal online payments and Skype Web telephone service [Auchard2006]. However, they can also be decoupled to create the foundation of a complete platform for new services, something close to the perception of the 'eBay universe'.*

4. Protect fundamentals. This is the great challenge, as the operation of half of all constructs (all instruments of control) are based on confidence in the proper operation of eBay. Such a challenge must be met not only by the continuous vigilance (that eBay has already claimed and implemented), but (recursively following the model) also by addressing trust and control that contribute to this fundamental confidence.

   (a) Trust in fundamentals. The smooth experience with eBay contributes to the overall trust in its operation, but not much more evidence is available. The foundation of *continuity*, the persistent memory, runs on unknown technology and (from the user perspective) has never been audited or cryptographically protected. The site seems to be *competent*, but the technology behind is entirely obscured. Finally, even though the technology is definitely delivering today, the *motivation* to provide the leading edge infrastructure is not substantiated by any evidence.

(b) Control. Evidence of control (user's control over eBay's operation as a whole) is less visible and it is believed that eBay can improve in this area. The perception of *influence* can be improved by better communication between users and eBay; *knowledge* can be addressed, e.g. by exposing key technical performance indicators while *assurance* in the longer run might be addressed, e.g. by voluntary regulation. Note that confidence in such control instruments depends (recursively, again) on the proper operation of eBay so that eBay is able to build the complete set of fundamentals rooted only in itself.

*Currently eBay is failing to improve its delivery on either account. Potentially, it is the most important risk factor that eBay is facing.*

5. *Improve the reputation-based system. Feedback, reputation and its scoring are the crown jewels of eBay, and with every feedback received the system becomes even more important for all its users. However, it is not (and never will be) perfect. Certain improvements that may further increase the value of the reputation are listed below:*

(a) *Expand the notion of PowerSeller to identify different types of sellers and buyers as there are different expectations regarding the behaviour (i.e. address differences in the similarity enabler). Currently the system is skewed towards the seller's reputation and buyer's feedback, i.e. there is no perceived value of buyer's reputation. The recent separation of buyer's reputation and seller's reputation is the first step in this direction.*

(b) *Create specialised thematic sub-groups with their own reputation (again, the similarity enabler). Currently reputation is attached to the abstracted activity of trading, without any specialisation. However, the reputation associated e.g. with selling cars may have little relevance to a reputation regarding antiques. It is not only the value of items, but mostly the expected competence in assessing and presenting goods as well as shipping and resolving problems. The problem here is the equivalent of brand over-stretch where the established brand may be perceived as incompetent (and therefore have a lower reputation) outside of its area of expertise.*

(c) *Differences in value of items traded may require different clustering of reputations, e.g. separating a reputation associated with low-value items and those that have been gained by trading high value ones. As buyers increasingly base their purchase decision on the reputation rating, it is important for them to receive clear indication about the perceived level of financial risk that past buyers have been facing.*

(d) *While evaluating and presenting the rating, cater for cultural dependencies, such as preferred types of trades (fixed price, auction) and the usage of different methods to build confidence (e.g. public communication vs. private communication) and enforce behaviour (e.g. group pressure, fear of losing 'face', etc.)*

(e) *Embed temporal characteristic of feedback into the reputation process, to allow a reputation discount in time. Deliver also certain forms of forgiveness. Specifically forgiveness, currently in its very crude form, is essential to prevent driving users who defaulted incidentally into assuming a new identity (hence improving on the identity enabler).*

## 13.5 Conclusion

We are witnessing the continuous rapid development of different forms of digitally-augmented communication, from e-mail to convergent communication systems to sophisticated workgroup support or projects in integrated healthcare systems. Successful adoption of those systems depends on whether we will be able to recreate and reinforce the confidence assessment and building process that is characteristic of face-to-face communication.

The complexity-based model of confidence provides a good framework that allows us to analyse the confidence. The model, applied to the operation of eBay, identifies a specific pattern, showing that eBay addresses (even though in several different ways) all the components of the confidence assessment process. This gives eBay a very strong position as a trading marketplace, as trading between relative strangers requires a good assessment of confidence.

Even though this chapter does not discuss the actual development strategy of eBay, it is already obvious that such a development tends to address all the components as specified by the model. Therefore, it is possible to outline certain expectations regarding future directions of eBay. As some components of the model are under-represented within eBay, such components are most likely to be addressed in the future.

We can see that there is a way to replicate the success of eBay in assessing, creating and maintaining confidence, also in other areas such as e-health or e-government. The complexity-based model of confidence provides the list and the characteristic of different areas that together form the confidence-reinforcing structure. By providing the components that fulfil the requirements of the framework (even though not necessarily by replicating and copying eBay's solutions), it should be possible to create systems that will deliver confidence – and will benefit from it.

What is the future of eBay? Knowing that prophecies in an ever-changing digital marketplace are particularly short-lived, it is unwise to claim certainty. However, it is obvious that the relationship between eBay and the majority of its users has developed well into the identity phase and that trust has been established. Several businesses and individuals use eBay as their major trading channel and have tied the future of their businesses to the future development of eBay.

Certainly eBay is not perfect. It occasionally fails, it attracts crooks, its reputation system can be deceived, etc. However, the current strength of the relationship between eBay and its users can easily withstand those problems. There must be something more dramatic that will upset the relationship. Knowing that eBay itself is not easy to be trusted (as it apparently does not care about evidence that may lead to increased trust and confidence), the greatest threat for eBay is eBay itself. Not external attacks, but potentially eBay's apparent negligence, significant abuse of its dominant position, large privacy violation or visibly skewed business model may cause users to abandon eBay.

They can easily flee to one of the existing several competitors of eBay that currently serve slightly different markets. Such competitors will still exist, even though they are likely to be small or local. They may specialise, may offer different trading models (e.g. with more focus on fixed prices sales), may introduce strong identity (e.g. by compulsory vetting of all traders), or may alter the reputation-based system to suit more balanced needs (e.g. better addressing the buyer's side of the equation). The number of strategies that allow them to survive is quite large, but as long as eBay does not make a major mistake, they will be only marginal players.

## Bibliography

[Auchard2006] Eric Auchard: EBay Sees Faster Growth Ahead, Investors Unmoved. Available at: http://yahoo.reuters.com/stocks/QuoteCompanyNewsArticle.aspx?storyID=urn: newsml:reuters.com: 20060505:MTFH07376_2006-05-05_01-05-25_N04268920&symbol=EB AY.O& rpc=44. 2006.

[Bolton2004] Gary E. Bolton, Elena Katok, Axel Ockenfels: How Effective are Electronic Reputation Mechanisms? An Experimental Investigation. *Management Science*, 50(11): 1587–1602. 2004.

[Bunnell2000] David Bunnell: *The eBay Phenomenon*. Chichester: John Wiley & Sons, Ltd. 2000.

[Chang2006] Elizabeth Chang, Tharam Dillion and Farook K. Hussain: *Trust and Reputation for Service-Oriented Environments: Technologies for Building Business Intelligence and Consumer Confidence*. Chichester: John Wiley & Sons, Ltd. 2006.

[Chen2005] Edward T. Chen: Leveraging IT for a Competitive Advantage: Case of eBay. Available at: http://www.iacis.org/iis/2005_IIS/PDFs/Chen.pdf. 2005.

[Cofta2006] Piotr Cofta: Confidence Creation Framework of eBay. In *Proc. of Networking and Electronic Commerce Research Conf. NAEC2006*. 2006.

[Dellarocas2003] Chrysanthos Dellarocas and Paul Resnick: Online Reputation Mechanisms: A Roadmap for Future Research. Available at: http://ccs.mit.edu/dell/papers/symposiumreport03.pdf. 2003.

[Egger2000] Florian N. Egger: From Interactions to Transactions: Designing the Trust Experience for Business-to-Consumer Electronic Commerce. PhD thesis, Eindhoven University of Technology, The Netherlands. ISBN 90-386-1778-X. 2000.

[Haines2004] Lester Haines: Spanish eBayers in Open Revolt. Available at: http://www.theregister.co.uk/2004/11/23/ebay_spain_strike/. 2004.

[Howard2004] Theresa Howard: Ads Pump up eBay Community with Good Feelings. Available at: http://www.usatoday.com/tech/webguide/internetlife/2004-10-17-ebay-community-ads_x.htm. 2004.

[Josang2002] Audun Josang and Roslan Ismail: The Beta Reputation System. In *15th Bled Electronic Commerce Conf.: e-Reality: Constructing the e-Economy*. Bled, Slovenia. 17–19 June. Available at: http://security.dstc.edu.au/papers/JI2002-Bled.pdf. 2002.

[Karvonen2000] Kristiina Karvonen: The Beauty of Simplicity. In *Proc. of the ACM Conf. on Universal Usability*. 2000.

[Lewicki1995] R.J. Lewicki and B.B. Bunker: Trust in Relationships: A Model of Development and Decline. In B.B. Bunker and Rubin (eds): *Conflict, Cooperation and Justice: Essays Inspired by Works of Morton Deutsch*. San Francisco: Jossey-Bass, pp. 133–173. 1995.

[Leyden2006] John Leyden: Skype Uses Peer Pressure Defense to Explain China Text Censorship. Available: http://www.theregister.co.uk/2006/04/20/skype_china_censorship_row/. 2006.

[Marsh1994] Stephen Marsh: Optimism and Pessimism in Trust. Technical Report CSM-117. In *Proc. IBERAMIA94/CNAISE'94*. Ed. J. Ramirez. New York: McGraw-Hill. 1994.

[McKnight2001] D. Harrison McKnight and Norman L. Chervany: Conceptualizing Trust: A Typology and E-Commerce Customer Relationships Model. In *Proc. of the 34th Hawaii Int. Conf. on System Sciences*. 2001.

[McKnight2003] D. Harrison McKnight, Chuck Kacmar and Vivek Choudhury: Whoops... Did I Use the Wrong Concept to Predict E-Commerce Trust? Modelling the Risk-Related Effects of Trust versus Distrust Concepts. In *Proc. of the 36th Hawaii Int. Conf. on System Sciences (HICSS'03)*. Available at: http://csdl2.computer.org/comp/Proc./hicss/2003/1874/07/187470182b.pdf. 2003.

[Miller2005] Ernest Miller: eBay to Law Enforcement: We're Here to Help. Available at: http://research.yale.edu/lawmeme/modules.php?name=News&file=article&sid=925. 2005.

[Mueller2004] John Paul Mueller: *eBay Web Services: Building Applications with the eBay API*. Sybex. 2004.

[Resnick2001] Paul Resnick and Richard Zeckhauser: Trust Among Strangers in Internet Transactions: Empirical Analysis of eBay's Reputation System. Available at: http://www-csag.ucsd.edu/teaching/cse225s04/Reading%20List/E-bay-Empirical-BodegaBay.pdf. 2001.

[Resnick2006] Paul Resnick: The Value of Reputation on eBay: A Controlled Experiment. *Experimental Economics*, 9(2): 79–101. Available at: http://www.si.umich.edu/~presnick/papers/postcards/PostcardsFinalPrePub.pdf. 2006.

[Tan2000] Yao-Hua Tan and Walter Thoen: Formal Aspects of a Generic Model of Trust for Electronic Commerce. In *Proc. of the 33rd Hawaii Int. Conf. on System Sciences*. 2000.

# 14

# Privacy and Authentication

> ... as we transfer our whole being to the data bank, privacy will become a ghost or echo of its former self.
>
> (Marshall McLuhan (1980))

## 14.1 Introduction

Privacy is our undeniable right and authentication is an unnecessary evil – or is it really so? We have accepted driver licences, credit cards, PIN codes, passwords and a myriad other authentication means, sometimes with pride, sometimes as a necessary evil, sometimes even as an improvement (specifically if the previous scheme has been painfully complex).

What we probably would like to have is an unobtrusive method to convince the other party that we are who we are claiming to be. However, we have learned – the hard way – not to tell too much about ourselves, specifically on the Internet, where the current (2007) rush towards social networking ends up too often in questionable marketing practices and PR stunts. But then, how can I convince you that I am who I claim to be (myself, driver, aged 21+, decent person, etc.) without telling you too much about myself? There should be a practical balance between our ability to convince the other party and our unwillingness to tell everything.

This chapter discusses the authentication framework that builds on the foundation of trust and confidence. This framework offers the improved authentication scheme where several parties can reassure each other about the identity of the entity without the need to endanger the privacy of such an entity that is being authenticated, by cross-linking or sharing the same identity among parties. In addition, such authentication can be constrained by a set maximum usability overhead.

The chapter is organised as follows. We start with clarifications regarding terminology, to understand what identity, privacy or authentication really mean. From there we will move to the description of a confidence-based authentication scheme that promises to protect privacy and keep complexity constrained at the same time. Living in a convergent world means that we may want to have a look at how existing technologies such as SIM can be used within this scheme. Finally, we will close the loop by discussing the impact of the improved authentication on confidence.

*Trust, Complexity and Control: Confidence in a Convergent World*   Piotr Cofta
© 2007 John Wiley & Sons, Ltd

Throughout this chapter the term 'party' refers to the authenticating party, i.e. the entity that validates the identity claim. This can be e.g. a service provider, a bank teller or a social security worker. As usual, Alice and her friends will play the role of such a party. The term 'entity' is used to identify the one that is being authenticated – whether a person or an agent. As usual, this role is played by Bob.

## 14.2 Terminology

We start with clarifications regarding terminology. We will explore three interleaving concepts: identity, privacy and authentication. The proper definition of those concepts (even though only within the scope of this chapter) will allow us to see that it is indeed possible to reach the proper balance between authentication and privacy.

### 14.2.1 Identity

The discussion about privacy should rightfully start with the discussion about identity. Identity theft is growing and we are becoming painfully aware what constitute our digital identity (specifically if it is too late to recover from the loss of it). Still, capturing the exact meaning of identity may be hard. We will not delve here, however, into different definitions of identity, but we will consider the simple model of the relationship between ourselves and the external world and from there we will derive a simple working definition of identity.

Let's assume (having a look at Figure 14.1) that everyone has something that we can call the self-identity, the primary perception of oneself. This true self, the self-identity can be defined [Giddens1991] as a self-reflective memory of oneself that should be continuously reconciled to construct the narrative of life. This definition, linking the self-identity with active personal memory, also encompasses the fact that one's self-identity is closely linked to a body and a physical presence – or rather with memories of it. Note also that several authors stress that unique properties (or properties in general) of an individual are foundations of his or her identity. Those two approaches are not conflicting and will be reconciled later in this chapter.

This self-identity is identified in Figure 14.1 by the Chinese character that can be translated as 'self'. This is (symbolically) what we really are (in a sense that it includes all our active memories) and this is (usually) what we would rather not present to the general public. For the purpose of our communication with the external world we construct presentations of ourselves, the specialised (and probably slightly improved) sub-sets of the self-identity, designed to do a particular job – representing us in different communication contexts. Some

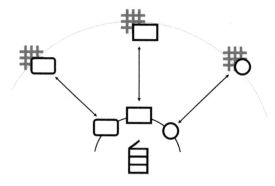

**Figure 14.1** Several identities of one self

elements of such presentations may overlap, but then some may be constructed (intentionally or not) for the purpose of one's presentation. They together form a multitude of presentation-identities, each representing only a part of the self-identity. The main function of each presentation-identity is to encapsulate the memory of interactions with particular 'others', e.g. the identity with a visible representation in a form of a credit card encapsulates memories of interactions related to shopping. Note that it is quite common for such identities to encapsulate elements that make them unique within a given interaction context and convenient to prove, e.g. in the form of a credit card and a PIN code.

For the purpose of this chapter, we will call each presentation-identity simply 'identity', as it is (also) designed to casually identify us in a particular context. Specifically, every identity should contain an arbitrary selected group of information that an entity (person, group, device, network) is willing to disclose to other parties (e.g. service providers) for the purpose of its unique identification and possibly personalisation. We can see that a single person (entity) can have and does have several identities, and that such a situation is quite popular, even though it does not always go to the extremes of having 300 SIM cards (which is actually a true story).

## Disembedding

Digital communication has been specifically damaging to the element of the self-identity that is not easily transferable over digital media: our body. The body represents an important part of self-identity as well as an important part of physical presentations. Digital communication created the real disembodiment [Giddens1991], by depriving us of our body and shifting the issue of identity entirely into the domain of the mind. Such disembodiment is a socially new experience and it is not yet integrated into the fabric of society.

More recently, digital communication allows us to freely create (and destroy) several forms of presentations such as log-in names, user profiles, etc. [Turkle1997]. In this process, each presentation becomes a specialised shield of the self, linked back to the true one, but designed to protect and defend it in case of any problems. For example, we may develop separate presentations regarding professional work and private life, in the expectation that this will prevent our private problems from damaging our professional career – or as a way to manage the work–life balance.

## Fundamental Tension

There is an underlying fundamental tension between the human desire to remain anonymous (not disclosing the self-identity) and the social desire to make other parts of life transparent, the tension that has to be managed somehow – otherwise our life will be unbearable. This tension has been usually managed by several social protocols and practices and more recently by legal regulations in the domain of privacy and data protection.

Such tension can be formulated as follows: an individual, for several reasons, tends to keep part of his life private and allows only partial access to it. Some memories are simply not to be shared, at least not with everyone. For that purpose, an individual may develop separate identities for different private and public parts of his life. This behaviour can be observed e.g. among politicians, famous artists, etc., even to the extent of creating entirely fake identities for public relations purposes. Even an average person may feel better e.g. with several independent financial identities (credit cards) so that the failure of one card does not negatively impact his presentation associated with other cards (e.g. with other banks). We can say that an individual may prefer to create several disjointed identities as a means to protect the true self from complete disclosure.

Society, however, has a vested interest in knowing as much as it can about its members, e.g. to make sure that they conform to social rules or simply to make sure that they are humans (so that they can be expected to behave in a human way). Society would like to see the self-identity and would like to deal with identities that are compatible with the self-identity (so that there is a relationship of identity between the self-identity and the identity-presentation). As there are several identities already present, society is pushing towards collecting and collating information about an individual in the hope of reconstructing the true self from this information. Specifically, if an individual can create several identities, society is willing to correlate or link those identities. This drives both gossip columns and the Total Information Awareness System, not to mention national identity card programmes.

## 14.2.2 Privacy

Privacy is an interesting thing. Depending on the person you are listening to, you may have an impression that we should not expect any privacy in the convergent world or that privacy is a defining element of our lives, the oxygen of the Internet. The often-quoted words of Scott McNeal: 'You have no privacy, get over it' are in stark contrast to the statement issued by Justice William O. Douglas: 'Privacy is the beginning of all freedom.'

The majority of users of communication systems (including users of the Internet) are one way or another concerned about their privacy. There can be even a quantifiable value assigned to the privacy, as demonstrated in [Buffett2004] for contact preferences and in [Huberman2005] for the disclosure of personal information. Interestingly, significantly fewer can clearly and coherently define what they mean by privacy. It seems that, whatever it is, we should be concerned if it is not here.

Let's return to our discussion about self-identity and presentation-identities. Within this context, privacy can be defined simply as the ability to restrict the cross-linking of one's presentation-identities. The disclosure of identity to the relevant party itself, being the inherent part of the communication, does not pose a threat to an entity's true self. In fact, the given identity has been designed to be disclosed, so that its disclosure cannot harm us. The main challenge is the disclosure of an identity that has not been designed for the particular relationship (disclosure out of context), or the cross-linking of more than one identity.

Note that the problem of potential identity abuse (e.g. identity theft) by the other party is not discussed here. We conveniently assume that the identity is designed in the way that it will withstand at least certain forms of abuse. For those really concerned about the potential of such abuse, anonymity or disposable identities remain viable options. The problem of privacy violation on the server side (e.g. unjustified extraction of data, data sharing, etc.) requires separate analysis using different tools. For example, [An2006] proposes the use of a Bayesian detection system to identify suspicious actions. Further, privacy standards together with supporting legislation [Robbins2006] are being developed.

We can see schematically in Figure 14.2 what privacy is – it is a wall between different identities engaged in different interactions. This perception of privacy as conceptual integrity was proposed in [Nissenbaum2004] and formalised in [Barth2006]. However, for the sake of the reasoning presented here the simplified approach is satisfactory. As long as nobody (except for ourselves) can correlate our different identities against our will, we can assume that privacy is preserved. If two service providers (or any other parties we deal with) can come to an agreement that may result in cross-linking our identities, then privacy can be compromised. Note that the desire for privacy may hamper our ability to build our reputation [Sartor2006] – there is always a price for the privilege of being left alone.

## 14.2.3 Authentication

Within an interaction, the entity may claim a certain identity, i.e. he or she may claim to be a certain person (agent). Certainly, the other party may have no desire to take such a claim for granted. Authentication is actually the verification of an identity claim within the context of

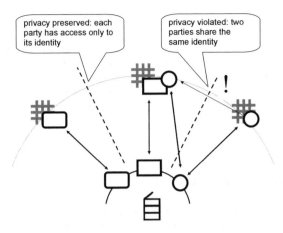

**Figure 14.2** Privacy and identities

an interaction, i.e. authentication responds to the question whether an entity is indeed what it claims to be, from the perspective of other parties involved. The concept of authentication is one of the well-defined constructs of access control and security (e.g. [Schneier2004]).

From the psychological perspective, authentication is a process of reconciliation between the expectation of one's identity and the actual identity. The reconciliation is performed by processing such elements of the history that are available to both parties. So, if Alice would like to authenticate Bob, she in fact reconciles that what she know currently about Bob matches her previous knowledge. Specifically, Alice would like to interact with Bob to verify some of her knowledge. Interestingly, this implies that Bob's presentation-identity becomes a part of Alice's self-identity. Alice may resort to a simple trick to augment her memory: she may issue Bob with a credential, a token that signifies the relationship, and then ask Bob later to present such a credential.

If Alice wants to authenticate Bob, she would like to make sure that he remained the same person (has the same self-identity) that she has experienced before, at least within the presentation that is available to her. However, as Bob's identity undergoes continuous small changes, Alice's reassurance about the stability of Bob's identity fades away in time. In order to make sure that Bob remains the same, she needs frequent interactions. Actually, if the intensity of interaction falls below a certain minimum level, Alice may start suspecting that Bob's identity is no longer the real one, but perhaps that Bob has abandoned this identity (e.g. passed away). In order to prevent this, both Alice and Bob should put some effort into their interaction.

The question of identity, expressed within a context of authentication, is therefore whether Alice is confident enough that Bob is indeed Bob. In a manner similar to the assessment of confidence that we use throughout this book, Alice is looking for evidence of Bob's identity. Bob sends some evidence involuntarily (e.g. his physical look), but can control the release of other evidence (e.g. what he knows). Authentication therefore becomes an interactive game where Bob is trying to convince Alice about being himself (or about being someone else, if he is actually impersonating another person). Alice's task is to determine to what extent Bob's identity meets her expectations.

Authentication happens through interaction and as such it places a burden on both sides. Following the concept of identity as memory, we can see that authentication supports the reconciliation of common memories. It is therefore not only the external party (e.g. Alice) who should put some work in authenticating Bob, but it is also Bob who is required to put some effort in authenticating himself to Alice. Hopefully, Bob may be interested in being authenticated by Alice.

The usability of the authentication process is currently seen as one of the major obstacles to the wide acceptance of security. The user's inability to manage several passwords has led to the concept of single sign-on schemes where the single set of credentials can be used to access several systems. However, this implies that the identity of the user must be shared among those systems, either through implicit action or with the explicit user's consent. This, in turn, may impair privacy.

## 14.3 Privacy and Authentication

Identity, as we understand it here, is a set of information known by Alice that allows Bob to be identified (preferably uniquely) in the given context by Alice. Authentication, in turn, is a process of verifying the identity claim, i.e. the process of determining by Alice her level of confidence regarding the relationship between the claimed and the actual identity.

The process of authentication is a source of irremovable conflict of interest between an individual (Bob) that attempts to restrict the usage of identity to its designated area and society that would like to capture the 'whole person' information across all the possible identities of the individual. This conflict is visible currently as a tension between the need for privacy and the demand for stronger authentication.

Authentication always disposes of certain privacy. One cannot expect to remain entirely anonymous and yet be recognised by others. However, there are several ways that decrease the scope of the potential damage to privacy. If rendered in technology (as contrary e.g. to a legal system), they are collectively referred to as privacy-preserving authentications, and are part of the larger collection of privacy-enhancing technologies.

Privacy-preserving authentication techniques all share the same approach as they attempt to restrict the disclosure of private information during the authentication process. However, they employ different ways to tackle this task.

The first group draws on foundations of trust management as it separates the identification from authentication by delegating authentication to specialised parties. Such an authenticating party deals with the authentication and then issues a token that can be redeemed for access to other parties' services (which interestingly resembles e.g. the ticketing scheme). A rather brief but useful overview of current trends in privacy-enhancing technologies can be found e.g. [Weis2006]. Some interesting works regarding ubiquitous computing are presented in [Al-Muhtadi2002]. In its simplest forms, Microsoft's Passport [Kormann2000] provides the centralised variant of this method while Liberty Framework [Liberty2003] defines the federated version. More universal meta-frameworks are promised by Bandit (http://www.bandit-project.org) and Higgins (http://www.eclipse.org/higgins/). Replacing individual tokens with group identities [Scavo2005] or with capabilities [Konidala2005] further improves certain properties of this group of authentication methods.

The second group retains the link between authentication and service (there is no intermediary token) but minimises the amount of information disclosed in the process. This perception of privacy as conceptual integrity has been proposed in [Nissenbaum2004] and formalised in [Barth2006]. Brand's selective disclosure [Brands2000] restricts the overall usage of private information while secret handshake [Tsudik2005] demonstrates that certain group-identifying properties can be authenticated entirely without disclosing individual identity.

The proposition discussed in this chapter accepts that the disclosure of one identity is the prerequisite for authentication and it assumes that an entity has several (possibly lightweight) identities – the concept discussed e.g. in [Ernst2006]. Therefore, the breach of privacy is associated not with the disclosure of one identity but with the ability to cross-correlate different identities that may lead to the disclosure of the 'true identity'.

Further, the proposition does not seek certainty in the authentication. Similarly to e.g. the concept of confidence described in [Al-Muhtadi2002], it accepts that all authentication methods are subject to fallibility and that the party seeks only sufficient confidence in the outcome of the authentication process, which can be potentially counterbalanced e.g. by business risk analysis.

## 14.4 Confidence-based Privacy-preserving Authentication

Referring to the complexity-based model of confidence, one can see the similarity between the creation of confidence and authentication. In both cases there is a desire to assess (and potentially increase) a level of confidence, but only within the limits of bearable increase in complexity. Authentication can be viewed as the process whereby one party (Alice, e.g. service provider) is ascertaining her confidence in the entity's (Bob's) claim regarding his identity. Drawing the similarity further, the choice of methods that Alice has at her disposal should be restricted by her affordable complexity.

Note, however, that authentication has an interesting property regarding complexity – it adds the burden of proof also onto the entity that is authenticated, not only on the party that desires such authentication. It is not only Alice that bears the complexity, but Bob as well. Only if an entity is identified without authentication, it is Alice alone that bears the whole complexity of the process. However, if Bob is being authenticated, he becomes part of the process, so that the more demanding Alice is, the more burden Bob bears. In the case of significantly asymmetric resources (e.g. a person being authenticated by the computer), it is the lower of those complexities that practically restricts the process.

As long as Alice stays within the identity designated for this relationship, Bob's privacy is of no concern. However, Alice may violate Bob's privacy by sharing his identity with others. It may happen either because others may perform additional authentication that reinforces the confidence in identity or because others are satisfied with the authentication that has been already made. In either case, the identity leaks beyond its original scope.

### 14.4.1 Privacy-preserving Authentication (PPA)

The privacy-preserving authentication defines the protocol that allows Alice to seek help from other parties without sharing Bob's identities, i.e. without violating the contextual integrity of his identity. Each party can authenticate Bob by using his identity that is known only to such a party while one party discloses to another party only its confidence in such authentication, not the identity. Such confidence is perceived by other parties as the evidence of control over the entity (evidence of knowledge), thus enhancing the strength of the authentication.

As parties do not share identities, it is important to make sure that all actually authenticate Bob. This is guaranteed by sharing the single one-time authentication token which is e.g. established by Alice and then used by others.

Note that one can rightfully point out that, as parties do not share any common identity, they are never certain whether they are authenticating the same Bob. However, this holds true for other authentication methods as well to the extent that it is included in legal statements supporting the authentication. For example, the small print on the credit card agreement says that one is liable for all transactions where the PIN has been correctly entered. If Bob discloses his PIN to his friend Dave, Dave can act as if he is Bob – and the bank will never know (and probably will never care).

It may be quite easy for Alice to ask other parties for additional authentication in order to seek even higher confidence. Therefore, the process of authentication may easily exceed

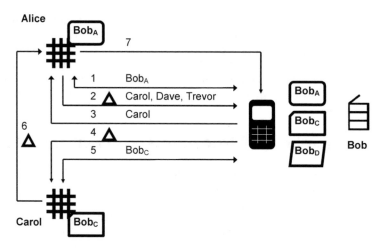

**Figure 14.3** Privacy-preserving authentication

Bob's affordable complexity, and he may refuse to co-operate. Throughout the process it is necessary to track the overall complexity burden and optimise it so that the maximum bearable complexity is not exceeded.

Even though it is Bob that primarily bears the complexity, the proposed model assumes that it is Alice who can define the limit of complexity, together with the required level of confidence. The reason for this is that Alice bears the risk of insufficient authentication as well as the risk of alienating Bob. She is also usually more experienced in balancing both complexity and confidence.

## 14.4.2 The PPA Protocol

In general, the proposed solution may work as follows (see also Figure 14.3). Alice is trying to authenticate Bob. Alice solicits the help of another party, Carol, to increase her confidence in Bob's identity. Note that Bob has three different presentation-identities ($Bob_A$, $Bob_C$ and $Bob_D$) established for the purpose of being authenticated by Alice, Carol and Dave, respectively. Bob presumably prefers to keep those identities separate.

1. Alice and Bob establish the connection for the purpose of authentication. Alice determines the desired level of confidence and the level of complexity associated with authentication. Alice also performs the authentication against $Bob_A$, i.e. the identity (and associated authentication scheme) that she is aware of. In this process she gain certain confidence at the expense of some complexity.

2. She may not be satisfied with it, so that she sends to Bob an authentication token and the list of entities she is willing to receive the assessment from, together with the outstanding balance of complexity.

3. Bob compares the list sent by Alice with his list of his identities. He decides to use Carol to vouch for his authenticity. Bob can use more than one agent to vouch for his authenticity if they match the common list. Bob reply to Alice telling that he has asked Carol to perform an additional assessment.

Example

**247**

4. Bob sends an authentication token to Carol. The token points to Alice as the source of authentication. Further, Bob forwards the level of complexity to Carol so that Carol will perform authentication only if it is no more complex than the allowed one.

5. Carol performs an authentication of Bob using $Bob_C$ identity and respective authentication scheme.

6. Carol sends the assessment of her confidence (but not the identity) to Alice, together with an authentication token and the outstanding complexity.

7. Alice consolidates the outcome from Carol and if she is satisfied, then she informs Bob about it. Otherwise, she may request an additional round of assessments.

8. Carol (as well as anyone else) may recursively invoke the scheme to satisfy her needs regarding her confidence.

## 14.5 Example

Let's consider the example where Alice would like to become confident in the identity claim of Bob. Due to the importance of the transaction Alice would like to reach the confidence of at least 0.9 and expects that Bob is willing to bear the burden of authentication expressed as 6 units. Note that 'complexity units' are used for illustration only and do not imply any specific measuring scale. A certain interpretation of the confidence value and complexity units is presented later. The example is illustrated in Figure 14.4 as the diagram captures the architecture, the message flow and the evaluation of confidence together.

Bob has established three different identifies $Bob_A$, $Bob_C$ and $Bob_D$ and Alice knows only the identity $Bob_A$. However, Alice can use Carol and Dave to further verify the identity of Bob,

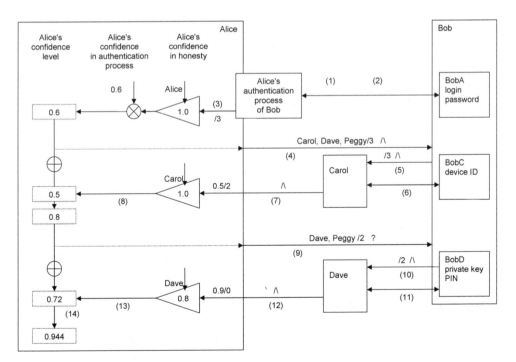

**Figure 14.4** Example of the message flow and calculation

this time using $Bob_C$ and $Bob_D$, respectively. Alice is fully confident in Carol's authentication but Alice's experience with Dave has been not perfect so that Alice is confident in Dave's authentication only at the level of 0.8.

Considering the control path from the model of confidence, Alice can use the authentication method that Bob must respond to. For example, Alice may ask Bob to provide the name and the password, previously established between them. If the password is correct, then Alice is confident in Bob's identity claim to the extent that is related to the overall quality of password authentication. Such an authentication process is the element of the control path as it is Alice who decides when it should be performed and it is Alice who knows the expected outcome of the process.

For the purpose of authentication, Alice establishes the connection with Bob (1) and performs authentication (2), requesting Bob to enter login name and password (attributes of $Bob_A$). The authentication is successful, but it is Alice's experience that the confidence of such authentication is only at the level of 0.6 and this form of authentication bears the expense of 3 in terms of complexity (3).

As the level of confidence is not sufficient and the complexity is not used up, Alice requests from Bob to become authenticated by Carol within the limit of 3 units (4). This request is passed by Bob to Carol (5). Carol performs such authentication (6) using e.g. the device identifier, already stored in Bob's terminal (as $Bob_C$). The cost of such authentication is 1 but the confidence is quite low, at the level of 0.5, and this is reported back to Alice (7).

Alice then calculates that she has reached the confidence of 0.8 (0.6+0.4*0.5) (8), at the expense of 4 units. As this confidence is still insufficient, Alice decides to ask Bob to perform additional authentication, this time with Dave at the expense of no more than 2 units (9). Bob forwards this request to Dave (10).

Dave performs the digital signature verification with $Bob_D$ (11), which requires Bob to enter his PIN – at the expense of exactly 2 units. The confidence associated with the digital signature is 0.9, and this is reported back to Alice (12). However, when Dave reports back to Alice, Alice weights her confidence in this authentication taking into account her confidence in Dave, having 0.72 as the confidence value (13). This finally allows Alice to reach the level of confidence of 0.944 (0.8+0.2*0.72) and proceed with the transaction (14).

The example above is using (for illustration only) the specific arbitrary formula (14.1.) to calculate a sum of confidences. This formula is equivalent to the confidence value proposed in [Al-Muhtadi2002], even though the underlying reasoning may differ. The formula assumes that the confidence is interpreted as a subjective probability of values between 0 to 1, inclusively. Further, the formula assumes that confidence is built gradually, by 'adding' the existing one and the new one. If there is an established confidence $c_A$ and the new independent source reports confidence $c_B$, then the new confidence is evaluated as a 'sum' of both:

$$c_A \oplus c_B = c_A + (1 - c_A) * c_B \tag{14.1}$$

Similarly the usability metrics is used for illustration only. It is assumed to express the complexity associated with authentication in 'units' where each unit represents the similar amount of complexity for the perspective of Bob. For an interesting overview of this subject, see e.g. [Sollie2005].

## 14.6 The Role of SIM

SIM (Subscriber Identity Module) is the small-format smart card that can be found in the majority of modern mobile phones. The main purpose of SIM is to hold the subscriber's identity that can be authenticated remotely by the mobile operator. In addition, SIM can hold several other identities, e.g. SWIM stores the PKI-based identity while ISIM extends the SIM identity for the purpose of multimedia connectivity.

The growing usage of mobile devices for access and authentication purposes makes SIM a very attractive element of the scheme. As different fragments of SIM can be used to authenticate the same device independently to different operators or service providers, it can be reasonably simple to consolidate evidences of such authentication. Even the simple ability to ask the operator for the 'second opinion' regarding the current authentication can greatly improve the confidence in such authentication.

The scheme may work as follows. Let's assume that Bob is known to both Alice and Carol. For example, Carol is Bob's network service provider while Alice is Bob's banker. Bob's identity known to Carol ($Bob_C$, Bob's phone number) may be different from Bob's identity ($Bob_A$, which is Bob's bank account number) known to Alice. It may be that even if Alice sends to Carol all information about Bob (something that she should not do), Carol will not be able to identify Bob on the basis of this information. It is only Bob that is able to reliably link $Bob_A$ and $Bob_C$ – but he may not be willing to do it, to protect his privacy.

Let's consider now two authentication techniques that are likely to reside in the mobile device. The first one is based on SIM authentication and works with Carol's infrastructure (identity and authentication for network purposes). The other may be the account number and password provided by Alice to enable Internet banking.

Let' assume that Bob performs the transaction with Alice over the Internet and that Alice would like to be confident enough that Bob is what he is pretending to be. Alice is likely to use Bob's device for this purpose. The device, for example, can prompt Bob to enter his bank account number and password. Using the scheme described above, Alice can also solicit help from Carol. Carol can query the device, perform SIM authentication and report back to Alice her confidence that such authentication is successful (e.g. that SIM has not been stolen, blocked, etc.). In brief, Alice can ask Carol for the second opinion about Bob's authenticity even though Bob operates under different, disconnected identities – as long as Bob uses the same device to communicate with both Alice and Carol.

# 14.7 Continuous Authentication

From Alice's perspective, the authentication provides confidence related to Bob's identity at a given moment in time. However, such confidence does not immediately disappear once the authentication is done. Rather, Alice has a reasonable expectation regarding the continuity of Bob's identity, so that she accepts further interaction without re-authenticating Bob – at least up to a certain point. For example, the Internet bank may not require re-authentication for up to half an hour assuming that there has been no more than one minute of inactivity. Similarly, Yahoo has recently introduced the feature not to re-authenticate its users for up to two weeks, if they connect from a given computer.

One may say that the confidence in identity generally decreases in time, specifically if not supported by fresh evidence. Generalising, the party has always certain confidence in the identity, even though in time such confidence may go down to almost zero.

If Alice is still confident about Bob's identity and another party (e.g. Carol) seeks the confirmation of Bob's identity, Alice may issue such confirmation without performing the actual authentication, thus significantly decreasing the complexity perceived by Bob. Even though the level of such confidence may not be very high, it may be sufficient for the planned action or it may be further increased by requesting similar action from other parties.

With enough parties being involved (e.g. in the ubiquitous computing environment), Bob is continuously authenticated by the group of parties, with a confidence level fluctuating in time, depending on recent updates. Assuming that parties can communicate and disclose their authentication capabilities and their current state, it may be possible to negotiate the mix of authentication methods that deliver at least the required level of confidence, always with the minimum complexity, thus greatly improving the usability of authentication.

## 14.8 Identity and Confidence

Up until now we have been considering the use of the model of confidence to ascertain the identity, i.e. to determine the level of confidence in one's identity through authentication. It seems that not only is it possible to increase confidence without endangering privacy, but also that such a process conforms to the model. However, the reverse question is also valid: if the identity is strong (i.e. it is assessed with higher confidence), how does it influence the process of assessing and building confidence in general? For example, let Alice depend equally on Bob and Carol and the only differentiating factor is that she can determine Bob's identity much better then Carol's. Does it make a difference to Alice's confidence regarding Bob and Carol? Is she more or less confident in Bob – or is she confident differently in Bob?

### 14.8.1 Identity and Enablers

The immediate answer, already discussed within the model, is that identity is the prerequisite of confidence – if there is a doubt regarding one's identity then evidence cannot be reasonably collected and processed over time so there will always be a higher level of uncertainty associated with such assessment. Consequently, Alice should be less certain regarding what she knows about Carol than what she knows about Bob. Potentially, this may lead to Alice being more confident about Bob (if evidence is generally positive), compared to Carol. So, strong identity lowers uncertainty – but for the moment it does not seem to increase confidence as such.

If we consider the way evidence is gathered, there seems to be a difference between evidence that is available directly from an agent (Bob, Carol) and evidence that is available from other sources (indirect evidence). For direct evidence, identity is indeed only a convenient way to collate evidence over time and to assign them to the same entity. Here, less confidence in identity translates in less certainty about evidence. However, for indirect evidence the situation is different. While direct evidence may come from direct observations of certain properties (hence may be hard to fake), indirect evidence is usually comes in a form of acts of speech (that can easily be faked at low cost). The most important property of a good observer is honesty – the willingness to tell the truth.

Assuming that what we say depends entirely on us (i.e. that the other party cannot easily tell whether we are lying or not), we should be willing to tell the truth according to our honesty, e.g. our inner willingness to support the person (by giving our opinion). Control actually does more harm then good here: under pressure, we can tell what the other person would like to hear, but it will not improve the position of such a person: he will be deprived of the truth, thus our evidence will not only be useless but also potentially misleading. Looking at large examples of such process, the collapse of the Soviet Union economy was partly attributed to the increasingly untrustworthy reporting system where the headquarters exercised pressure on its subordinates to deliver optimistic data and – not surprisingly – received such data, thus getting further and further from the real state of the economy.

### 14.8.2 Identity and the Perception of Truth

It is known that indirect evidence from known sources of established identity is regarded higher than evidence from sources that prefer to preserve anonymity. If Bob is unafraid of speaking openly, then it is expected that Bob has reasons to believe that what he says will withstand scrutiny of questioning. Personal delivery of the message implies personal involvement, implies putting one's reputation on the line to support the message. Bob's perception of truth may not be the one that is shared by his recipients, but it at least alleviates worries regarding his beliefs.

This behaviour is controlled by a social mechanism that binds identity, reputation and the strength of one's evidence. Openly and sparsely used reputation adds value to the message and reinforces reputation. Those that abuse such a mechanism (e.g. Aesop's 'cry wolf' boy) are ignored, punished or ostracised. For as long as identities are permanent (or at least long living), truth can be eventually discovered, and there is an expectation of repeat interactions, the mechanism works. However, the Internet guarantees none of those.

One must be also aware of the possible downside of strong personal evidence. In [Goffee2005] it is noted that the desire for privacy sometimes goes against the desire for the acceptance of a message. A person who wants to deliver the message to the public must choose either secure relative anonymity and weaken the message or disclose identity to strengthen the message. However, as disclosing the 'true self' is not beneficial in the long run, even authentic leaders should exercise caution and construct several presentations. This strategy will not only preserve the true self, but it will also facilitate the delivery of the message to different groups of recipients, as the presentation can be 'authentic' yet still shaped to increase the local acceptance.

So, Alice may not be more confident in Bob because she can identify him better, but she may be more confident in evidence provided by Bob if she is confident about his identity. Bob's identity makes little difference to evidence about Bob, but may make a significant difference if he is a source of indirect evidence about someone else. The amplifying effect of Bob's strong identity may be less important for Bob, but may be more important for others.

Societies that consider themselves free have developed several methods to guarantee the delivery of the truth. Constitutional guarantee of free speech, existence of free media, etc. are all instruments that can be used for the benefit of truth, to build confidence in others. They are built on the general premise that Bob cannot be punished (and even enjoys certain forms of protection) if he is willing to disclose his identity. Such identity can be further withheld (e.g. as a form of protection, e.g. a journalist may not disclose her sources of information), but in such case it is the journalist and the newspaper that put their identity and their reputation behind the message.

Not discussing casual aberrations of such provisions (such as the abuse of free speech or the biased reporting by the media), it is worth considering what the possible impact of 'citizen journalism' is on confidence. The Internet allowed low-cost publishing to become available to millions of people across the world. However, available tools do not usually cater for a strong identity, but allow the near-anonymity of postings. While this encourages participation, it may not positively influence the perception of honesty in such postings. While the majority of posters are benevolent (even though possibly not always objective), the availability of popular postings can easily be hijacked by the unscrupulous few that may turn it into the threat to online reputations or may be used to disguise themselves for financial gain, e.g. in marketing. Should this happen (and it already has), there may be a general loss of confidence in information published on the Internet.

Where does this lead us? We cannot guarantee the identity of the source of information, but we can ascertain our confidence in such an identity, possibly without endangering our privacy. Further, we do not know whether the statement from such a source is true or not, whether an agent is honest or not. What we can do is to use our confidence in the identity as a measure of our confidence (regarding the ability to tell the truth) in a source. Following observed social behaviour, we can assume that those who are not afraid to be identified (specifically not afraid to be identified in a permanent way) are more willing to tell the truth than those who prefer to remain anonymous.

This is the crude assessment of honesty and trustworthiness indeed, but it is also the attractive one. First, it is supported by social observations. Second, it seems to be fair: if I cannot ascertain someone's identity, I could not be able to assess his trustworthiness anyway, so that if I assess such trustworthiness on assumption that the identity is correct, then I should discount it to include such an assumption. It is also attractively simple: there is no need to engage in any evaluation whether the other party is willing to tell the truth – if one can present the solid proof of his identity, he is believed to tell the truth. Finally, this encourages people to behave in a trustworthy manner if they want to be heard: permanent identity (that gives a high rating with regard to the truth) is also a guarantee of considerate behaviour in the presence of social reputation schemes (as one cannot easily replace his identity, but must live with it).

### 14.8.3 Shared Identity and Confidence

Finally, yet another take on the subject of identity and confidence. The notion of identity goes beyond that of an individual and encompasses also group identity, the construct where individual agents develop a shared and common identity, potentially through shared memories and experiences. Even though such an identity is no more than a collective superset of individual presentations (as individuals may have presentations that do not contribute to the group), group identity plays an important role in the development of a group.

Considering the long-term dynamics of confidence (rather than the transactional assessment of confidence), we can see that confidence can be built within the relationship through phenomena such as reciprocity, experimentation, etc. It has already been stated in this book that the relationship usually grows in distinctive stages, from a control-based one (where the support of external institutions is needed) through knowledge-based (where habitual behaviour of participants is discovered and accepted) towards identity-based (where participants identify a common interest and develop a shared identity).

Once the identity-based stage is reached, the identification and confidence become intrinsically interconnected, not because one's identity is simply known but because one's identity is to certain extent entangled with others' identities – so that it is intimately known. If Alice and Bob are members of the same group, Alice implicitly trusts Bob, exactly because Alice's identity and Bob's identity are interconnected (at least within a group) so that Alice trusts Bob partly because she is sharing some of her identity with him.

Trust and identity here become self-propelling and self-fulfilling. Alice develops shared identity with Bob because she is interacting with him. Alice is interacting with Bob because she trusts him. She trusts him because they share the identity. The loop continues. Unless Bob significantly abuses such a relationship, they both are embedded in it – and actually they may be happy with it.

## 14.9 Conclusion

We took the circular route. We started from the observation that identity can be interpreted in terms of confidence as the outcome of an authentication process. We then saw that the ability to authenticate is one of the elements of confidence building. Whichever way we go, it seems that confidence is the key element of the story. We saw that a strong identity makes confidence in one's message higher, even though such a strong identity may harm the privacy. The mechanism that allows for strong authentication without compromising privacy is an ideal balanced proposition, addressing social and individual needs.

This chapter concentrates on the proposition of such a mechanism, presenting an interesting approach to authentication. The mechanism uses concepts derived from social sciences (such as trust and confidence) to be built on top of existing authentication methods so that the contextual integrity of privacy can be protected, i.e. identities assigned for different interactions cannot be cross-linked.

The mechanism allows for the authentication to be delegated and confidence combined from several sources, thus enabling the flexible and extensible framework that can cater for several different configurations and authentication methods, as long as several authentication methods are available for the same entity.

The proposed method demonstrates significant potential for addressing the problem of least-complexity continuous authentication in ubiquitous environment which may lead to the highly-usable authentication, significantly improving user experience and increasing the acceptability of information coming from the entity, eventually leading to the growth in confidence.

# Bibliography

[Al-Muhtadi2002] J. Al-Muhtadi et al: A Flexible, Privacy-Preserving Authentication Framework for Ubiquitous Computing Environments. In *Proc. Int. Workshop on Smart Appliances and Wearable Computing*, Vienna, Austria. July. 2002.

[An2006] Xiangdong An, Dawn Jutla and Nick Cercone: Privacy Intrusion Detection Using Dynamic Bayesian Networks. In *Proc. of ICEC'06*. Fredericton, Canada. 14–16 August. 2006.

[Barth2006] Adam Barth et al.: Privacy and Contextual Integrity: Framework and Applications. In *Proc. of the 27th IEEE Symposium on Security and Privacy, IEEE Computer Society*. Available: at http://www.adambarth.org/papers/barth-datta-mitchell-nissenbaum-2006.pdf. 2006.

[Brands2000] Stefan Brands: *Rethinking Public Key Infrastructures and Digital Certificates: Building in Privacy*. Cambridge, MA: MIT Press. 2000.

[Buffett2004] Scott Buffett et al: Determining Internet Users' Values for Private Information. In *Second Annual Conf. on Privacy, Security and Trust (PST'04)*. pp. 79–88. Fredericton, New Brunswick, Canada. 14–15 October. 2004.

[Ernst2006] Johannes Ernst: Light-Weight Identity 2.0. White Paper. Available at: http://lid.netmesh.org/docs/NetMesh-LID.pdf. 2006.

[Giddens1991] Anthony Giddens: *Modernity and Self-identity: Self and Society in the Late Modern Age*. Cambridge: Polity Press. 1991.

[Goffee2005] Rob Goffee and Gareth Jones: Managing Authenticity: The Paradox of Great Leadership. *Harvard Business Review*, December: 87–94. 2005.

[Huberman2005] Berardo A. Huberman, Etyan Adar and Leslie R. Fine: Valuating Privacy. *IEEE Security & Privacy*, 3(5): 22–25. 2005.

[Konidala2005] Divyan M. Konidala, Dang N. Duc, Dongman Leey and Kwangjo Kim: A Capability-based Privacy-preserving Scheme for Pervasive Computing Environments. In *Third IEEE Int. Conf. on Pervasive Computing and Communications Workshops (PERCOMW'05)*, pp. 136–140. 2005.

[Kormann2000] David P. Kormann and Aviel D. Rubin: Risks of the Passport Single Signon Protocol. *Computer Networks*, 33: 51–58. Available at: http://avirubin.com/passport.html. 2000.

[Liberty2003] Liberty Alliance: Introduction to the Liberty Alliance Liberty Architecture. Available at: https://www.projectliberty.org/resources/whitepapers/LAP%20Identity%20Architecture%20Whitepaper%20Final.pdf. 2003.

[Nissenbaum2004] Helen Nissenbaum: Privacy as Contextual Integrity. *Washington Law Review*, 17: 101–139. Available at: http://crypto.stanford.edu/portia/papers/RevnissenbaumDTP31.pdf. 2004.

[Robbins2006] Jim Robbins: Managing Information Privacy. *IEEE Security & Privacy*, 4(4): 92–95. 2006.

[Sartor2006] Giovanni Sartor: Privacy, Reputation and Trust: Some Implications for Data Protection. In K. Stolen et al. (eds): *iTrust 2006*, Berlin: Springer. LNCS 3986, pp. 354–366. 2006.

[Scavo2005] Tom Scavo and Scott Cantor (eds): Shibboleth Architecture Technical Overview. Available: http://shibboleth.internet2.edu/docs/draft-mace-shibboleth-tech-overview-latest.pdf. 2005.

[Schneier2004] Bruce Schneier: *Secrets and Lies: Digital Security in a Networked World*. Chichester: Wiley Publishing Inc. 2004.

[Sollie2005] Roar S. Sollie: Security and Usability Assessment of Several Authentication Technologies. Master's thesis, Gjovik University College. Available at: http://www.hig.no/imt/file.php?id=1039. 2005.

[Tsudik2005] Gene Tsudik and Shouhuai Xu: A Flexible Framework for Secret Handshakes. *ACM Conf. on Principles of Distributed Computing (PODC'05)*. Available at: http://eprint.iacr.org/2005/034.pdf. 2005.

[Turkle1997] Sherry Turkle: *Life on the Screen: Identity in the Age of the Internet*. New York: Simon & Schuster. 1997.

[Weis2006] Weis: Privacy-Enhancing Technologies. *IEEE Security & Privacy* 4(5): 59. 2006.

# 15

# Trust Management

We fortify in paper and in figures,
Using the names of men instead of men,
Like one that draws the model of an house
Beyond his power to build.

(William Shakespeare)

## 15.1 Introduction

As usual with everything about trust, the concept of trust management is also confusing. The basic definition of trust management defines it as a set of instruments that help replicate the relationship between people in the relationship between their digital agents. Extending this definition, we can position trust management as the middle tier between authentication and authorisation [Blaze2003], where such replication is combined with certain forms of delegation.

The way trust management is implemented today is usually as a policy-based credential management linked with authentication and authorisation capabilities. Such a scheme does not conduct, build or maximise trust but actually decreases the need for trust – it is essentially an element of control. As such, despite its name, it has little to do with trust and even less to do with managing it. Even if we restrict ourselves to very narrow expectations regarding trust, we can see that current solutions are not up to challenges of the future.

This chapter starts by exploring the traditional notion of trust management to see how it must evolve to address future needs. By identifying problems and extrapolating trends we can see that there is a new case for trust management – interestingly in managing actual trust. In a gradual shift, trust management should move from control and increasingly embrace evidence of trust to assess the extent of trust and to communicate trustworthiness.

## 15.2 Trust Management

Can we manage trust? Should we manage trust? Is trust manageable? Intuitively, 'trust' and 'management' are worlds apart. Management invokes images of supervision, enforcement

*Trust, Complexity and Control: Confidence in a Convergent World*   Piotr Cofta
© 2007 John Wiley & Sons, Ltd

and regulation while trust seems to be none of those. Therefore, first and foremost, clarification is required as to whether trust management is about managing trust.

Trust management is understood here as a set of instruments that are used to replicate in the technical domain the relationship that exists between agents in the human domain. This definition captures quite well what is currently associated with the concept of trust management, whether in literature or in practice.

If Alice and Bob are both humans, and they decide to deploy their respective agents (Alice) and (Bob) into the technical domain, they may wish to replicate the relationship that exists between them by endowing both technical agents with a proper set of instruments so that such agents behave towards each other in a way that reflects the relationship that exists in the human world (e.g. [Gerck1999], [Nikander2001]).

### 15.2.1 Example

Let's consider a popular experience of a mobile phone fitted with SIM card, as shown in Figure 15.1. Making a call requires cooperation between two parties: the user – Bob and the operator – Alice. Alice and Bob formalise their relationship in the form of a contract and Bob receives from Alice a token that confirms the existence of such relationship – a SIM card. Alice-operator controls her agent (Alice) that is a mobile network.

By modifying her user's register, she instructs her network to extend the service to any device that demonstrates that it has valid SIM card fitted. Bob deploys his agent (Bob) in a form of a mobile phone and inserts the SIM into such a phone so that the phone can demonstrate the existence of the relationship. When Bob wants to make a call, (Bob) presents SIM to (Alice) and is granted access to the network.

The first, most immediate observation (yet the very important one) is that this scheme has nothing to do with trust. Alice gave Bob a token (SIM) so that she can identify his device whenever he wants to use her network. In this process SIM became a part of Bob's identity that is known to Alice. It is also a credential that can be conveniently verified remotely by digital means. Once Bob's request is authorised, Alice can decide whether she want to grant him access, deliver service, etc.

This scheme is structurally familiar to credit cards, ATM cards, passwords and several other authentication schemes that pervade our everyday life. Essentially it is an authentication and authorisation process that can be found in several security textbooks. The only difference (yet the important one) is the separation of authentication and authorisation phases. The SIM (or

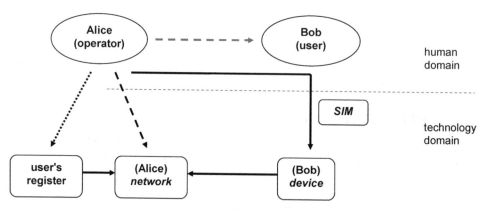

**Figure 15.1**   A simple trust management scheme

any other token) acts as a credential that is given to Bob, but can be used by whoever has it in their possession.

The only relationship to trust that we can see is that the design structure of the system indeed may replicate relationships of trust between different agents [Clark2007]. If Alice trusts Bob (e.g. she believes that he will pay his bills on time), then the SIM becomes a token of trust. However, if Alice does not trust Bob, she can still provide him with a service (e.g. in a pre-paid form) and give him a SIM.

This may lead sometimes to conflicts when the single device (specifically the mobile device) may be forced to implement conflicting trust relationships [Pfitzmann1997]. For example, the same device can have a relationship with an operator via a SIM card, with a software manufacturer via trusted computing platform, etc. Possible conflicts may lead to the situation where one device will have more than one trusted agents, each one trying to fulfil its agenda. The full impact of this observation on trust management goes beyond the scope of this book, but in general the conflict of relationships may be damaging to the user's confidence in the device and to the adoption of a scheme.

So, trust management does not in fact manage trust. It has nothing to do with trust. Trust management deploys credentials, verifies requests and implements policies – nothing more and nothing less. If parties want to instruct their agents to cooperate, then we may think of a certain behaviour that evokes the perception of trust. However, trust is not managed here – it is neither created nor destroyed, neither assigned nor removed. Alice and Bob are free to instruct their agents in any way they like, also creating the fake perception of a happy relationship where in fact one does not exist or the perception of hostility that is not true.

Trust management is about delegating responsibilities and letting agents do as they are told. Certainly, such delegation requires confidence (hence may require some trust). Both Alice and Bob must be confident about their agents: the operation of the network, the working of the mobile device, etc. However, Alice and Bob do not have to trust each other to use trust management (it is enough that Alice is confident about Bob, e.g. by controlling him) and they are unlikely to trust each other only because Alice gave Bob a credential that granted him rights to some of her services. Alice-operator does not trust Bob-user: she diversifies her business risk by serving millions of users and cuts off Bob whenever she thinks that his actions may undermine her profit.

### 15.2.2 Alternatives

There is also an alternative approach to trust management [Josang2005] where the concept of trust management is associated with the inter-human trust that is maintained through digital means. The main goal of such trust management seems to be the enablement of a proper assessment of the extent of trust one person has in the other. This approach is close to the concept of Trust-Enhancing Technologies, presented in this book, but unfortunately it is not what is usually called 'trust management'.

Yet another approach links trust management with the ability to induce trust regardless whether the other party deserves such trust [Elgesem2006]. This 'marketing approach' assumes that trust can be induced in Alice by cleverly managing Bob's presentation, e.g. by appropriately managing design elements of a website.

### 15.2.3 Generic Trust Management

Generalising from the previous example, we can now draw a structure of a generic trust management scheme. It has two distinctive operational stages and a structure shown in

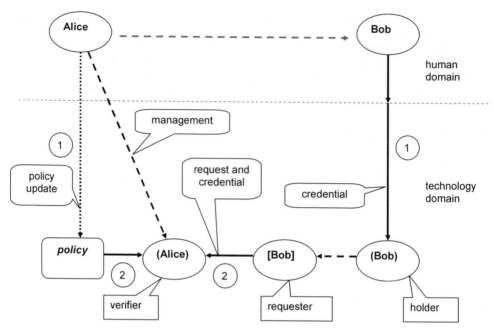

**Figure 15.2** Generic trust management

Figure 15.2. The prerequisite to any trust management is the real-life relationship of confidence between agents, e.g. Alice is willing to be confident ('trust') about Bob with regard to access to her services.

Trust management starts with provisioning. Here, Bob's agent (Bob) is issued with the credential that reflects Alice's confidence in Bob and (Bob) becomes a holder of such a credential. At the same time Alice's agent (Alice) may have her policies updated to reflect the existence of new credentials, i.e. reflecting the existence of a relationship between Alice and Bob. Note that Alice has also some means to manage (Alice), e.g. to stop her, to monitor her, etc, but those are for the moment less important here. We can simply assume that (Alice) is operational and that she follows policies.

In the operational phase, the requester [Bob] uses credentials to issue requests to Alice's agent and such credentials are used by (Alice) to verify whether the request can be fulfilled, using the policy. Note that there is an intentional gap between the provisioning and the operational stage, as the requester is not necessary a holder. (Alice) is willing to deliver service to whoever holds credentials – e.g. to [Bob], not necessary to (Bob). In fact, the holder (Bob) is not even required to retain or use a credential but can transfer it (e.g. to the requester [Bob]) or dispose of it.

The disconnection of authentication and authorisation is a strong point for trust management systems – it caters for much more diverse configurations where the authority to request a service can be freely delegated. For example, the typical ticketing system cannot be reasonably modelled within the limitations of a combined authentication and authorisation, but it can easily be expressed in terms of trust management. Tickets in general are excellent examples of transactional tokens that – once acquired by one party – can be transferred to another party who can fully benefit from access to the service.

## 15.3 Research Overview

There are several operational trust management systems and there is a large body of research work dedicated to this subject. Most of trust management systems (even WS-Trust) trace back their roots to KeyNote, but there is also a new group of systems that derive their operation from social observations. The following is a short overview.

Of different trust management systems, KeyNote [Blaze2003] that developed from the original PolicyMaker is a well-known example of a trust management system that follows the generic operational principle described above. KeyNote concentrates on the operation of a trust management engine, the decision-making algorithm that is used by the verifier (Alice) to validate the set of credentials in the request against the policy. The decision is made by asserting whether credentials are satisfactory, assuming that certain basic assertions (root of trust) are true. Further, KeyNote defines a standardised policy description language, built on assertions, that can be used to express a large number of policies.

KeyNote facilitated the development of different trust management systems and its architecture has been re-used in many practical implementations. For example, Fidelis [Yao2003] provides certain improvements to KeyNote, by separating credentials and policies and allowing local semantics with greater expressiveness of the policy language.

Web Services Trust Language (WS-Trust) [Anderson2005] brings trust management into the environment of web services. The core proposition remain generally unchanged: the Web Service (verifier) accepts a request from the Requestor only if the request contains proofs of claims (credentials) that satisfy the policy of a Web Service. The novelty comes mostly from well-defined protocols, data formats, etc. and from the clear separation of an issuer (Security Token Service) within the scheme. Specifically, it is possible to construct requests where credentials are dynamically delivered by the issuer within a scope of a request (e.g. as one-time credentials).

The step beyond simple representations of Alice and Bob is to allow their agents to alter their policies as the relationship progress. Such changes can be substantiated by the observable behaviour of another party so that Alice can collect information about Bob from several sources (the way e.g. credit card industry operates) and make transactional decisions to let Bob access the service or not. Such systems are collectively called Trust-Based Access Control (TBAC) [Dimmock2005] and their applicability has been studied in several different application areas (e.g. [Adams2005]).

An interesting deviation from the strict rules described here is trust management that employs initial blind trust where trust is expressed (and 'locked in') towards the first agent contacted. Alice-agent, once started, checks the presence of other agents and trusts the first agent that replies to her. If Bob happens to be such as agent, Alice decides to trust Bob from this moment. This interesting trust management mechanism is actually modelled after the behaviour of animals and – to certain extent – follows the development of trust between a child and a mother throughout early childhood.

## 15.4 Problems

Even though trust management has been used for quite a long time, there are still several problems that hamper its adoption. Those problems are attributable to the concept of trust management, to the way it captures the real-world relationship as well as to its operational environment. Providing solutions to those problems is one of the important drivers of current developments in trust management.

### 15.4.1 Reinstatement

The flexibility of the use of credentials can sometimes be considered to be a limiting factor. This problem is clearly visible in situations when the credential has been involuntarily transferred to an unauthorised party. Let's consider a situation where a SIM card (a credential) has been stolen from Bob (possibly together with his mobile phone). Alice (operator) is willing to serve whoever presents the SIM, not necessarily Bob – and that's exactly what she does, even though the contractual agreement is between her and Bob, not between her and the current holder of the SIM. The disparity between the operation of trust management and the contract can be only restored by a process of SIM revocation.

Suddenly the more traditional authentication looks attractive again – if only Alice can determine that it is always Bob that she serves, thieves will have no use for the stolen phone (and presumably will not be interested in stealing it). The conflict between flexibility (provided by transferable credentials) and the ability to reinstate the original relationship with Bob can be practically resolved, as can be experienced by anyone who has had to stop his credit card or a phone's subscription. However, methods that are used to manage such reinstatement go beyond the original premise of trust management.

### 15.4.2 Revocation

The premise of revocation is simple: if there is a termination of the relationship between Alice and Bob, both agents should be informed about such change so that the future relationship between agents will remain in accord with the real-life relationship. Unfortunately, the revocation is never immediate as it requires either Alice to push updates to all her agents or all agents to continuously update their policies. Nor is feasible e.g. in mobile, low-power networks where devices tend to conserve their resources and the communication may be slow or erratic.

The most immediate aspect of the revocation is as follows: if Alice stops being confident about Bob, she should have the ability to alter her policy so that Bob's request is no longer served. However, there is a deeper problem: should Alice not only revoke Bob's rights but also revoke all actions and processes he is involved in? Should she interrupt the call, terminate the transaction, evict him from the theatre? If the revocation is not instant (as is usually the case), should Alice revoke (or re-evaluate) all past transactions where Bob has been involved? To what extent should others that operated through Bob be affected? There are no simple answers to those questions as the restitution is never perfect.

The ability to 'turn back time' and provide more complete restitution can be of significant importance not only to the integrity of trust management systems, but also to the general adoption of technology [Lacohee2006].

### 15.4.3 Reliable Agents

It sounds obvious that Alice should be confident about her agents – otherwise the whole exercise of trust management will be meaningless. For trust management, the most important property of an agent is its reliability and dependability – the agent must do what it said. Such a 'trusted' agent (the use of the word 'trust' in this context is of course confusing yet popular) does not seem to be fully attainable.

The problem of an agent's reliability is beyond the scope of this chapter, but it rightfully belongs to the discussion of confidence (specifically trust) that people have in technology in general. There are too many things that can go wrong with modern complex agents: from confusing user interface to careless software to environmental effects and intentional attacks. Alice may never be sure that an agent will exactly follow her order (which may be quite frustrating), but she should have at least a reasonable level of assurance that an agent does what it said.

Trusted computing can provide some support here. While it cannot guarantee that Alice's agent will operate flawlessly, it can at least ensure that it has not been altered and – for as long as the hardware has not been tampered with – it is in fact Alice's agent.

### 15.4.4 Secure Communication

It seems obvious that Alice should have some means to communicate with her agent, but it is not obvious if the requirements of such communication channel are well understood. The concept of a 'trusted path' (yet another confusing use of the word 'trust') requires that there is a secure communication path all the way between Alice and her agent.

While it is reasonably easy to provide cryptographic protection for a digital communication channel or authenticated communication, it is surprisingly complicated to authenticate Alice (not the card, the password, etc., but actual person or institution) and to make sure that no undesired intermediary interferes with or monitors the interaction. Even the simple act of typing in the password on the computer screen can be tampered with or intercepted in so many different ways (hostile scripts, phishing websites, key, video recording . . . ) that it is still challenging to guarantee that Alice's agent really interacts with Alice (and vice versa).

Surprisingly, help may come from an unexpected direction: for the purpose of content management, modern operating systems introduce the concept of 'protected path', i.e. the hardware components that only authenticated software can use. This, applied to e.g. monitors and keyboards, may guarantee that there is no intermediary between Alice and her agent.

### 15.4.5 Monitoring and Reporting

The endowment in trust management works only one way – Alice tells her agent how to behave. Even if an agent is able to alter its behaviour, this updated knowledge about the relationship stays with the agent. In principle, nothing prevents two-way communication so that Alice can be informed about certain events and can enquire about the current decisions of her agent and about the level of confidence that her agent may have regarding Bob's agent.

In fact, such communication has been implemented in some agents. For example, the browser (that acts as Alice's agent) tells Alice whether it believes that a given page can be 'trusted' and (sometimes in an annoying manner) asks for support when its policy disallows automatic decisions.

This behaviour of an agent, even though potentially useful, is not standardised. Therefore the experience that Alice has with one type of agents cannot easily be transferred to another type of agent. Similarly, experience of different agents cannot easily be collected or processed. The ability to discuss with an agent will become increasingly important with the growth in the autonomy of agents. While simple agents were mostly on the receiving side, agents that can observe, enquire and reason may have something interesting to tell.

### 15.4.6 Expressiveness

The expressiveness of the policy language restricts what Alice can command, what credentials she may request and how detailed a description she may provide. It seems that the majority (if not all) of policy languages suffer at least in their inability to handle negative permissions, i.e. cases where a particular credential should prevent Alice from delivering the service. While it is possible to create a list of negative certificates (e.g. the certificate revocation list) as a kind of add-on solution, it is hard to embed it into the fabric of the policy.

The most important advantage of positive permissions is the lack of conflict. If there is the possibility to express positive and negative permissions, then the policy verifier (Alice's agent) must make a decision how to resolve the conflict, which is both complicated and problematic. However, the lack of negative permissions significantly restricts the expressiveness of policy in the real world. For example, the policy to allow service to everybody except for the listed few cannot be expressed at all. As the list of those that are allowed is potentially indefinite, then it is not possible to substitute the negative permission with a list of positive ones.

The inclusion of obligations provides another interesting development path to increase expressiveness of policies (e.g. [Brown2000]). Obligations and prohibitions request agents to undertake (or respectively not to undertake) certain actions. The inclusion of obligations and prohibitions allows for policies such 'Alice grants rights to the service on condition that Bob will pay for it later' to be correctly specified. The reasoning (and the verification) become potentially quite complicated (not to mention that it includes elements of temporal logic) but the expressiveness is definitely improving.

### 15.4.7 Identity

Identity-related problems emerge if trust management starts monitoring and assessing agents. If, for example, Bob (the user of the mobile phone) tends to make a single call a day to a local number but suddenly a burst of calls to some exotic locations shows up, Alice (the operator) faces a tough question: it may be that Bob has radically changed his lifestyle or it may be that his phone has radically changed hands. Should Alice refuse to serve such requests?

From the perspective of the model of confidence, Alice has good reasons to refuse, as an important enabler seems to disappear. Note, that Alice is contractually bound to serve Bob – not an impostor. Alice is continuously aware that the relationship between Bob and the SIM may change, whether voluntary or involuntary. Similarly, she is aware that Bob may change. She may perceive changes in a call profile as a change in Bob's identity, as seen though his call log – 'old' Bob makes local calls while 'new' Bob goes global.

Such a drastic change in identity may (from Alice's perspective) suspend or invalidate the original agreement – this is not the Bob that she used to know. Therefore, even though the relationship may not change, the perception of identity changes. Apparently she has a right to refuse a service and validate whether Bob is still the one she knows, i.e. to reconcile his identity.

## 15.5 Autonomy

One of the most important developments in trust management is the increase in autonomy of agents. The relationship between Alice (or Bob) and their agents may be quite straightforward but it may also grow to be complicated. Therefore, before we embark on the discussion of the future of trust management, we should investigate different aspects of agents' autonomy. For the purpose of this chapter, the autonomy of an agent can be structured along two dimensions (Figure 15.3.).

The growth of autonomy is driven by the overall growth in the number of agents, combined with the growing detachment of agents. As they are growing in numbers and are further away from both Alice and Bob, it is necessary to grant them more autonomy, otherwise they will not be able to perform their tasks. What we can expect is that the development in trust management should follow this challenge and it will move from managing relatively non-autonomous agents towards managing those that are autonomous.

The discussion below structures the area of autonomy. By determining how autonomous agents should behave, we will be able to see what are the main challenges to trust management and how trust management should respond to those challenges.

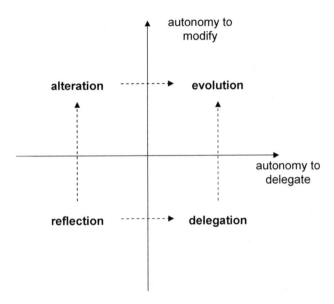

**Figure 15.3** Dimensions of autonomy

### 15.5.1 Reflection

Reflection is the simplest type of endowment, where agents have no autonomy. Alice's agent is instructed how they should behave when they encounter Bob (to be exact: any agent that presents credentials that have been originally issued to Bob) and such instructions are binding on those agents. Similarly, Bob's agent is instructed to present a given credential to Alice in order to receive services.

Note that because the simple reflection precludes Bob's agent from transferring the credential to another agent, Bob's agents are usually endowed with at least delegation autonomy. However, the problem of revocation (mentioned earlier) demonstrates that there is a hidden assumption that even though there is a delegation in place, this is not fully acknowledged and the relationship is expected to be of a reflection kind.

Note that the reflection itself is not a simple or static relationship. It should cater for the whole richness and possible dynamics of the interaction: Alice may stop being confident about Bob, Alice's agents may require additional confirmation, they may experience problems with Bob's agents, Alice may be confident about Bob only if Carol is also confident about him, she may be confident with regard to one service but not with another, etc. The reflection assumes that there is at least a working two-way communication channel between Alice and her agents, so that Alice can revoke her trust at any time and agents are instructed to check with her on all suspicious cases. Presumably, there is a similar two-way channel between Bob and his agent.

### 15.5.2 Delegation

The autonomy to delegate implies that an agent can employ other agents to perform its task, potentially according to a subset of a policy or a limited set of credentials. Delegation seems to be close to the spirit of the original trust management, specifically on Bob's side. Here, Bob's agent (Bob) can delegate his responsibility (to access Alice's service), together with a credential to [Bob]. Nothing (except its own rules) prevents (Bob) from delegating it to several agents or [Bob] from forwarding it to yet another agent.

On Alice's side the delegation has not been originally included in the concept, but practical implementations often distribute policies to several agents or allow several agents to share the same policy, effectively delegating activity to several agents. On both sides delegation re-defines the sense of identity: instead of talking of (Alice) and (Bob), we can talk about Alice's trusted domain and Bob's trusted domain – all agents that share the same policy or the same credential.

Delegation is the most obvious source of detachment: with every delegation, the agent acting on behalf of Alice or Bob is moving away from the original relationship of confidence. It has been already discussed that delegation makes revocation doubtful, as Bob cannot always reach immediately all agents that act on his behalf. Delegation makes also management more problematic.

For example, let's consider an example of delegation shown in Figure 15.4. Alice has delegated responsibilities to (Alice) and (Alice) has delegated them to [Alice1] and [Alice2]. (Alice) has also decided to endow them with different policies, *policy1* and *policy2*. Note that [Alice1] and [Alice2] are managed by (Alice), not directly by Alice. If this is the scheme, questions are whether Alice should be aware of such an arrangement and whether she should be able to directly communicate with [Alice1] and [Alice2], e.g. in case (Alice) stops responding to Alice's calls (the line with the question mark).

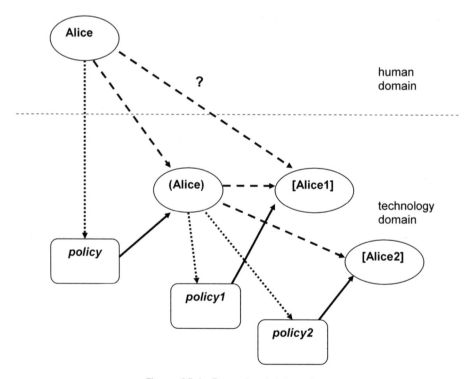

**Figure 15.4**  Example of delegation

### 15.5.3 Alteration

Alice is not always the best person to turn to in order to make the final decision. Agents can operate across a wide range of environments, they can work non-stop and they may engage with a large number of other agents. If an agent is unable to make an unattended decision that alters its policies then Alice may become overwhelmed with the support that she is expected to provide. This is probably the everyday experience of an Internet user that has a highly suspicious browser: dialogue boxes pop up quite often asking for verification and confirmation of myriads of situations in which an agent (the browser) is uncertain how to proceed – and the user is unlikely to be an expert.

As agents become more detached, even the sheer idea of Alice's support becomes problematic. Not only Alice has no expertise and no time for it, but the communication between Alice and her agents may be erratic: agents are quite often on their own as they travel across the network to meet other agents. Returning to an example presented in Figure 15.4, [Alice1] is probably not expected to bother Alice and all. The viable solution is therefore to let agents to alter their policies – to let them respond to their changing environments with agreed alterations to their policies.

For example, Figure 15.5 shows the alteration scheme. Even though Alice has endowed (Alice) with the initial set of policies, (Alice) has rights to alter her own policy depending on circumstances.

Such alteration does not change the foundation of an endowment, as (Alice) is still instructed by Alice what to do, but it definitely extends the scope of it. Alice may include in the endowed policy instructions of how (Alice) should react to changes in its environment, so that initial settings regarding agent's confidence are only a starting point, not the rule that cannot be changed. Alice may even tell her agents how to act upon meeting strangers that she is not aware of at the time of endowment, so that an agent can actually develop a complete relationship with other agents.

The ability to alter behaviour has been studied as part of already mentioned Trust-Based Access Control systems (e.g. [Dimmock2005]), where access rights are granted on the basis of observed behaviour as well as on the basis of initial policy. Further, this alteration is characteristic to autonomous agents where by interacting with other agents, they can change

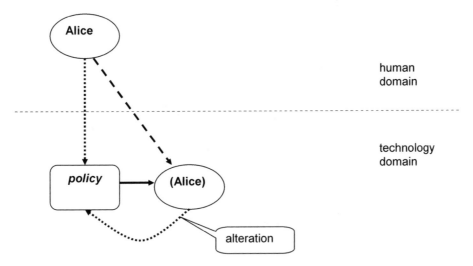

**Figure 15.5** Example of alteration

their assessment of them (see [Luck2005] for an overview). Current research is dominated by assessment that is essentially performance-based and built on the concept of knowledge (hence using control) and it seems to be concentrated on improving the overall welfare of the system (the total value of 'trust') rather than the ability of an individual agent to perform the correct assessment of its confidence.

### 15.5.4 Evolution

Potentially the ultimate endowment is to let an agent evolve by developing itself into new agents that may or may not resemble the original one, depending on the environment it operates in. Specifically, agents may specialise, replicate, etc. to solve problems that an individual agent is unable to handle and to act and move faster. They can alter themselves so that they may be better suited to work in a particular environment. There is even the potential of genetic-like evolution of agents where natural selection results in agents that are able to operate on problems and in environments that Alice has never even considered.

An example of replication is shown in Figure 15.6. Alice has endowed (Alice) with the autonomy to evolve. (Alice) has been developing her policy for some time, but then she decided to create a new agent, [Alice] and to endow her with a newest versions of her own policy. Once she has done it, (Alice) can terminate itself and delegate the primary connection from Alice to the newly created [Alice]. It is likely that [Alice] can replicate this process, altering her own policy and then creating another agent.

This combination of delegation and alteration allows agents to develop into a new form (or forms) that will better suit Alice in serving Bob, specifically if the environment is changing. However, such endowment comes with a caveat. Alice relinquishes all her remaining control over agents and relies only on the ability of her directly reporting agents to manage their respective offspring – and to tell Alice what has happened.

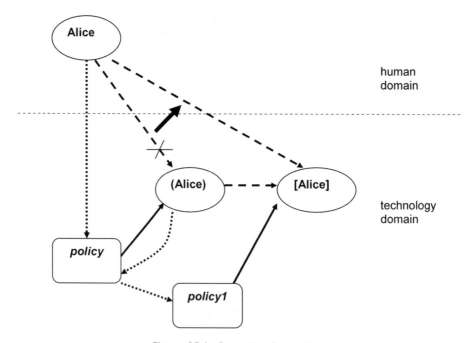

**Figure 15.6** Example of evolution

As agents replicate, mutate and alter their policies, the scheme can easily grow beyond reasonable control, even if Alice retains certain fundamental rights to manage agents (such as e.g. terminating some of them). The limit of this scheme is defined not by technology but by Alice's bounded rationality, her inability to control what she does not understand. In the end, Alice may be in possession of a system that is doing something and delivers services to Bob, but she will be unable to comprehend why and how it is been done.

## 15.6 Challenges to Trust Management

As time progresses, we are increasingly immersed in the digital communication. Currently we are usually dealing with the digital space through a perceived single layer of intermediaries: we operate devices that are performing requested actions. However, the proliferation of software agents, web services or even the proliferation of personal websites or blogs means that we will delegate increasingly important tasks to agents that are detached from us. Should this happen, the question of trust management become a central one: how can we, in an increasingly complex and connected world, manage the plethora of agents that work for us (and those that work against us)? Below is a list of challenges that go beyond improvements to existing schemes, to concentrate on the future.

### 15.6.1 Autonomy

The growing autonomy of agents is unstoppable. Considering the taxonomy of autonomy from the perspective of trust management, we can see that it will lead to agents that can evolve, replicate themselves, alter their policies, solicit support and delegate responsibilities. However, as autonomy grows, Alice's ability to manage and even simply understand her agents becomes a limiting factor.

Alice cannot control her own autonomous agents – either because there are too many of them or because their policies have grown beyond her comprehension or maybe because she has lost contact with the majority of them. However, Alice may be able to trust them – if she has reason to believe that they are trustworthy. By trusting her own agents, Alice can relate to them and can accept the outcome of their actions even though she would not always be able to understand how and why they performed a given action. Trust can also decrease the complexity that Alice must face, to the level that is possible for her.

### 15.6.2 Identity

As the number of intermediate digital agents grow, Alice is less assessed by the technical skills of those agents and more by the overall 'image' that such agents deliver – whether her website is consistent with her blog, whether her e-commerce shop conveniently links to a discussion forum, etc. The integrity of Alice's identity no longer depends on her ability to control every detail of her presentation, but on her ability to deliver the generally consistent image of herself. From 'hard' expectations regarding mastering and obedience, it is shifting towards 'soft' perception of consistency and continuity – from control to trust.

It is interesting to see whether the increase in the number of agents will also lead to the diffusion of the sense of identity, i.e. whether there will be several autonomous bodies that, once started, run relatively independently, unchecked and unmanaged. The current expectation is that it will probably not, even though the perception of identity may change. The individual identity of an agent will gradually be less important as agents will be easily replaceable and will be available in abundance. However, the group identity, the allegiance of such agents may become even more important, as the management will gradually move towards the same direction – from managing individual agents to managing a group of them.

We can manage only what we can understand and the new perception of the identity may be modelled on a loyal trusting crowd, an obedient army where an individual agent is important only if it can be singled out as no longer acting as a loyal servant to the higher cause.

### 15.6.3 Complexity

Complexity has been visible throughout the discussion as a driver and a limiting factor. The growth in the number of devices, agents, sites, information, content, etc. is extremely fast and it will probably continue to be so. For as long as an increment in efficiency or convenience can justify investment, we will increase the complexity of our digital environment. The ugly side of complexity shows up all the time: as the number of agents grow, it becomes hard to control them in the traditional sense of this word: issuing commands, receiving responses, altering policies, understanding what they do, etc.

We may alleviate the problem by grouping agents and constructing hierarchical management structure, but it will only delay the moment when such a structure will again be unmanageable. Following the visible change in organisational management of companies, this may eventually lead to increased reliance on trust, not on control. This may be considered a surrender, but it may be the only way to have at least a certain level of management in place.

## 15.7 The Future of Trust Management

The complexity-based model of confidence provides clues on how to respond to the challenges presented above. The growing complexity, increased autonomy, soft identity, uncertainty – all are addressed by trust better than by control. If agents can trust each other (rather than replicate real-life confidence through control instruments), then the future will look interesting (even though still challenging). If they must resort to control, then we are reaching the limits of what can be achieved.

The way trust management is being practised today leaves little if no space for trust: it is all about faithful representation of relationships and such representation can be better attained through control, not through trust. Modern trust management is therefore managing control to replicate trust. However, the future requires a reliance on trust that will increase interest in the tools and methods to communicate, assess and embed. Should this happen, trust management will be about rules of trust and trustworthiness.

It seems that the future of trust management is – paradoxically – in managing trust. If this is true, it is essential that trust management address two key challenges of its own:

1. whether trust can be communicated and assessed between agents; and
2. whether agents can reliably report their true trustworthiness.

A positive answer to the first question may lead to systems where signals that may lead to trust (i.e. signals of trustworthiness) will be delivered and accessed correctly, thus leading to the appropriate adjustment of the level of trust. Answering positively to the second question may lead to the development of a new class of agents: agents where trustworthiness is their design feature, non-negotiable by an agent and externally verifiable by other agents.

## 15.8 The Assessment of Trust

Trust, evidences of trust, levels of trust, are all hard to express in digital terms (contrary to control). Of the three classes of trust-related evidences, the hardest one is continuity:

expectation that certain forms of behaviour will continue into the future. Others (competence and motivation) can be more readily translated into the digital domain.

For reference, we earlier identified certain technical properties of protocols between agents that can deliver evidence of trust. Those were: for continuity: standards and norms, for motivation: inter-dependent processing and for competence: performance indicators. Here, we will look deeper into those classes of evidences, this time from the perspective of trust management.

## 15.8.1 Continuity

Continuity, the 'shadow of the future' is the class of evidence that builds an expectation that rules that bind Bob's behaviour will last into the future and that Alice understands and approves of such rules. In this book we have already discussed standards as a means to communicate the perception of a continuity, within the context of a single protocol. This discussion concentrates on norms.

The construct of continuity assumes that Bob is assessed as a 'whole person', so that Alice can look at his behaviour across several contexts. For example, Alice-customer can see Bob-businessman during the dinner or maybe at the theatre to understand how Bob will behave during the business negotiations. Alice can reasonably infer Bob's business behaviour from his table manners only if she can expect that both are governed by the same set of norms that provide Bob with general guidance about the proper behaviour across different situations. Once such norms are discovered, Alice receives certain assurance about Bob's future behaviour, as he – despite occasional hiccups – is likely to adhere to such norms.

The construct of norms offers a new view on Bob (see [Elgesem2006] for an interesting discussion on norms and trust and [Barwise1996] for basic concepts of information flow). While confidence and control are essentially contextual (Alice can reasonably talk about her confidence in Bob only within the context of certain transaction), norms are inter-contextual and they are supposed to be visible through Bob's actions and statements across several contexts. Whether Bob is a helpful person or not shows equally well at a golf course and during business negotiations, but the same cannot be said about his performance: he may be an expert businessman but a very poor golfer. There is a difference between his will to act in an appropriate way and the way he acts (where his capability may affect his actual performance).

That's all very well for Bob-person, but is it possible to think of Bob-agent that demonstrates adherence to a certain norm or Bob-agent that has been developed so that he always observes certain norms? Norm is essentially an isomorphism of behaviours where Bob's activity across several domains exhibits a similar meta-pattern. Finding such a meta-pattern requires unfortunately more than just running simple comparison or statistics across domains, as semantics of different actions may invalidate the reasoning. However, to talk about norms, it requires Bob to have several contexts and it requires actions within those contexts to be semantically comparable.

Possibly not surprisingly, design meta-patterns have been around for some time. The most visible location of such patterns in the desktop of an average computer. Since IBM published its SAA/CUA (System Applications Architecture/Common User Access) around 1990, the majority of desktop applications follow the same set of generic principles of how to organise user interactions. Those norms are now so ingrained in us through everyday experience that every user (hopefully) understands the meaning of 'OK' and 'Cancel' buttons or expects that he is always in control of the application. The strong belief in the continuity of those norms is one of the elements that builds trust in computers and is specifically visible when certain applications do not adhere to them. It is yet to be seen whether such design meta-norms can be reliably inferred from observations of Bob's behaviour alone.

## 15.8.2 Motivation

The aspects of motivation as an inter-dependent processing have been already discussed in this book. Here, we concentrate on another aspect of motivation, that implies a more holistic approach to Bob: his desires.

Motivation can be understood as a driving desire, a high-level intention that shapes particular intentions that act as strategies to fulfil the desire [Griffiths2003]. It is possible to model agents so that they are driven by certain motives and desires (such as perceived hunger, desire for light, etc.) but inferring motives from actual behaviour may be more problematic. As motivation is able to override intentions (e.g. hungry agent will not deliver a service), motivation can be inferred from behavioural traits that are visible across contexts, specifically in situations of conflict. While continuity caters for normal behaviour and can be inferred from observations, motivations caters for extreme behaviour and can be inferred from experiments.

The classic evidence of motivation that Alice may seek calls for proof that Bob's interest encapsulates Alice's interest [Hardin2002], i.e. that while acting at his best interest he actually satisfies hers. This can be demonstrated if Bob supports Alice in times of distress and insufficient resources – something that may not be encountered throughout the normal course of actions. The reasoning is as follows: distress reduces Bob's intentions to those that are strongly driven by his motivation; if he still supports Alice in such a situation, then apparently supporting Alice is strongly encapsulated in his motivation. Therefore, even though Alice may not know exactly why Bob supports her, she may be certain that somehow her interest (i.e. being supported) is encapsulated in Bob's motivation.

Let's consider an example where Bob is an agent that is providing a storage space for other agents. Alice would like to be sure that she can trust Bob regarding availability of the space. For that purpose she constructs a stress test where she pushes requests up to the limit (within a contract but way above other's requirements) and she cooperates with Carol so that Carol also places significant requests. Bob, presumably under stress, may decline to serve some of the requests. If he refuses Alice more often then Carol, Alice can believe that Bob is not motivated. If Carol receives more refusals or if both are served at the similar level, Alice receives certain evidence that Bob is motivated to support her (or at least he is not motivated not to support her).

## 15.8.3 Competence

Competence seems to be more contextual as it is concerned with capabilities rather than with willingness: a good driver may make a bad surgeon and vice versa. As such, the proof from past performance within a given context or given set of contexts (possibly extrapolated and adjusted) may work better to determine the level of competence. For example, the server may demonstrate its capability by listing services that it is able to run and by providing information about its relative performance such as relative processor load, percentage of memory in use, current throughput in transactions per seconds, etc.).

Such solutions, even though they are relatively simple to implement, deliver good indicators of competence. They have been implemented e.g. for grid computing [Dragovic2003] to monitor the fulfilment of contractually guaranteed quality of service. Note that even though competence is contextual, the evidence of competence may again require a more holistic approach to Bob. It may be necessary e.g. to disclose by Bob not only his workload within a context of an interaction, but also his general performance, across several interactions to demonstrate that his current performance is as good as it can be.

## 15.9 Embedded Trustworthiness

It is now time to investigate the second question: is it possible to build (Bob) in a way that he will have no choice but always honestly disclose his own trustworthiness in a simple yet reliable way? We would like (Bob) to be built in a way that precludes him from cheating about his own trustworthiness – a task that unfortunately is not attainable in our society, but may be possible among agents. We will explore here two different methods, embedding this functionality into agent's environment and into the language that has been used to create (Bob).

The first method is to embed norms in an agent's immediate environment that is used to run such agent. Such an environment will continuously monitor what (Bob) is doing (across a wide range of his activities) and it will assess his trustworthiness – in the sense of (Bob's) adherence to norms, fulfilment and performance. Note that such an environment is not preventing (Bob) from behaving in an untrustworthy manner – it is more relaxed then e.g. a security-driven sandbox. While the sandbox will prevent Bob from doing actual harm, the environment will monitor whether he acts from goodwill.

Once we have an environment that is watching (Bob), other agents may ask such environment about him and receive its best assessment of his trustworthiness. As other agents cannot possibly differentiate between (Bob) and his immediate environment, they will perceive this trustworthiness assessment as coming from (Bob). At the same time, because such environment is outside of (Bob's) influence, the assessment is guaranteed to be honest.

Another way is to alter the language that has been used to program (Bob's) behaviour. If the language allows for defining behavioural meta-templates then they can be used as a way to express rules of behaviour that will be consequently reflected across all contexts of (Bob). Such meta-norms can then be disclosed to other agents to demonstrate foundations of (Bob's) operation. Interestingly, programming language concepts such as templates or patterns (not to mention XML) seem to have already noticed the need for such approach or at least point in such a direction.

There is yet another alternative: to vouch for (Bob) by someone who knows him intimately and is trusted. For example, if (Bob) has been programmed by Carol and Alice is confident about Carol, then Alice may be satisfied with knowing that it was Carol indeed who made (Bob). This is quite a popular solution that can be implemented with e.g. code signing, specifically if combined with trusted computing platform. However, this solution does not answer the challenge – it merely shifts it from Alice being confident in (Bob's) behaviour into Alice being confident about Carol.

## 15.10 Trusted Third Party

It has been already mentioned earlier in this book that trusted third party (TTP) does not contribute to the creation of a trusted relationship. It does not mean that there is no place for the TTP – even though TTP does not build trust, it helps managing the relationship. Potential roles of trusted third party is extensively studied in [Rea2001], in a context of a model that combines human and technical domains. Trusted third parties are believed to be a main facilitator of several aspects of the creation and maintenance of confidence.

Within the scope of current trust management (and within the understanding that such management has nothing to do with trust), there is a valid and important place for TTP: to act as a trusted intermediary between human and technical domains, to issue credentials, manage policy updates, etc. Trust that is vested in TTP by all parties is a prerequisite to accept policies and credentials issued by TTP. Figure 15.7 shows the position and the role of

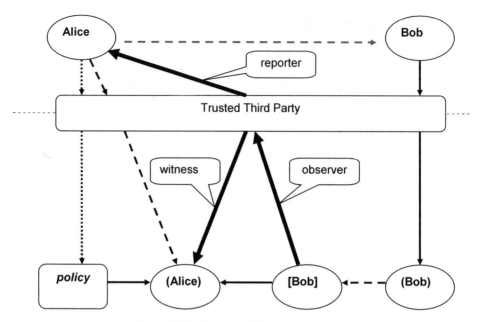

**Figure 15.7** The role of Trusted Third Party

TTP within the generic trust management. Note that bold solid lines refer to new roles, as discussed below.

This role of TTP is well understood and some parties have already assumed such roles. Fore example, Certificate Authorities (CA) serve as TTP for the public key infrastructure (PKI), approving the relationship between a certificate and a real-life entity that owns the matching private key. Notwithstanding some operational hurdles (it is sometimes surprisingly easy to generate a certificate for an arbitrary institution without proper credentials), the existence of such a TTP enables and facilitates the development of trust management. Similarly, financial institutions and mobile operators are trying to act as TTP for other industries, leveraging in this process their existing infrastructure (EMV cards or SIM cards, respectively). Further, some governments act as TTP in issuing digital identity cards, entitlement cards or similar solutions.

The natural extension of this role is for TTP to act as a trust bridge between domains of trust [Yan2003]. Such role requires the TTP to be trusted by at least two parties that are otherwise unable to establish the relationship of trust, e.g. because of their conflicting interest. Again, this role has already been assumed by various institutions such as credit card issuers (that intermediate trust between customer and merchants), quite often combined with certain forms of insurance-based protection. This role can be also reflected in a system design where system components that are not intended to trust each other can establish a meaningful communication with the help of another party [Clark2007]. Note that none of those roles actually creates trust, but they facilitate the creation of relationship by introducing the element of mutually accepted control.

It is worth, however, considering what the future of TTP might be, beyond what has been already incorporated into trust management. The new role of TTP (that may develop on top of the existing ones) is closely associated with the expected development of trust management – towards managing trust. TTP should assume the role of a neutral (trusted) witness of an agent's behaviour, delivering individual observations as well as aggregated ratings, justified opinions, etc. Such a TTP will become a holder of evidence, observer of norms, tester of motivation and collector of capabilities, shifting its interest towards evidences that support trust relationship. Figure 15.7 shows with bold solid lines those potentially new roles.

Some parties seem to understand this fundamental shift. Reputation-based systems, as they grow in complexity and sophistication, are gradually becoming conducive to evidence of trust and are increasingly willing to export their aggregated opinion beyond their original domain of interest. It is yet to be seen whether a good e-commerce trader can be trusted when it comes to governmental posts (as currently the tattered credit rating does not seem to preclude business or political career), but the direction has been set: there is a need for the back-translation of trust: from the technical domain into the human one.

## 15.11 Conclusion

Not surprisingly, trust management, as we know it, has nothing to do with trust. Surprisingly, the future of trust management lies in managing trust. From its current position as an instrument of control and policies, trust management will eventually evolve into an instrument to manage trust assessment and trustworthy behaviour of a multitude of agents that will be available around us.

It is interesting that the concept of trust management here has gone through the whole loop. By uncovering the primary relationship that trust management is supposed to handle, it turned out to be no more than a toolbox of control, aimed at faithfully reproducing (but not managing) trust. The deconstruction of the concept is, however, followed by the construction of it: trust management finds its new role in managing trust, driven by the increase in **number** and complexity of agents.

## Bibliography

[Adams2005] William J. Adams and Nathaniel J. Davis IV: Towards a Decentralized Trust-Based Access Control System for Dynamic Collaboration. In *Proc. of the 2005 IEEE Workshop on Information Assurance and Security*. 2005.

[Anderson2005] Steve Anderson et al.: Web Services Trust Language (WS-Trust). Available at: http://specs.xmlsoap.org/ws/2005/02/trust/WS-Trust.pdf. 2005.

[Barwise1996] J. Barwise and J. Seligman: Information flow. *The Logic of Distributed Systems*. Cambridge: Cambridge University Press. 1996.

[Blaze2003] Matt Blaze et al.: Experience with the KeyNote Trust Management System: Applications and Future Directions. In *Proc. of First Int. Conf. on Trust Management iTrust 2003*. Berlin: Springer. *LNCS 2692*, pp. 284–300. 2003.

[Brown2000] Mark A. Brown: Conditional Obligation and Positive Permission for Agents in Time. *Nordic J. of Philosophical Logic*, 5(2). 83–112. 2000.

[Clark2007] David Clark: The End-to-end Argument in Today's World: Do We Have to Destroy the Principle to Save it? Presentation, in BT UK. 28 Feb. Suffolk. 2007.

[Dimmock2005] Nathan Dimmock, Jean Bacon, David Ingram, and Ken Moody: Risk Models for Trust-Based Access Control (TBAC). In P. Herrmann (ed.): *iTrust2005*, Berlin: Springer. LNCS 3477, pp. 364–371. 2005.

[Dragovic2003] Boris Dragovic, Steven Hand, Tim Harris, Evengelos Kotsovinos and Andrew Twigg: Managing Trust and Reputation in the XenoServer Open Platform. In *Proc. of iTrust 2003*. In P. Nixon, and S. Terzis (eds): *Trust Management 2003*, Berlin: Springer. LNCS 2692, pp. 59–74. 2003.

[Elgesem2006] Dag Elgesem: Normative Structures in Trust Management. In K. Stolen et al. (eds): *iTrust 2006*, Berlin: Springer. LNCS 3986, pp. 48–61. 2006.

[Gerck1999] Ed Gerck: Overview of Certification Systems: X.509, PKIX, CA, PGP & SKIP. Black Hat Conf., Las Vegas, NV. Available at: http://nma.com/papers/certover.pdf. 1999.

[Griffiths2003] Nathan Griffiths, Michael Luck and Mark d'Inverno: Annotating Cooperative Plans with Trusted Agents. In: R. Falcone et al. (eds): *AAMAS 2002 Workshop on Trust, Reputation* . . . LNAI 2631, pp. 87–107. 2003.

[Hardin2002] Russel Hardin: *Trust and Trustworthiness*. New York: Russell Sage Foundation. 2002.

[Josang2005] Audun Josang, Claudia Keser and Theo Dimitrakos: Can We Manage Trust? In: P. Herrmann et al. (eds): *iTrust 2005*, Berlin: Springer. LNCS 3477, pp. 93–107. 2005.

[Lacohee2006] Hazel Lacohee, Stephen Crane, and Andy Phippen: *Trustguide*. Available at: http:www.trustguide.org. 2006.

[Luck2005] Michael Luck et al: Agent Technology: Computing as Interaction. A Roadmap for Agent-based Computing. *AgentLink 2005*. 2005.

[Nikander2001] Pekka Nikander, Kristiina Karvonen: Users and Trust in Cyberspace. In *Security Protocols : 8th Int. Workshops*. Cambridge, UK. 3–5, April. *Revised Papers*. LNCS 2133/2001. pp. 24–35. 2001.

[Pfitzmann1997] Andreas Pfitzmann et al: Trusting Mobile User Devices and Security Modules. *IEEE Computer*. February: 61–68. 1997.

[Rea2001] Tim Rea: Engendering Trust in Electronic Environments. Roles for a Trusted Third Party. In Cristiano Castelfranchi and Yao-Hua Tan (eds): *Trust and Deception in Virtual Societies* Pordrecht: Kluwer Academic Publishers. pp. 221–236. 2001.

[Yan2003] Zheng Yan and Piotr Cofta: Methodology to Bridge Different Domains of Trust in Mobile Communications. In *The First International Conference on Trust Management (iTrust'03)*, Berlin: Springer. LNCS 2692/2003, pp. 211–224, Greece, May. 2003.

[Yao2003] Walt Teh-Ming Yao: Fidelis: A Policy-Driven Trust Management Framework. In *Proc. of First Int. Conf. on Trust Management iTrust 2003*. Berlin: Springer. LNCS 2692, pp. 301–317. 2003.

# 16

# Economy of Confidence

Our names are labels, plainly printed on the bottled essence of our past behaviour.

(Logan Pearsall Smith)

## 16.1 Introduction

Confidence, trust and reputation are in a short supply while they are increasingly desired around the world. Commerce cannot grow with them, governments would like to gain them while businesses try to grow them. Apparently, if there is scarcity of particular goods, there should be a market for them and economy of confidence should flourish. It should be that confidence becomes generated, packaged and commoditised, price for the reputation should be set by the free market and stock exchanges should trade trust derivatives. However, this is not exactly what we can see happening.

Confidence is not a durable good so that there may be the problem of commoditising it in the way gold and frozen orange juice have been. Confidence is not a service either, even though we may consider it an essential part of every service and its ultimate market differentiator. Confidence cannot be produced and consumed in an anonymous way characteristic of modern trading. It is the property of an individual relationship and is not directly transferable. If Alice is confident in Bob, she believes that in need she can solicit Bob's favourite behaviour. Her belief does not transfer to her relationship with Carol and Bob's willingness to support her does not necessarily apply to requests coming from Dave.

However, this should not hamper the market for confidence. In economy, market imperfections tend to drive transactions, not to inhibit them. If there is a demand and supply, the question is price, as the appropriate conveyance mechanism can be always found. Therefore we can see that Alice's confidence in Bob should be tradeable, should have its market valuation and there should be instruments by which such confidence can be conveyed from one relationship to another. The fact that they may not be immediately recognisable is attributable to the specific nature of confidence, but it does not inhibit their existence.

Of the two components of confidence (control and trust), trust seems to be the more elusive and more valuable one. While control can be acquired through several different mechanisms

---

*Trust, Complexity and Control: Confidence in a Convergent World*   Piotr Cofta
© 2007 John Wiley & Sons, Ltd

(such as legal enforcement, risk assessment, insurance, etc.), trust apparently escapes easy instrumentalisation. At the same time, companies and individuals see greater benefits in trust than in control – it is more cost-efficient and there are expectations of greater returns. Increasing the offset between trust and control (which is a measure of trustworthiness) seems to be of particular value.

This chapter discusses different manifestations of the concept of 'economy of confidence' (or 'economy of trust' as it can be sometimes called). It starts from looking at four different approaches to what the relationship between economy and confidence really is. From there, it demonstrates how the economy of confidence can operate, with a focus on the means of conveyance that allows confidence to be transferred between relationships. The discussion of additional business opportunities in confidence follows. The chapter continues with a description of an interesting proposition where confidence is used as a currency to trade secrets, linking it back to considerations about convergent communication and its security. Finally, it closes with a slightly unexpected valuation of trust.

## 16.2 Confidence and Economy

The relationship between confidence and economy is obvious yet multi-faceted. Below, we explore four different aspects of this relationship, highlighting their importance in the convergent world. Of all the elements of confidence, this chapter heavily concentrates on trustworthiness and on reputation. We can see reputation as the socially embedded, objectified metrics of one's expected trustworthy behaviour, linked to one's trustworthiness. For simplification, we assume that such a reputation is globally acquired and globally visible and as such it is the foundation of confidence-based decisions in the marketplace.

### 16.2.1 Confidence as Facilitator

The first approach is to treat confidence as a facilitator for economic relationships and transactions (see [Frankema2005] for a brief overview). Similar to the lubricant in a gearbox, confidence may not have a significant value in itself, but the lack of confidence significantly impairs the machinery of the economy, thus increasing the cost of doing business. Here, the substitutive relationship between trust and control is fully visible. While the facilitation can be achieved either by control (e.g. legal framework) or trust, trust is preferred on the ground of its low cost (in terms of complexity but also in terms of accrued expenditures). It is worth noting, however, that it is either trust or control (and usually the mix of both) that is usually used to facilitate the trade.

The value of confidence can be assessed here only through the negative estimate, e.g. by the cost incurred if confidence is not present. The recognition of additional cost of control (management) of contracts can be formalised in the context of game theory [Raub2001]. Specifically, the substitute relationship of trust and control here permits assessment of the offset between trust and control (trustworthiness), to support the case that trust is the less expensive of those two.

Subcontracting and outsourcing can be used as an excellent example, as both are likely to be accelerated by a convergent communication. For the sake of efficiency (but not necessary the effectiveness), companies willingly break the relationship of trust and try to develop the relationship of control with external subcontractors. Intra-company ties of loyalty and mutual support are replaced by inter-company contractual responsibility. It is therefore surprising that while making a giant step back in the relationship with their contractors, companies demand trust from them and try to develop the relationship of trust, possibly without understanding that the relationship is a two-way street: one cannot request trust at one end and impose control on the other.

It is not surprising that companies soon realise that the cost of managing the control-based relationship is higher than the trust-based one. Sources [Willcocks2006] report savings of 20–40 per cent in management cost if the relationship is built on trust rather than on control

Considering cultural differences, the picture is even a darker than expected. The problem is not only that in some countries people are 'bowling alone' [Putnam2000]. Considering industries in the US and Japan that rely heavily on subcontracting and outsourcing [Sako1998], we can see that in case of Japan the relationship evolves in time in an expected way: towards increased trust and shared identity while in the US the relationship develops in a reverse direction: from initial trust to increasing usage of control-based instrument of contract, as if the initial endowment of trust is used up in the relationship.

## 16.2.2 Confidence as Social Capital

Another approach is to treat confidence (again, mostly trustworthiness) as social capital. Even though the notion of social capital is not always clearly defined (compare e.g. [Putnam2000] and [Fukuyama1996]), it is usually associated with the existence of a network of relationships (often informal ones) where people justifiably trust each other (rather than control each other) and are voluntarily bound by certain norms. Members of such networks gain additional benefits, ranging from the perception of ontological security, to more pragmatic ones such as access to limited resources, priority of contacts, better terms of trade or the ability to influence strategic decisions.

If we use the concept of trustworthiness as an 'offset' between the extent of trust and the extent of control within an individual relationship, then the volume of social capital can be expressed as the size of network multiplied by such offset (where the 'size of network' incorporates whatever 'network effect' may be present [Briscoe2006]). This leads to the immediate observation that growing the network is not sufficient to grow the capital. If several relationships lessen control (as happens quite often in the early stages of such relationship), then they actually decrease the overall capital. Thus convergent network itself does not contribute to the growth of social capital unless it provides a means to build a trusted relationship.

The concept of confidence (trust, reputation) as social capital has several proponents. The ability to manipulate confidence through beliefs [Castelfranchi2006] should increase, decrease and transfer confidence as if it is a capital indeed. In a cautionary note, some of those manipulations can be classified as soft forms of attacks (or maybe just marketing tactics). The fact that groups and individuals can 'spend' such capital in time of distress [Mishra1996] (even though the social capital does not necessary decrease in this process) well reflects reality of e.g. transformational change within the company (e.g. [Rodgers2006]) and further reinforces the perception of confidence as a capital.

Further, the concept of social capital can explain the enthusiasm behind Web 2.0 (e.g. [O'Reilly2005]), concentrated on developing socially embedded applications, in the expectation that they will increase the 'offset', shifting relationship towards trust. This similar (yet more technology-oriented) line of thought is presented in this book while discussing Trust-Enhancing Technologies.

Social capital has been also criticised (see e.g. [Sobel2002] for a critical review or [Tonkiss2004]). It has been pointed out that the social cohesion of the group may not always be in line with the norms and goals of the society it is embedded in. The 'network effect' can lead to closed cliques aimed at supporting their members despite or even against social norms as a whole, thus hampering overall social development. Further, the 'fuzziness' of the concept may lead to the over-rating of it: social capital is not a panacea to all problems, as it may decrease operational efficiency and the responsiveness of organisations. Trust and cooperation may arise from reasons that do not adhere to concepts of the civic community [Durlauf2002]. Finally, from the more pessimistic perspective, stable social trust is not believed to be created by any spontaneous action by the society itself, but it is likely to disintegrate if left to such actions [Hawthorn2000].

Castelfranchi [Castelfranchi2006] offers an insight into a potential conflict between Alice's individual interest and social welfare, something that has been noted several times throughout this book. While society may be interested in evenly distributed average level of trust, Alice may be more interested in being in the concentration spot of trust, where she may enjoy additional benefits.

## 16.2.3 Confidence as Differentiator

Confidence can be also interpreted as a market differentiator, so that while Alice is facing a choice on the marketplace, she may prefer to deal with an agent she is confident in. It is quite common in this approach to equal individual confidence with reputation, i.e. the shared perception of the past performance of an agent, usually being used as a good approximation of agent's intentions and future performance. This equivalence is particularly valid for global markets (facilitated by the convergence) where traditional customer loyalty may be harder to find.

Considering the competition on the modern market, the ability to maintain reputation is an important differentiating asset, that may win a significant amount of business and potential profit. On top of a higher volume, agents that enjoy high reputation can also expect a better profit margin. The reputation (hence confidence) significantly influences on-line auction prices, specifically for higher valued items, as reported in [Mui2002]. Experiments with eBay [Resnick2006] demonstrated the 8.1 per cent premium associated with good reputation – the value that no trader can ignore. Reports regarding perceived brand value (which has a strong correlation with reputation) indicate that for some companies such a reputation constitutes a significant fraction of their market value.

Confidence can be also used at the more technical level, e.g. in trust-based routing ([Twigg2003]) where more trusted peers attract increased traffic. Such traffic, even though technically it represents an additional burden, may be converged into new business opportunities in market analysis and advertising, if only there are means to monetise this advantage. Examples of large search engines that were able to monetise their reputation as the 'most trusted' places on the Internet can serve as a good example.

The fact that confidence is a differentiator is particularly important when the convergent, global market is considered. One of the aspects of convergence is a brand virtualisation, where customer confidence is increasingly attached to portable brands and – as such – it can be transferred between technology solutions to create market differentiation.

## 16.2.4 Confidence as Currency

This interesting approach somehow reverses the trend: instead of asking what is a monetary value of confidence, it attempts to use confidence as a form of currency on its own, currency that can be used in other economical relationships. The concept of using confidence as a currency is relatively new, but it has been already demonstrated, e.g. to improve resilience against certain forms of attacks. The following is an overview of a few propositions.

An interesting step towards using reputation as a currency is presented in [Kerr2006]. What on the surface looks like a conventional reputation scheme, differs by requesting sellers to put an amount of their reputation (measured in 'trunits') into an escrow. Such amount is proportional to the monetary value of transaction and is lost if the transactions are deemed unsatisfactory by the buyer. Otherwise it is returned to the seller, with a fractional bonus.

By expressing reputation ratings in units, the scheme prevents sellers from engaging into too many transactions at the same time or from rapidly moving from low-value to high-value transactions. Even though the system has its drawbacks (risk of dishonest buyers, need for the market operator), it is an interesting attempt to monetise confidence.

Another proposition is to spend agents' own reputation (trust) on providing recommendations to other agents [Seigneur2005]. Reputation here is used as a quasi-currency, as it is not centrally stored but can be passed between agents. Agents are passing a fraction of their own reputation to those agents that they recommend, as a proof of recommendation. If Alice is interested in knowing Bob's reputation, Bob can simply demonstrate how much reputation he has received from others. This scheme has been originally designed as a protection against Sybil attack [Douceur2002], and is similar to the concept of certification discussed later in this chapter.

An intriguing question is what can be bought with such confidence-based currency. The majority of proposals assume that it is the reputation that can be paid for with another reputation. The different approach is presented later in this chapter, where reputation is used to buy secrets.

## 16.3 Economy of Confidence

It is worth discussing the economy of confidence (specifically the economy of trust) only if there are indicators that such an economy may exist. Essentially, to have the economy we need scarce goods, supply and demand and means of conveying goods. Then we can sit back, relax, and see how the market sets the price, whether in its monetary or in barter formats.

Scarcity can easily be observed and has been acknowledged by several sources. There is a continuous supply of confidence, as every relationship can produce it. Due to its economic value, there is also continuous demand. The only missing elements are means of conveyance. Conveying confidence between agents seems to be a complex task. It has been already mentioned that confidence is a relational property, i.e. it does not yield itself to the process of packaging and shipping. However, economic activities know several sophisticated other forms of conveyance (e.g. title, share, etc.) that do away with packaging so that we definitely can find something that suits confidence.

As an illustrative example let's consider the situation that happened between Alice, Bob and Carol (Figure 16.1). Alice has certain confidence in Bob and some (presumably higher) confidence in Carol. If we have a common rating system (e.g. central reputation storage), the

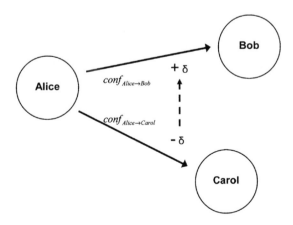

**Figure 16.1** The question of means of conveyance

difference between $conf_{Alice \to Bob}$ and $conf_{Alice \to Carol}$ can be objectively visible (otherwise we may want to rely on their respective perception of the difference).

Let's assume that we are interested in increasing $conf_{Alice \to Bob}$, by a certain value of $\delta$. Following the model of confidence, we know that this can be done in several different ways. First, Alice can seek instruments that improve her assessment (or her confidence) or second – Bob may behave well so that Alice will alter her belief in his intentions. Even though both options are interesting as business opportunities (and are discussed briefly later in this chapter), they are not – strictly speaking – a part of the economy of confidence. It is only when Bob or Alice spots Carol and decide to transfer $\delta$ of $conf_{Alice \to Carol}$ to increase $conf_{Alice \to Bob}$ (and possibly pay for it) that the economy of confidence develops.

While we may discuss whether such transfer is feasible in a strict sense, we can observe that the market has already developed some methods to perform it. Currently, there are three main schemes that can be used to transfer: certification, representation and branding.

### 16.3.1 Certification

The process of certification is aimed at Alice improving her confidence in Bob's intentions (e.g. regarding the delivery of the service). Knowing that Alice is confident in Carol, Bob can ask Carol to vouch for him, i.e. he can ask her to deliver evidence to Alice that is favourable to him. Assuming that Alice did not have access to Carol's opinion, and because of the presumed high confidence Alice has in Carol, the relative weight of such evidence may have significant and immediate impact on $conf_{Alice \to Bob}$ (Figure 16.2). Carol can actual tune her recommendation so that Alice will experience an expected gain of $\delta$.

Note that this scheme works only if Alice values Carol's honesty and her ability to judge others, not her intentions regarding the service that she asks from Bob. Alice's confidence in Carol should therefore be of a different nature than Alice's confidence in Bob. In fact, Carol's ability to deliver the service is not relevant here, unless Alice's approach to Carol is that of the 'whole person'. Carol may be also losing a bit of her reputation in this process, at least from Alice's perspective: if Carol strongly vouches for everybody, then the value of such recommendation (that is Alice's confidence in Carol's honesty) decreases.

Interestingly, Carol may see this difference between two forms of confidence as a business opportunity. By specialising in delivering ratings and certification, she may convert honesty and accuracy of her ratings into her primary offering – and charge for it. Actually, if Carol is willing to specialise and is good at rating others, Alice's confidence in her can grow with

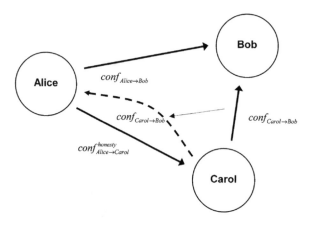

**Figure 16.2** Certification

every successful certification, potentially allowing Carol to generate higher profit from her certification business.

Several practical solutions already exist in this area, such as certification centres, public notary, trusted parties, etc. An interesting example of commercialisation of such usage of confidence is the 'e-mail' tax' planned to be introduced by some large Internet service providers in 2006. Senders of bulk e-mail (usually corporations) will be able to bypass ISP's spam filters by paying the small fee per each e-mail (approx. 1 cent). Such mail will be additionally marked as 'certified' and the delivery will be guaranteed. Other bulk mail is likely to end up being marked as 'spam' and even outright rejected. The scheme is interesting as it provides an instrument that increases confidence in mail delivery and in the acceptance of such mail by using the certification on the basis of existing confidence in the ISP. Note that the scheme is not addressed to mail recipients. Should they want to increase their confidence in e-mail system, they can resort to spam filtering packages (that may come at a price as well).

The ability to pay for a certification service has been already discussed in [Jurca2006] and the distributed mechanism to collect and deliver such evidences has been proposed in [Obreiter2004]. Further, if we are willing to use reputation as a currency to pay for reputation, [Seigneur2005] proposes a scheme where Carol can issue a certificate only at the expense of her reputation.

## 16.3.2 Representation

In this scheme, Bob appoints Carol to represent him to Alice in the expectation that Alice's low confidence in him will not endanger this new relationship. Carol becomes a part of every communication between Alice and Bob and she is entitled to certain compensation for her work – and possibly even greater compensation for using her confidence. The compensation should also cover the risk that Carol is facing: she may end up with her reputation being damaged if Bob does not perform (Figure 16.3).

This is a popular scheme and there are several examples of representation in electronic commerce. Carol can become an integrator, representing several agents under her single reputation. Companies such as Amazon integrate smaller suppliers through their marketplaces (as well as through its logistics system) and use their own reputation as a means to provide higher visibility, increased differentiation and ultimately better pricing. In a similar

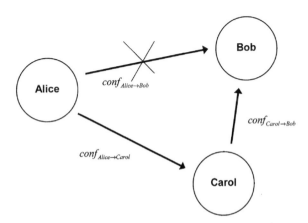

**Figure 16.3** Representation

vein, eBay's trading posts collect sell orders from individuals and trade wares under their own name, enjoying additional profit arising from their higher reputation.

There may be several versions of how the relationship between Carol and Bob develops: whether Carol is only a thin front or whether she significantly supplements Bob's functionality with her own. Similarly, there are different means to manage the risk that Carol is facing: from entirely dealing under her own name to disclosing Bob as an original provider.

### 16.3.3 Branding

Branding can be seen as a logical extreme of representation: Carol's contribution to Bob's operation is getting gradually thinner, to the extent that it is only her name (brand) that Carol sells to Bob. While Alice believes that she is dealing with Carol, she in fact is dealing directly with Bob, but her confidence in Bob is substituted by her confidence in Carol (Figure 16.4.).

Brands is the encapsulation of market identity and associated reputation, but without attachment to actual operation. Brand is a well-understood method to capture and trade confidence. Indeed, the process of determining brand equity is related to the assessment of how much the brand itself is worth, without the underlying operation. Note, however, that branding can be also viewed as an attack on confidence, as in fact Alice is deceived regarding the real identity of an agent she is dealing with.

Convergence is particularly conducive to the notion of brand-based trading, as it virtualises market propositions on top of standardised technology. From this perspective, the future of such a market is important. Tadelis [Tadelis1999] investigates how the reputation itself can be traded, possibly in a form of a brand name. However, the outcome of the ability to trade brands is not all positive, as there may appear a discrepancy between the agenda of agents that own the brand and the reputation associated with the brand – 'bad guys' are buying good brands without informing other agents about the change in ownership. Eventually, such brands lose their reputation and the process starts again. If allowed, this trading is highly attractive in the short term, creating an equilibrium on brand values. In the long run, the outcome is less attractive, as brands (with attached reputation) lose their meaning as the differentiator if reputation.

Interestingly, the reputation market works as the iterative version of the deceitful two-stage strategy for the game of trust. Such strategy operates by developing the relationship and subsequently (in the last move) to capitalise on it, by betraying the partner. The repeatability

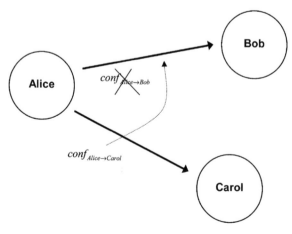

**Figure 16.4** Branding

of such strategy is limited by the unchangeable identity of agents, where the deceived partner may learn and not cooperate at the next attempt to build reputation. If, however, the brand (which is the visible identity of the agent) can be refreshed with one of the good reputation, the deceitful strategy may be repeated.

## 16.4 Opportunities in Confidence

Apart from the concept of directly trading confidence (already discussed above), there are several other market opportunities associated with the relationship between Alice and Bob. Below there are some brief scenarios where revenue can be made by working with confidence. As usual, opportunities are where there is a market imperfection: information does not flow, confidence is unevenly distributed or is of a different kind.

*Facilitation* improves the flow of evidences without interfering with them, so that evidences are not produced, rejected or processed. Facilitation operates on the level of enablers (mostly availability). Both Alice and Bob may be interested in such a service, even though possibly for different reasons: Alice may benefit from facilitation to achieve a more accurate assessment of her confidence in Bob while Bob may expect that an improved flow of evidences may eventually improve Alice's confidence in him. Depending on the expected outcome, either Alice or Bob (or both) may be willing to pay for the service. As the expectations are about improving flow, actual Alice's confidence in Bob may not be affected.

Directory services, search engines and advertising are simple examples of facilitation, and some of them are quite successful. Convergence will increase the need for such services, as the number of agents will grow. Even currently, the relative confidence in Web-based information can quite often be attributed more to its availability than to its quality.

*Instrumentalisation.* While the economy of confidence has generally been concerned with the transition from control to trust, instrumentalisation strongly plays on the control side. Instrumentalisation delivers instruments of control that are built on top of confidence in the provider of such instrument, and then potentially charge for its usage. It has been already discussed that there is a visible trend to instrumentalise our relationships and convergence accelerates this trend.

Instrumentalisation is already the foundation of a huge computer security market, where tools (security appliances, security software, etc.) are being used to increase confidence in computer systems. The value of such instruments of control quite often depends on the level of confidence that Alice has in their provider. For example, if Alice wants to improve her confidence in Bob she may want to use an instrument of control (e.g. the firewall) that is provided by Carol. The price Alice is willing to pay for such a firewall is related to her confidence in Carol as a provider of such a firewall.

*Agency* decreases Alice's complexity in her dealings with Bob. Here, Carol herself becomes an instrument of Alice that can be used to deal with Bob. Agency changes the nature of a relationship: Alice can build her confidence in Carol on trust, yet Carol may be more likely to use control to deal with Bob. Because of this Alice is facing lower complexity (using trust instead of control) and it is expected that Alice may be willing to pay Carol for being her agent.

*Delegation* is similar to agency in that Alice uses Carol to deal with Bob. However, there is a difference in operation: Carol is not Alice's instrument but Carol is taking over the whole relationship with Bob, acting as Alice's sole agent. Alice is benefiting from her higher confidence in Carol in that she can manage such a relationship with lower complexity, so that Alice may be willing to compensate Carol for her efforts.

The existing market for agencies and delegation follows the general trend to institutionalise our relationships. The number as well as the variety of agencies is growing (e.g.

representation, enforcement, etc.). The concept of agency and delegation is fundamental to the convergent world, where it covers not only human agencies but also hardware and particularly software agents that can travel the Internet. An interesting review of an agent future can be found e.g. in [Luck2005].

## 16.5 Fences of the Internet

The ability to pay with confidence (i.e. to use confidence as a form of currency) brings the question of what can be bought with such currency. Below is the proposition (intentionally futuristic) of the scheme where confidence can be used to buy secrets, in the expectation that this will lead to a more balanced approach to the Internet security. The metaphor of 'fences of the Internet' is used here to describe the security associated with digital communication. Convergence, with its proliferation of communication means, makes the problem of adequate security even more important.

Let's start from a simple observation. The traditional way to protect one's place is to put up a fence around it. Depending on the perceived risk, such a fence can be a low hedge or a patrolled concrete wall, but the purpose is the same: to demarcate what's ours, to provide protection against casual snoopers and to make life harder for an attacker. No fence is entirely unbreakable – an attacker with sufficient resources in his hands can jump over the hedge or dig under the wall, if needed.

Regarding data communication and its protection, what we are witnessing today on the Internet is an anomaly, a deviation from our everyday experience. By using modern cryptography, for free, we can erect fences to information that are so tall and impenetrable that nobody can get through them. Even we cannot jump the fence if we forget our password. Certainly, we are tempted to hide behind such technology, in an understandable avoidance of any risk. Encrypted communication, content management, digital vaults – all explore the premise of digital impenetrable fences. What digital fences have done, however, was to direct crooks to less protected areas of security. Instead of cracking the communication, they concentrate on the password. Instead of attacking the secret code they chop off the finger of the culprit.

In fact, impenetrable fences are a kind of excess – and not only because the law enforcement officers prefer to snoop at private communication from time to time. They effectively deprive Internet communication of any kind of social supervision and disable whatever social protocols can be developed. In real life, if Alice's neighbour (or child) Bob behaves strangely, she can call for a help. On the Internet, as long as he uses encrypted communication, Alice has no idea what he is talking about, whether this is casual gossip (that he may even want her to join in) or a secretive ploy. Should he be able to build fences of different height, he could have rated his desire for protection, and they both would have been able to develop an appropriate social protocol.

The impenetrable fence of modern cryptography is built on the premise that the cryptographic key cannot be cracked. This may change when quantum computing and quantum cryptography realise their potential [DiVincenzo1995]. The promise of breaking every cryptographic protection in a reasonable time is combined with the promise of the new wave of cryptography that is resistant to such attacks. Should quantum computing be an overly elusive target, mandatory key escrow, such as the ill-fated Clipper can essentially offer the similar proposition: two-level cryptography: a breakable one for all and an unbreakable one for a select few.

Considering the complexity and cost of the quantum computational equipment we may expect that for the reasonably long period of time quantum computing will be confined to few centres, similarly to early mainframe computing. Let's assume that in a true spirit of free market, those quantum-centres will offer cryptographic services to the rest of the agents community (whether human or automated). Such services may be of the following kind:

- providing traditional (breakable) encryption;
- breaking traditional (breakable) encryption;
- running the permanent register with associated registration.

Now we have tools to build fences of different heights, at least for the majority of us. We can have the fence for free, built with traditional security. Such a fence can be broken at a certain expense (those centres will presumably charge for the service). We can also build the fence that is truly impenetrable (at least within limits of quantum technology) – potentially at a high price.

Now, how to set the price for those services? The possible 'price list' is shown in Table 16.1 but only as an example, to illustrate the concept presented here. The underlying concept is as follows: breaking traditional encryption should depend on whether the encryption has been provided by the centre or not. It should be reasonably cheap to break something that has not been provided by the centre. Encryption provided by the centre can also be broken by the centre at a price that depends on how much the user has paid to establish such encryption.

We are allowing for fences of different heights and respective difficulties. While competitor's fences are cheap to break, an agent is attracted to acquire a fence from the centre – and decide about its height. A tall fence is more expensive but they do provide more protection.

The price list is an interesting one. Breaking the encryption that Alice gets for free bears only the nominal cost, say 10 EUR – quite cheap. Breaking the encryption that has been ordered from the centre is more expensive, but it is still at a level one can afford, say, it should be fixed at 100 EUR. Finally, the costs of providing encryption vary, from 0.1 EUR to 10 EUR. Note that the encryption Alice receives from the centre can be always broken at the expense of 100 EUR, whether she has paid 0.1 or 10 EUR for it. It seems unfair and discouraging, but we will get back to it. For the moment it is important to remember that Alice can get a certain encryption, but she cannot get the perfect one.

The centre runs one service that has not been yet discussed – the permanent register. Such a register is a public list of services that it has provided – all the facts that encryption has been provided and (what is even more important) all encryption breaks that it has done. Every record is tagged with the undeniable identity (protected by quantum cryptography) of

**Table 16.1**  The hypothetical 'price list'

| Service | Price | |
|---|---|---|
| | Money | Reputation |
| Setting up encryption | 0.1–10 depending on the required strength | 1–100 depending on the strength |
| Breaking encryption if set up by centre | 100 | 1 of set by oneself 100*monetary price paid for it if set by others |
| Breaking encryption if not set up by centre | 10 | None |

the person who has ordered the service. Everybody knows that Alice is using the fence and everybody knows that somebody has broken her fence.

This is not the key escrow scheme. The encryption that Alice orders from the centre is not 'hers'. Even though she is the owner of the particular encryption key, she is not the sole master of it. Everybody can break her security by requesting the service from the centre. She or her enemy, they can pay the centre and have the encryption broken.

It seems like the answer to the problem of fences. Alice can order a fence of an arbitrary height, depending on her perception of risk and her need for protection. What she does is public and visible. What she can do in secret can be disclosed so cheaply that it does not provide any permanent protection. If she needs, she can break her own fence. Others can break her fence as well, but what they do will be visible to her as well as to others.

Now we are getting to the issue of reputation and confidence. Money is not the best currency to pay for the encryption (or encryption breaking) service – or rather it is not the only currency that is needed here. It seems to be unfair to assume that rich people can learn the secrets of the poor while the poor cannot learn theirs. Building and breaking digital fences should incur cost in reputation. Money is used here mostly to discourage trivial cases, not to shape the market.

Dealing with the centre is not free, it costs some 'reputation points', but the way the price list is structured is quite different. Alice incurs little cost (in 'reputation points') in breaking her own encryption, but she is not getting it for free – the one who habitually breaks her own lock does not seem to be trustworthy anyway. Breaking an other's encryption is expensive – in proportion to the money the owner has paid for the encryption. Breaking Bob's 0.1 EUR encryption can be done from time to time, but breaking the 10 EUR encryption costs all the reputation Alice can acquire. It must be a desperate measure for her to order such a service.

So, if Alice is not concerned much about her encryption then Bob can take a casual peek into her data from time to time. Bob cannot do it all the time (as his reputation budget will not allow for it) and Alice will know that he has done it. If an organisation is very concerned about its data, the determined individual can still look at them, but this will cost him all his reputation – the digital equivalent of a determined whistle-blowing activist.

Buying encryption should also cost some reputation points, in addition to money. Certainly, excessive secrecy does not inspire confidence so that higher-grade encryption should bear some penalty in the area of reputation. On the other hand, casual encryption (e.g. at 0.1 EUR) should not incur any significant loss in reputation, as it is more of a border marking than actual fencing.

There is an open question how an agent can acquire reputation points that are needed for this scheme. The similar question related to money has not been answered here – we have simply assumed that agents have money from somewhere outside of the scheme to pay for services. We can answer the reputation-related question in a similar manner – that reputation has to be earned somewhere else.

## 16.6 Trust in a Bottle

Finally, there is a price set on trust by the pharmacy business: 25 US$ for two weeks' supply of liquid trust (volume discounts apply). That's not a joke – the substance called oxytocin is available in a ready to apply (spray) packages to create the atmosphere of trust.

Studies of MRI brain activity suggest that Alice's decision about her confidence in Bob (specifically her decision to trust Bob) is both rational and emotional, and that those two parts of the decision are processed in different parts of the brain [Adolphs2002]. Being exposed to evidence of potential untrustworthiness activates amygdale that feeds its perception of the lack of trust to the whole decision-making process (interestingly, amygdale responds also to situations of fear and unfamiliarity).

Oxytocin is a hormone that suppresses the activity of amygdale [Kosfeld2005]. Oxytocin is usually released in situation related to social recognition and bonding and plays an important role in the social behaviour of mammals. If applied externally, oxytocin suppresses fear and then perception of untrustworthiness, effectively making Alice ignore evidence that speaks against Bob – thus assigning more importance to evidence that is positive.

So, oxytocin is maybe not the 'trust potion', but it definitely facilitates early contact and the creation of swift trust – something that every salesman will appreciate. If Bob behaves reasonably, his actions will be viewed in a positive light by Alice. The only problem Bob faces is the proper application – so that he does not inhale it. Otherwise, they both will start trusting each other – and this is not something that every salesman will appreciate.

## 16.7 Conclusion

The relationship between confidence and economy has several aspects. The economic interest is generally in increasing the trust component while reducing control, in the expectation that this will have a positive impact on the economy as a whole, or a least for certain groups of agents. However, this approach does not address the ability to commoditise confidence and to transfer it between agents, being satisfied with the traditional means of relationship-building, amplified by the network effect.

Alternatively, confidence can be seen as a commodity that can be acquired and spent, presumably in an exchange for money. This approach addresses the emerging economy of confidence, further fuelled by the brand virtualisation aspect of convergence. It is also possible to use confidence as an equivalent of currency, and use it as an intermediary to trade other goods. An interesting attempt to trade security on the basis of confidence is proposed in this chapter, in an attempt to re-define traditional relationships between those two concepts.

Convergence accelerates and facilitates all the trends that have been discussed in this chapter, so that different aspects of the 'economy of confidence' become increasingly relevant to the coming convergent world.

## Bibliography

[Adolphs2002] Ralph Adolphs: Trust in the Brain. *Nature Neuroscience*. 5(3): 8–9. 2002.

[Briscoe2006] Bob Briscoe, Andrew Odlyzko and Benjamin Tilly: Metcalfe's Law is Wrong. *IEEE Spectrum*. July. 2006.

[Castelfranchi2006] Cristiano Castelfranchi, Rino Falcone and Francesca Marzo: Being Trusted in a Social Network: Trust as Relational Capital. In K. Stolen et al. (eds): *iTrust 2006*, Berlin: Springer. LNCS 3986, pp. 19–32. 2006.

[DiVincenzo1995] David P. DiVincenzo: Quantum Computation. *Science* 270(5234): 255–261. 1995.

[Douceur2002] John R. Douceur: The Sybil Attack. In *Proc. of the First Int. Workshop on Peer-to-Peer Systems, IPTPS02 Workshop*. Cambridge, MA. March. 2002.

[Durlauf2002] Steven N. Durlauf: On the Empirics of Social Capital. *Economic J.*, 112(483): 459–479. 2002.

[Frankema2005] Katinka Bijlsma-Frankema and Ana Cristina Costa: Understanding the Trust-Control Nexus. *Int. Sociology*. 20(30): 259–282. 2005.

[Fukuyama1996] Francis Fukuyama: *Trust: The Social Virtues and the Creation of Prosperity*. New York: Touchstone Books. 1996.

[Hawthorn2000] Geoffrey Hawthorn: Three Ironies in Trust. In Diego Gambetta, (ed.): *Trust: Making and Breaking Cooperative Relations*, electronic edition, Department of Sociology, University of Oxford, Chapter 7, pp. 111–126. Available at: http://www.sociology.ox.ac.uk/papers/hawthorn111–126.pdf. 2000.

[Jurca2006] Radu Jurca and Boi Faltings: Using CHI Scores to Reward Honest Feedback from Repeated Interactions. In *Proc. of AAMAS'06*. 8–12 May. Hakodate, Japan. 2006.

[Kerr2006] Reid Kerr and Robin Cohen: Trunits: A Monetary Approach to Modelling Trust in Electronic Marketplaces. In *Proc. of Fifth Int. Conf. on Autonomous Agents and Multiagent Systems AAMAS-06*. Hakodate, Japan. 2006.

[Kosfeld2005] Michael Kosfeld et al.: Oxytocin Increases Trust in Humans. *Nature* 435: 673–676. 2005.

[Luck2005] Michael Luck et al: Agent Technology: Computing as Interaction. A Roadmap for Agent-based Computing. *AgentLink*. 2005.

[Mishra1996] Aneil K. Mishra: Organizational Responses to Crisis. The Centrality of Trust. In: Roderick M. Kramer and Tom R. Tyler: *Trust in Organizations*. London: Sage Publications. 1996.

[Mui2002] Lik Mui et al.: A Compotational Model of Trust and *Reputation*. In *Proc. of 35th Hawaii Int. Conf. on System Science* (HICSS). IEEE Computer Society Press, USA. 2002.

[O'Reilly2005] Tim O'Reilly: What is Web 2.0.?: Design Patterns and Business Models for the Next Generation of Software. Available at: http://www.oreillynet.com/pub/a/oreilly/tim/news/2005/09/30/what-is-web-20.html. 2005.

[Obreiter2004] Philipp Obreiter: A Case for Evidence-Aware Distributed Reputation Systems: Overcoming the Limitations of Plausibility Considerations. In C. D. Jensen et al. (eds.): *iTrust 2004*, Berlin. Springer. *LNCS 2995*, pp. 33–47. 2004.

[Putnam2000] Robert D. Putnam: *Bowling Alone: The Collapse and Revival of American Community*. New York: Simon & Schuster. 2000.

[Raub2001] Werner Raub and Chris Snijden: A Reluctant Match: Models for the Analysis of Trust in Durable Two Party Relations. In Cristiano Castelfranchi and Yao-Hua Tan (eds): *Trust and Deception in Virtual Societies*. Dordrecht: Kluwer Academic Publishers. pp. 195–220. 2001.

[Resnick2006] Paul Resnick: The Value of Reputation on eBay: A Controlled Experiment. *Experimental Economics*, 9(2): 79–101. Available at: http://www.si.umich.edu/~presnick/papers/postcards/PostcardsFinalPrePub.pdf. 2006.

[Rodgers2006] Chris Rodgers: *Informal Coalitions: Mastering the Hidden Dynamics of Organizational Change*. Basingstoke: Palgrave Macmillan. 2006.

[Sako1998] Mari Sako and Susan Helper: Determinants of Trust in Supplier Relations: Evidence from the Automotive Industry in Japan and the United States. *J. of Economic Behavior and Organization*. 34: 387–417. 1998.

[Seigneur2005] Jean-Marc Seigneur, Alan Gray and Christian Damsgaard Jensen: Trust Transfer: Encouraging Self-recommendations without Sybil Attack. In P. Herrmann et al. (eds): *iTrust 2005, LNCS 3477*, pp. 321–337. 2005.

[Sobel2002] Joel Sobel: Can We Trust Social Capital? *J. of Econ. Literature*. XL (March): 139–154. 2002.

[Tadelis1999] Steven Tadelis: What's in a Name?: Reputation as a Tradeable Asset. *American Economic Review*. 89(3): 548–563. 1999.

[Tonkiss2004] Fran Tonkiss: Trust and Social Capital. In Jane Franklin (ed.): *Politics, Trust and Networks: Social Capital in Critical Perspective*. London: London South Bank University. 2004.

[Twigg2003] Andrew Twigg: A Subjective Approach to Routing in P2P and Ad Hoc Networks. In *Proc. of First Int. Conf. on Trust Management iTrust 2003. LNCS 2692*, pp. 225–238. 2003.

[Willcocks2006] Leslie P. Willcocks and Sara Cullen: The Outsourcing Enterprise: The Power of Relationships. Available at: http://www.logicacmg.com/pSecured/admin/countries/assets/serve_asset.asp?id=3252. 2006.

# Index